The frontispiece depicts the basic cyclic particle system, a process with random dynamics. This complex spatial dynamic has one of the simplest imaginable local update rules. The states are coded as n colors labeled 0, ..., n−1, thought of as a color wheel. The instructions for each cell of the array are: choose one of 4 nearest neighbors at random and update to that color if it is your successor. On small arrays there is little indication of pattern formation since the process will either grind to a halt or interfere with itself. But on sufficiently large arrays, like the one shown, a remarkably stable spiral-laden stochastic equilibrium emerges. See Rick Durrett's article in this volume for more examples of self-organizing random systems.

Probability and
Stochastics Series

Topics in

Contemporary

Probability and

Its Applications

Probability and
Stochastics Series

Topics in

Contemporary Probability and Its Applications

Edited by
J. Laurie Snell

CRC Press
Boca Raton   New York   London   Tokyo

6-11-96-19 22531

Library of Congress Cataloging-in-Publication Data

Topics in contemporary probability and its application / [compiled
  by] J. Laurie Snell
      p.   cm. -- (Probability and stochastics series)
    Includes bibliographical references and index.
    ISBN 0-8493-8073-1 (alk. paper).
    1. Stochastic processes.   2. Probabilities.   I. Snell, J. Laurie
(James Laurie), 1925—   . II. Series.
    QA274.T65   1994 ✓
    519.2--dc20

                                                                94-23472
                                                                CIP

PREFACE

There was a time when most of probability theory could be put into a single book. Indeed, that is what J. L. Doob did in his book *Stochastic Processes* published in 1953. Any attempt to do this today would lead to an enormous encyclopedia. Still, it is useful to give some idea of the current trends in probability, and that is the purpose of this book. We asked a number of well-known probabilists to discuss their current work in a form that could be understood by graduate students, mathematicians in other areas, or scientists in other areas who use probability in their own work.

We found a continued interest in random walks and Brownian motion, though now random walks are sometimes on arbitrary graphs and Brownian motion might be on a manifold. As in the beginnings of probability, probabilists still find interesting problems suggested by the study of games and still enjoy looking at paradoxes.

Our first group of papers deals with random walks and Brownian motion. Robin Pemantle tells us what we can learn about random walks on graphs by answering the question "what does a typical spanning tree look like?" In the process he finds all kinds of interesting connections between random walks and electric networks, discrete harmonic functions and stationary Markov Chains. Gregory Lawler discusses random walks on the integer lattice. Lawler compares unrestricted random walks with the restricted self-avoiding random walks that arose from chemistry in the study of polymer chains. Lawler provides a number of fascinating conjectures, generally believed to be true, but whose rigorous proofs have yet to be developed. Ruth Williams explains Itô's stochastic calculus and uses the calculus to study the classical Dirichlet problem, the Schrödinger equation, and Laplace's equation with oblique derivative boundary conditions in a quadrant. She uses the latter to study a two-station queueing system in heavy traffic. Mark Pinsky asks the question: can you tell if a Riemann manifold is essentially a Euclidean space by looking at certain functionals of the Brownian motion on the manifold? He shows that for dimension n less than six, if the exit times for spheres have distributions agreeing with those for Euclidean space, the manifold is essentially Euclidean space. This need not be true for dimensions n greater than six.

Classical stochastic process theory as described in Doob's book dealt primarily with random variables indexed by time. Modern probability theory often studies random variables indexed by points in n dimensional space or even abstract spaces. These generalizations came from the study of the Ising model in physics and from biological problems that involve interaction between particles. Rick Durrett shows us the beautiful pictures and challenging mathematics problems that are suggested by computer simulations of epidemics, forest fires and other systems in which spatially distributed particles interact locally. Andrei Toom gives us a series of problems and examples relating to discrete time interacting particle systems. In the process he shows the relation between theoretical work and computer simulation. The problems range from simple exercises often described as "left to the reader" to research level problems.

The difficulty in obtaining explicit solutions for the Ising model led to the development of algorithms to obtain results by simulations. These algorithms, suitably generalized, have found fruitful applications in a variety of fields including image recognition, combinatorial optimization and statistics. Basilis Gidas describes the mathematical framework of these simulation algorithms and presents the main results concerning convergence and other properties of the algorithms.

Joel Cohen shows us that the concept of random graphs goes back at least to Charles Darwin's *Origin of Species* (1895). In ecology, random digraphs, called food webs, describe which species eat which species. Cohen demonstrates that stochastic models of food webs can explain and unify some of the empirical patterns one sees when a large number of real food webs are viewed as an ensemble.

The next group of papers all relate one way or the other to games. The question of how many times you should shuffle a deck of cards made the *New York Times* when Bayer and Diaconis, extending the work of Jim Reeds, showed that seven shuffles generally suffices. Brad Mann provides a self-contained exposition of their demonstration of this result.

Maitra and Sudderth consider sequential games in a quite general setting. Two players make a sequence of plays each of which leads to a particular state. When they are in a state, they select actions that determine the conditional probability of being at the various states at the next step. One player wants to reach a certain set of states or reach it repeatedly, while the other player wants to reach the complimentary set. Techniques for solving such a game are developed and applied to a number of examples.

Chris Thron tells us about the "bandit problem". This is one of a class of problems that has wide applications including medical drug testing. In the simplest description you have a choice between playing two slot machines that pay off with different unknown probabilities. How do you decide which machine to play to maximize your long range winning? Thron derives Gittins' elegant solution to this problem and shows how to implement the solution with a computer program.

The book ends with a discussion by Snell and Vanderbei of three bewitching switching paradoxes, one of which, the Monty Hall problem, was immortalized by the discussions of the problem in Marilyn vos Savant's columns in Parade Magazine.

There are several ways that this book can be used: as supplementary material for an undergraduate or graduate probability course, as the basis of a seminar in probability or applied mathematics, for students interested in independent work in probability and for scientists in other areas who want to see how probability theory might be used in their work.

For an undergraduate probability course we would recommend the articles by Durrett, Pemantle, Lawler, Cohen, Mann, and the article by Snell and Vanderbei.

Any of the articles would be appropriate to supplement a graduate course in probability and the choice will depend upon the topics emphasized in the course. For example, if random walk is such a topic the articles by Pemantle and Lawler could be used, if the course includes a discussion of Brownian motion the articles by Williams and Pinsky would be appropriate. If the course emphasizes applications, the articles by Durrett, Cohen, Thron and the article by Maitra and Sudderth would be appropriate.

The article by Pemantle is ideally suited for a seminar on the popular topic of random walks and electric networks and the article by Gidas could be the basis for a seminar on annealing. The article by Maitra and Sudderth could serve as the basis for seminar on stochastic game theory.

For independent work for students in probability, we would suggest the article by Toom with its rich collection of problems.

All of the articles have contacts with other disciplines but we should single out the articles by Durrett and Cohen as especially interesting to biologists, the article by Lawler to chemists, and the article by Gidas to computer scientists

We would like to thank the authors for their patience in seeing this book materialize. While TeX certainly makes it easy for an author to write a single paper, the use of special macros and different versions of TeX have made putting papers together like solving a giant jigsaw puzzle. Without the tireless efforts of Fuxing Hou this puzzle would still be unsolved. We thank her for her fine work.

CONTRIBUTORS

Joel E. Cohen
Rockefeller University

Rick Durrett
Department of Mathematics, Cornell University

Basilis Gidas
Division of Applied Mathematics, Brown University

Gregory F. Lawler
Department of Mathematics, Duke University

Ashok Maitra
School of Statistics, University of Minnesota

Brad Mann
Department of Mathematics, Harvard University

Robin Pemantle
Department of Mathematics, University of Wisconsin

Mark A. Pinsky
Department of Mathematics, Northwestern University

J. Laurie Snell
Department of Mathematics and Computer Science, Dartmouth College

William Sudderth
School of Statistics, University of Minnesota

Christopher P. Thron
Department of Mathematics and Physics, King College

Andrei Toom
Incarnate Word College

Robert Vanderbei
Department of Civil Engineering and Operations Research, Princeton University

Ruth J. Williams
Department of Mathematics, University of California, San Diego

CONTENTS

Chapter 1

UNIFORM RANDOM SPANNING TREES

Robin Pemantle
Department of Mathematics
University of Wisconsin

ABSTRACT

A spanning tree of a graph is what you get when you erase enough of the edges so that no cycles remain, but leave enough edges intact so that it is still possible to get from any vertex to any other. This chapter answers the question: If you pick at random one of the many spanning trees of a graph, what will it look like? The analysis brings in many related topics: random walks on graphs, electrical networks, discrete harmonic functions, eigenvalues for the discrete Laplace operator, and some theorems on distributional convergence based on convergence of moments.

1. INTRODUCTION

There are several good reasons you might want to read about uniform spanning trees, one being that spanning trees are useful combinatorial objects. Not only are they fundamental in algebraic graph theory and combinatorial geometry, but they predate both of these subjects, having been used by Kirchoff in the study of resistor networks. This article addresses the question about spanning trees most natural to anyone in probability theory, namely what does a typical spanning tree look like?

Some readers will be happy to know that understanding the basic questions requires no background knowledge or technical expertise. While the model is elementary, the answers are surprisingly rich. The combination of a simple question and a structurally complex answer is sometimes taken to be the quintessential mathematical experience. This nonwithstanding, I think the best reason to set out on a mathematical odyssey is to enjoy the ride. Answering the basic questions about spanning trees depends on a sort of vertical integration of techniques and results from diverse realms of probability theory and discrete mathematics. Some of the topics encountered en route are random walks, resistor networks, discrete harmonic analysis, stationary Markov chains, circulant matrices, inclusion-exclusion, branching processes, and the method of moments. Also touched on are characters of abelian groups, entropy and the infamous incipient infinite cluster.

The introductory section defines the model and previews some of the connections to these other topics. The remaining sections develop these at length. Explanations of jargon and results borrowed from other fields are provided whenever possible. Complete proofs are given in most cases, as appropriate.

1.1   Defining the model

Begin with a finite graph G. That means a finite collection $V(G)$ of *vertices* along with a finite collection $E(G)$ of *edges*. Each edge either connects two vertices v and $w \in V(G)$ or else is a *self-edge*, connecting some $v \in V(G)$ to itself. There may be more than one edge connecting a pair of vertices. Edges are said to be *incident* to the vertices they connect. To make the notation less cumbersome we will write $v \in G$ and $e \in G$ instead of $v \in V(G)$ and $e \in E(G)$. For $v, w \in G$ say v is a *neighbor* of w, written $v \sim w$ if and only if some edge connects v and w. Here is an example of a graph G_1 which will serve often as an illustration.

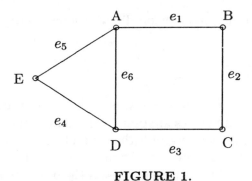

FIGURE 1.

Its vertex set is $\{A, B, C, D, E\}$ and it has six edges $e_1, \ldots, e_6$, none of which is a self-edge.

A subgraph of a graph G will mean a graph with the same vertex set but only a subset of the edges. (This differs from standard usage that allows the vertex set to be a subset as well.) Since G_1 has 6 edges, there are $2^6 = 64$ possible different subgraphs of G_1. A subgraph $H \subseteq G$ is said to be a *forest* if there are no cycles, i.e., you cannot find a sequence of vertices $v_1, \ldots, v_k$ for which there are edges in H connecting v_i to v_{i+1} for each $i < k$ and an edge connecting v_k to v_1. In particular ($k = 1$) there are no self-edges in a forest. A *tree* is a forest that is connected, i.e., for any v and w there is a path of edges that connects them. The *components* of a graph are the maximal connected subgraphs, so, for example, the components of a forest are trees. A *spanning* forest is a forest in which every vertex has at least one incident edge; a *spanning tree* is a tree in which every vertex has at least one incident edge. If G is connected (and all our graphs will be) then a spanning tree is just a subgraph with no cycles such that the addition of any other edge would create a cycle. From this it is easy to see that every connected graph has at least one spanning tree.

Now if G is any finite connected graph, imagine listing all of its spanning trees (there are only finitely many) and then choosing one of them at random with an equal probability of choosing any one. Call this random choice $\mathbf{T}$ and say that $\mathbf{T}$ is a *uniform random spanning tree* for G. In the above example there are eleven spanning trees for G_1 given (in the obvious notation) as follows:

$$e_1e_2e_3e_4 \quad e_1e_2e_3e_5 \quad e_1e_2e_4e_5 \quad e_1e_3e_4e_5$$

$$e_2e_3e_4e_5 \quad e_1e_2e_4e_6 \quad e_1e_3e_4e_6 \quad e_2e_3e_4e_6$$

$$e_1e_2e_5e_6 \quad e_1e_3e_5e_6 \quad e_2e_3e_5e_6$$

In this case, $\mathbf{T}$ is just one of these eleven trees, picked with uniform probability. The model is so simple, you may wonder what there is to say about it! One answer is that the model has some properties that are easy to state but hard to prove; these are introduced in the coming subsections. Another answer is that the definition of a uniform random spanning tree does not give us a way of readily computing local characteristics of the random tree. To phrase this as a question, can you compute probabilities of events local to a small set of edges, such as $\mathbf{P}(e_1 \in \mathbf{T})$ or $\mathbf{P}(e_1, e_4 \in \mathbf{T})$ without actually enumerating all of the spanning trees of G? In a sense, most of the article is devoted to answering this question. Events such as e_1 being in the tree are called local, in contrast to a global event such as the tree having diameter (longest path between two vertices) at most three.

1.2   Uniform spanning trees have negative correlations

Continuing the example in figure 1, suppose I calculate the probability that $e_1 \in \mathbf{T}$. That's easy: there are 8 spanning trees containing e_1, so

$$\mathbf{P}(e_1 \in \mathbf{T}) = \frac{8}{11}.$$

Similarly there are 7 spanning trees containing e_4 so

$$\mathbf{P}(e_4 \in \mathbf{T}) = \frac{7}{11}.$$

There are only 4 spanning trees containing both e_1 and e_4, so

$$\mathbf{P}(e_1 \in \mathbf{T} \text{ and } e_4 \in \mathbf{T}) = \frac{4}{11}.$$

Compare the probability of both of these edges being in the tree with the product of the probabilities of each of the edges being in the tree:

$$\frac{8}{11} \cdot \frac{7}{11} = \frac{56}{121} > \frac{4}{11}.$$

Thus

$$\mathbf{P}(e_1 \in \mathbf{T} \,|\, e_4 \in \mathbf{T}) = \frac{\mathbf{P}(e_1 \in \mathbf{T} \text{ and } e_4 \in \mathbf{T})}{\mathbf{P}(e_4 \in \mathbf{T})} < \mathbf{P}(e_1 \in \mathbf{T})$$

or, in words, the conditional probability of e_1 being in the tree if you know that e_4 is in the tree is less than the original unconditional probability. This negative correlation of edges holds in general, with the inequality not necessarily strict.

Theorem 1.1 *For any finite connected graph G, let $\mathbf{T}$ be a uniform spanning tree. If e and f are distinct edges, then $\mathbf{P}(e, f \in \mathbf{T}) \leq \mathbf{P}(e \in \mathbf{T})\mathbf{P}(f \in \mathbf{T})$.*

Any spanning tree of an n-vertx graph contains $n-1$ edges, so it should seem intuitively plausible — even obvious — that if one edge is forced to be in the tree then any other edge is less likely to be needed. Two proofs will be given later, but neither is straightforward, and in fact the only proofs I know involve elaborate connections between spanning trees, random walks, and electrical networks. Sections 1.4 and 2.4 will be occupied with the elucidation of these connections. The connection between random walks and electrical networks will be given more briefly, since an excellent treatment is available [8].

As an indication that the previous theorem is not trivial, here is a slightly stronger statement, the truth or falsity of which is unknown. Think of the distribution of $\mathbf{T}$ as a probability distribution on the outcome space Ω consisting of all the $2^{|E(G)|}$ subgraphs of G that just happens to give probability zero to any subgraph that is not a spanning tree. An event A (i.e., any subset of the outcome space) is called an *up-event* — short for *upwardly closed* — if whenever a subgraph H of G has a further subgraph K and $K \in A$, then $H \in A$. An example of an up-event is the event of containing at least two of the three edges e_1, e_3 and e_5. Say an event A ignores an edge e if for every H, $H \in A \Leftrightarrow H \cup e \in A$.

Conjecture 1 *For any finite connected graph G, let $\mathbf{T}$ be a uniform spanning tree. Let e be any edge and A be any up-event that ignores e. Then*

$$\mathbf{P}(A \text{ and } e \in \mathbf{T}) \leq \mathbf{P}(A)\mathbf{P}(e \in \mathbf{T}).$$

Theorem 1.1 is a special case of this when A is the event of f being in the tree. The conjecture is known to be true for *series-parallel* graphs and it is also known to be true in the case when A is an *elementary cylinder event*, i.e., the event of containing some fixed $e_1, \ldots, e_k$. On the negative side, there are natural generalizations of graphs and spanning trees, namely *matroids* and *bases* (see [19] for definitions), and both Theorem 1.1 and Conjecture 1 fail to generalize to this setting. If you're interested in seeing the counterexample, look at the end of [15].

1.3   The transfer-impedance matrix

The next two paragraphs discuss a theorem that computes probabilities such as $\mathbf{P}(e, f \in \mathbf{T})$. These computations alone would render the theorem useful, but it appears even more powerful in the context of how strongly it constrains the probability measure governing $\mathbf{T}$. Let me elaborate.

Fix a subset $S = \{e_1, \ldots, e_k\}$ of the edges of a finite connected graph G. If $\mathbf{T}$ is a uniform random spanning tree of G then the knowledge of whether $e_i \in \mathbf{T}$ for each i partitions the space into 2^k possible outcomes. (Some of these may have probability zero if S contains cycles, but if not, all 2^k may be possible.) In any case, choosing $\mathbf{T}$ from the uniform distribution on spanning trees of G induces a probability distribution on Ω, the space of these 2^k outcomes. There are many possible probability distributions on Ω: the ways of choosing 2^k nonnegative numbers summing to one are a $(2^k - 1)$-dimensional space. Theorem 1.1 shows that the actual measure induced by $\mathbf{T}$ satisfies certain inequalities, so not all probability distributions on Ω can be gotten in this way. But the set of probability distributions on Ω satisfying these inequalities is still $(2^k - 1)$-dimensional. It turns out, however, that the set of probability distributions on Ω that arise as induced distributions of uniform spanning trees on subsets of k edges actually has at most the much smaller dimension $k(k+1)/2$. This is a consequence of the following theorem which is the bulwark of our entire discussion of spanning trees:

Theorem 1.2 (transfer-impedance theorem) *Let G be any finite connected graph. There is a symmetric function $H(e, f)$ on pairs of edges in G such that for any $e_1, \ldots, e_r \in G$,*

$$\mathbf{P}(e_1, \ldots, e_r \in \mathbf{T}) = \det M(e_1, \ldots, e_r)$$

where $M(e_1, \ldots, e_r)$ is the r by r matrix whose i, j-entry is $H(e_i, e_j)$.

By inclusion-exclusion, the probability of any event in Ω may be determined from the probabilities of $\mathbf{P}(e_{j_1}, \ldots, e_{j_r} \in \mathbf{T})$ as $e_{j_1}, \ldots, e_{j_r}$ vary over all subsets of $e_1, \ldots, e_k$. The theorem says that these are all determined by the $k(k+1)/2$ numbers $\{H(e_i, e_j) : i, j \leq k\}$, which shows that there are indeed only $k(k+1)/2$ degrees of freedom in determining the measure on Ω.

Another way of saying this is that the measure is almost completely determined by its two-dimensional marginals, i.e., from the values of $\mathbf{P}(e, f \in \mathbf{T})$ as e and f vary over pairs of (not necessarily distinct) edges. To see this, calculate the values of $H(e, f)$. The values of $H(e, e)$ in the theorem must be equal to $\mathbf{P}(e \in \mathbf{T})$ since $\mathbf{P}(e, e) = \det M(e) = H(e, e)$. To see what $H(e, f)$ is for $e \neq f$, write

$$\mathbf{P}(e, f \in \mathbf{T}) \quad = \quad \det M(e, f)$$

$$= \quad H(e, e)H(e, f) - H(e, f)^2$$

$$= \ \mathbf{P}(e \in \mathbf{T})\mathbf{P}(f \in \mathbf{T}) - H(e,f)^2$$

and hence

$$H(e,f) = \pm\sqrt{\mathbf{P}(e \in \mathbf{T})\mathbf{P}(f \in \mathbf{T}) - \mathbf{P}(e,f \in \mathbf{T})}.$$

Thus the two-dimensional marginals determine H up to sign, and H determines the measure. Note that the above square root is always real, since by Theorem 1.1 the quantity under the radical is nonnegative. Section 4 will be devoted to proving Theorem 1.2, the proof depending heavily on the connections to random walks and electrical networks developed in Sections 1.4 and 2.4.

1.4   Applications of transfer-impedance to limit theorems

Let K_n denote the complete graph on n vertices, i.e., there are no self-edges and precisely one edge connecting each pair of distinct vertices. Imagine picking a uniform random spanning tree of K_n and letting n grow to infinity. What kind of limit theorem might we expect? Since a spanning tree of K_n has only $n-1$ edges, each of the $n(n-1)/2$ edges should have probability $2/n$ of being in the tree (by symmetry) and is hence decreasingly likely to be included as $n \to \infty$. On the other hand, the number of edges incident to each vertex is increasing. Say we fix a particular vertex v_n in each K_n and look at the number of edges incident to v_n that are included in the tree. Each of $n-1$ incident edges has probability $2/n$ of being included, so the expected number of of such edges is $2(n-1)/n$, which is evidently converging to 2. If the inclusion of each of these $n-1$ edges in the tree were independent of each other, then the number of edges incident to v_n in $\mathbf{T}$ would be a binomial random variable with parameters $(n-1, 2/n)$. The well-known Poisson limit theorem would then say that the random variable $D_{\mathbf{T}}(v_n)$, counting how many edges incident to v_n are in $\mathbf{T}$, converged as $n \to \infty$ to a Poisson distribution with mean two. [A quick explanation: integer-valued random variables X_n are said to converge to X in distribution if $\mathbf{P}(X_n = k) \to \mathbf{P}(X = k)$ for all integers k. In this instance, convergence of $D_{\mathbf{T}}(v_n)$ to a Poisson of mean two would mean that $\mathbf{P}(D_{\mathbf{T}}(v_n) = k) \to e^{-2}k^2/2$ as $n \to \infty$ for each integer k.] Unfortunately this can't be true because a Poisson(2) is sometimes zero, whereas $D_{\mathbf{T}}(v_n)$ can never be zero. It has however been shown [2] that $D_{\mathbf{T}}(v_n)$ converges in distribution to the next simplest thing: one plus a Poisson of mean one.

To show you why this really is the next best thing, let me point out a property of the mean one Poisson distribution. Pretend that, if you picked a family in the United States at random, then the number of children in the family would have a Poisson distribution with mean one (population control having apparently succeeded). Now imagine picking a child at random instead of picking a family at random, and asking how many children in the family. You would certainly get a different distribution, since you couldn't ever get the answer zero. In fact you would get one plus a Poisson of mean one. (Poisson distributions are the only ones with this property.) Thus a Poisson-plus-one distribution is a more natural distribution than it looks at first. At any rate, the convergence theorem is

Theorem 1.3 *Let $D_{\mathbf{T}}(v_n)$ be the random degree of the vertex v_n in a uniform spanning tree of K_n. Then, as $n \to \infty$, $D_{\mathbf{T}}(v_n)$ converges in distribution to X, where X is one plus a Poisson of mean one.*

Consider now the n-cube B_n. Its vertices are defined to be all strings of zeros and ones of length n, where two vertices are connected by an edge if and only if they differ in precisely one location. Fix a vertex $v_n \in B_n$ and play the same game: choose a uniform random spanning tree and let $D_{\mathbf{T}}(v_n)$ be the random degree of v_n in the tree. It is not hard to see again that the expected value, $\mathbf{E}D$, converges to 2 as $n \to \infty$. Indeed, for any graph the number of edges in a spanning tree is one less than the number of vertices and, since each edge has two endpoints the average degree of the vertices will be ≈ 2, if the graph is symmetric, each vertex will then have the same expected degree that must be 2. One could expect Theorem 1.3 to hold for B_n as well as K_n and in fact it does. A proof of this for a class of sequences of graphs that includes both K_n and B_n and does not use transfer-impedances appears in [2], along with the conjecture that the result should hold for more general sequences of graphs. This can indeed be established, and in Section 5 we will discuss the proof of Theorem 1.3 via transfer-impedances which can be extended to more general sequences of graphs.

The convergence in distribution of $D_{\mathbf{T}}(v_n)$ in these theorems is actually a special case of a stronger kind of convergence. To begin discussing this stronger kind of convergence, imagine that we pick a uniform random spanning tree of a graph, say K_n, and want to write down what it looks like "near v_n". Interpret "near v_n" to mean within a distance of r of v_n, where r is some arbitrary positive integer. The answer will be a rooted tree of height r. (A *rooted* tree is a tree plus a choice of one of its vertices, called the root. The *height* of a rooted tree is the maximum distance of any vertex from the root.) The rooted tree representing $\mathbf{T}$ near v_n will be the tree you get by picking up $\mathbf{T}$, dangling it from v_n, and ignoring everything more than r levels below the top.

Call this the *r-truncation* of $\mathbf{T}$, written $\mathbf{T} \wedge_{v_n} r$ or just $\mathbf{T} \wedge r$ when the choice of v_n is obvious. For example, suppose $r = 2$, v_n has 2 neighbors w_1 and w_2 in $\mathbf{T}$, w_1 has 3 neighbors other than v_n in $\mathbf{T}$, and w_2 has none. This information is encoded in the following picture. The picture could also have been drawn with left and right reversed, since we consider this to be the same abstract tree, no matter how it is drawn.

When $r = 1$, the only information in $\mathbf{T} \wedge r$ is the number of children of the root, i.e., $D_{\mathbf{T}}(v_n)$. Thus the previous theorem asserts the convergence in distribution of $\mathbf{T} \wedge_{v_n} 1$ to a root with a (1+Poisson) number of vertices. Generalizing this is the following theorem, proved in Section 5.

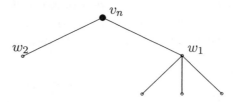

FIGURE 2.

Theorem 1.4 *For any $r \geq 1$, as $n \to \infty$, $\mathbf{T} \wedge_{v_n} r$ converges in distribution to a particular random tree, $\mathcal{P}_1 \wedge r$ to be defined later.*

Convergence in distribution means that, for any fixed tree t of height at most r, $\mathbf{P}(\mathbf{T} \wedge_{v_n} r = t)$ converges as $n \to \infty$ to the probability of the random tree $\mathcal{P}_1 \wedge r$ equalling t. As the notation indicates, the random tree $\mathcal{P}_1 \wedge r$ is the r-truncation of an infinite random tree. It is in fact the tree of a Poisson(1) branching process conditioned to live forever, but these terms will be defined later, in Section 5. The theorem is stated here only for the sequence K_n, but is in fact true for a more general class of sequences, which includes B_n.

2. SPANNING TREES AND RANDOM WALKS

Unless G is a very small graph, it is virtually impossible to list all of its spanning trees. For example, if $G = K_n$ is the complete graph on n vertices, then the number of spanning trees is n^{n-2} according to the well-known Prüfer bijection [17]. If n is much bigger than say 20, this is too many to be enumerated even by the snazziest computer that ever will be. Luckily, there are shortcuts that enable us to compute probabilities such as $\mathbf{P}(e \in \mathbf{T})$ without actually enumerating all spanning trees and counting the proportion containing e. The shortcuts are based on a close correspondence between spanning trees and random walks, which is the subject of this section.

2.1   Simple random walk

Begin by defining a simple random walk on G. To avoid obscuring the issue, we will place extra assumptions on the graph G and later indicate how to remove these. In particular, in addition to assuming that G is finite and connected, we will often suppose that it is D-regular for some positive integer D, which means that every vertex has precisely D edges incident to it. Also suppose that G is simple, i.e., it has no self-edges or parallel edges (different edges connecting the same pair of vertices). For any vertex $x \in G$, define a simple random walk on G starting at x, written SRW_x^G, intuitively as follows. Imagine a particle beginning at time 0 at the vertex x. At each future time $1, 2, 3, \ldots$, it moves along some edge, always choosing among the D edges incident to the vertex it is currently at with equal probability. When G is not D-regular, the definition will be the same: each of the edges leading away from the current position will be chosen with probability $1/\text{degree}(v)$. This defines a sequence of random positions $SRW_x^G(0), SRW_x^G(1), SRW_x^G(2), \ldots$ which is thus a random function SRW_x^G (or just SRW if x and G may be understood without ambiguity) from the nonnegative integers to the vertices of G. Formally, this random function may be defined by its finite-dimensional marginals which are given by $\mathbf{P}(SRW_x^G(0) = y_0, SRW_x^G(1) = y_1, \ldots, SRW_x^G(k) = y_k) = D^{-k}$ if $y_0 = x$ and for all $i = 1, \ldots, k$ there is an edge from y_{i-1} to y_i, and zero otherwise. For an illustration of this definition, let G be the following 3-regular simple graph.

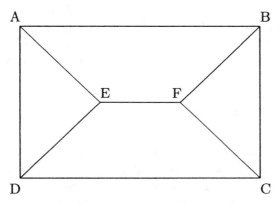

FIGURE 3.

Consider a simple random walk SRW_A^G starting at the vertex A. The probability of a particular beginning, say $SRW(1) = B$ and $SRW(2) = F$ is just $(1/3)^2$. The random position at time 2, $SRW(2)$, is then equal to F with probability $2/9$, since each of the two ways, ABF and AEF, of getting to F in two steps has probability $1/9$.

Another variant of random walk we will need is the *stationary Markov chain* corresponding to a simple random walk on G. I will preface this definition with a quick explanation of Markov chains; since I cannot do justice to this large topic in two paragraphs, the reader is referred to [11], [9], or any other favorite introductory probability text for further details.

A (time-homogeneous) *Markov chain* on a finite state space S is a sequence of random variables $\{X_i\}$ taking values in S, indexed by either the integers or the nonnegative integers and having the Markov property: there is a set of transition probabilities $\{p(x, y) : x, y \in S\}$ so that the probability of X_{i+1} being y, conditional upon $X_i = x$, is always equal to $p(x, y)$ regardless of how much more information about the past you have. (Formally, this means $\mathbf{P}(X_{i+1} = y \,|\, X_i = x$ and any values of X_j for $j < i$) is still $p(x,y)$.) An example of this is SRW_x^G, where S is the set of vertices of G and $p(x, y) = D^{-1}$

if $x \sim y$ and 0 otherwise (recall that $x \sim y$ means x is a neighbor of y). The values $p(x, y)$ must satisfy $\sum_y p(x, y) = 1$ for every x in order to be legitimate conditional probabilities. If in addition they satisfy $\sum_x p(x, y) = 1$ for every y, the Markov chain is said to be *doubly stochastic*. It will be useful later to know that the Markov property is time-reversible, meaning if $\{X_i\}$ is a Markov chain then so is the sequence $\{\tilde{X}_i = X_{-i}\}$, and there are backward transition probabilities $\tilde{p}(x, y)$ for which $\mathbf{P}(X_{i-1} = y \mid X_i = x) = \tilde{p}(x, y)$.

If it is possible eventually to get from every state in S to every other, then there is a unique *stationary distribution* that is a set of probabilities $\{\pi(x) : x \in S\}$ summing to one and having the property that $\sum_x \pi(x)p(x, y) = \pi(y)$ for all y. Intuitively, this means that, if we build a Markov chain with transition probabilities $p(x, y)$ and start it by randomizing X_0 so that $\mathbf{P}(X_0 = x) = \pi(x)$, then it will also be true that $\mathbf{P}(X_i = x) = \pi(x)$ for every $i > 0$. A *stationary* Markov chain is one indexed by the integers (as opposed to just the positive integers), in which $\mathbf{P}(X_i = x) = \pi(x)$ for some, hence every i. If a Markov chain is doubly stochastic, it is easy to check that the uniform distribution U is stationary:

$$\sum_x U(x)p(x, y) = \sum_x |S|^{-1}p(x, y) = |S|^{-1} = U(y).$$

The stationary distribution π is unique (assuming every state can get to every other) and is hence uniform over all states.

Now we define a stationary simple random walk on G to be a stationary Markov chain with state space $V(G)$ and transition probabilities $p(x, y) = D^{-1}$ if $x \sim y$ and 0 otherwise. Intuitively this can be built by choosing X_0 at random uniformly over $V(G)$, then choosing the X_i for $i > 0$ by walking randomly from X_0 along the edges and choosing the X_i for $i < 0$ also by walking randomly from X_0, thinking of this latter walk as going backward in time. (For SRW, $p(x, y) = p(y, x) = \tilde{p}(x, y)$ so the walk looks the same backward as forward.)

2.2   Random walk construction of uniform spanning trees

Now we are ready for the random walk construction of uniform random spanning trees. What we will actually get is a directed spanning tree, which is a spanning tree together with a choice of vertex called the root and an orientation on each edge (an arrow pointing along the edge in one of the two possible directions) such that following the arrows always leads to the root. Of course a directed spanning tree yields an ordinary spanning tree if you ignore the arrows and the root. Here is an algorithm to generate directed trees from random walks.

GROUNDSKEEPER'S ALGORITHM

Let G be a finite, connected, D-regular, simple graph and let x be any vertex of G. Imagine that we send the groundskeeper from the local baseball diamond on a walk along the edges of G starting from x; later we will take this to be the walk SRW_x^G. She brings with her the wheelbarrow full of chalk used to draw in lines. This groundskeeper is so eager to choose a spanning tree for G that she wants to chalk a line over each edge she walks along. Of course if that edge, along with the edges she's already chalked, would form a cycle (or is already chalked), she is not allowed to chalk it. In this case she continues walking that edge but temporarily — and reluctantly — shuts off the flow of chalk. Every time she chalks a new edge she inscribes an arrow pointing from the new vertex back to the old.

Eventually every vertex is connected to every other by a chalked path, so no more can be added without forming a cycle and the chalking is complete. It is easy to see that the subgraph consisting of chalked edges is always a single connected component. The first time the walk reaches a vertex y, the edge just travelled cannot form a cycle with the other chalked edges. Conversely, if the walk moves from z to some y that has been reached before, then y is connected to z already by some chalked path, so adding the edge zy would create a cycle and is not permitted. Also it is clear that following the arrows leads always to vertices that were visited previously, and hence eventually back to the root. Furthermore, every vertex except x has exactly one oriented edge leading out of it, namely the edge along which the vertex was first reached.

Putting this all together, we have defined a function — say τ — from walks on G (infinite sequences of vertices, each consecutive pair connected by an edge) to directed spanning trees of G. Formally $\tau(y_0, y_1, y_2, \ldots)$ is the subgraph $H \subseteq G$ such that if e is an oriented edge from w to z then

$$e \in H \Leftrightarrow \text{ for some } k > 0, \ y_k = z, y_{k-1} = w, \text{ and there is no } j < k$$
$$\text{such that } y_j = z.$$

As an example, suppose SRW_A^G in figure 3 begins ABFBCDAE. Then applying τ gives the tree with edges BA, FB, CB, DC and EA.

To be completely formal, I should admit that the groundskeeper's algorithm never stops if there is a vertex that the walk fails to hit in finite time. This is not a problem since we are going to apply τ to the path of a SRW, and this hits every vertex with probability one. As hinted earlier, the importance of this construction is the following equivalence.

Theorem 2.1 *Let G be any finite, connected, D-regular, simple graph and let x be any vertex of G. Run a simple random walk SRW_x^G and let* **T** *be the* *random spanning tree gotten by ignoring the arrows and root of the random directed spanning tree* $\tau(SRW_x^G)$. *Then* **T** *has the distribution of a uniform random spanning tree.*

To prove this it is necessary to consider a stationary simple random walk on G (called $SSRW^G$). It will be easy to get back to a SRW_x^G because the sites visited in positive time by a $SSRW^G$ conditioned on being at x at time zero form a SRW_x^G. Let T_n be the tree $\tau(SSRW(n), SSRW(n+1), \ldots)$; in other words, T_n is the directed tree gotten by applying the groundskeeper's algorithm to the portion of the stationary simple random walk from time n onward. The first goal is to show that the random collection of directed trees T_n forms a time-homogeneous Markov chain as n ranges over all integers.

Showing this is pretty straightforward because the transition probabilities are easy to see. First note that, if t and u are any two directed trees on disjoint sets of vertices, rooted respectively at v and w, then adding any arrow from v to a vertex in u combines them into a single tree rooted at w. Now define two operations on directed spanning trees of G as follows.

Operation F$(\mathbf{t}, \mathbf{x})$: *Start with a directed tree t rooted at v. Choose one of the the D neighbors of v in G, say x. Take away the edge in t that leads out of x, separating t into two trees, rooted at v and x. Now add an edge from v to x, resulting in a single tree $F(t, x)$.*

Operation F$^{-1}(\mathbf{t}, \mathbf{w})$: *Start with a directed tree t rooted at x. Choose one of the the D neighbors of x in G, say w. Follow the path from w to x in t and let v be the last vertex on this path before x. Take away the edge in t that leads out of v, separating t into two trees, rooted at x and v. Now add an edge from x to w, resulting in a single directed tree $F^{-1}(t, w)$.*

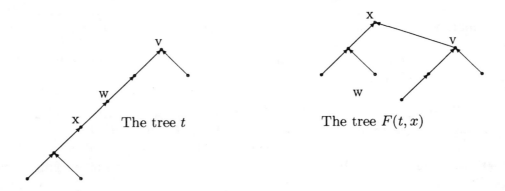

The tree t The tree $F(t, x)$

FIGURE 4.

It is easy to see that these operations really are inverse to each other, i.e., if t is rooted at v then $F^{-1}(F(t,x),w) = t$ for any $x \sim v$, where w is the other endpoint of the edge leading out of x in t. Here is a pictorial example.

I claim that for any directed trees t and u the backward transition probability $\tilde{p}(t,u)$ is equal to D^{-1} if $u = F(t,x)$ for some x and zero otherwise. To see this, it is just a matter of realizing where the operation F comes from. Remember that T_n is just $\tau(SSRW(n), SSRW(n+1), \ldots)$, so in particular the root of T_n is $SSRW(n)$. Now $SSRW$ really is a Markov chain. We already know that $\mathbf{P}(SSRW(n-1) = x \mid SSRW(n) = v)$ is D^{-1} if $x \sim v$ and zero otherwise. Also, this is unaffected by knowledge of $SSRW(j)$ for any $j > n$. Suppose it turns out that $SSRW(n-1) = x$. Then knowing only T_n and x (but not the values of $SSRW(j)$ for $j > n$) it is possible to work out what T_{n-1} is. Remember that T_n and T_{n-1} come from applying τ to the respective sequences $SSRW(n), SSRW(n+1), \ldots$ and $SSRW(n-1), SSRW(n), \ldots$ whose only difference is that the second of these has an extra x tacked on the beginning. Every time the first sequence reaches a vertex for the first time, so does the second, unless that vertex happens to be x. So the T_{n-1} has all the oriented edges of T_n except the one out of x. What it has instead is an oriented edge from v to x, chalked in by the groundskeeper at her very first step. Adding in the edge from v to some neighbor x and erasing the edge out of x yields precisely $F(t,x)$. So we have shown that $T_{n-1} = F(T_n, SSRW(n-1))$. But $SSRW(n-1)$ is uniformly distributed among the neighbors of $SSRW(n)$ no matter what other information we know about the future. This proves the claim and the time-homogeneous Markov property.

The next thing to show is that the stationary distribution is uniform over all directed trees. As we've seen, this would follow if we knew that $\{T_n\}$ was doubly stochastic. Since $p(t,u)$ is D^{-1} whenever $u = F(t,x)$ for some x and zero otherwise, this would be true if for every tree u there are precisely D trees t for which $F(t,x) = u$ for some x. But the trees t for which $F(t,x) = u$ for some x are precisely the trees $F^{-1}(u,x)$ for some neighbor x of the root of u, hence there are D such trees and the transition probabilities for $SSRW$ are doubly stochastic.

Now that the stationary distribution for $\{T_n\}$ has been shown to be uniform, the proof of Theorem 2.1 is almost done. Note that the event $SSRW(0) = x$ is the same as the event of $\tau(SSRW(0), SSRW(1), \ldots)$ being rooted at x. Since SRW_x^G is just $SSRW$ conditioned on $SSRW(0) = x$, $T_0(SRW_x^G)$ is distributed as a uniform directed spanning tree conditioned on being rooted at x. That is to say, $T_0(SRW_x^G)$ is uniformly distributed over all directed spanning trees rooted at x. But ordinary spanning trees are in a one to one correspondence with directed spanning trees rooted at a fixed vertex x, the correspondence being that to get from the ordinary tree to the directed tree you name x as the root and add arrows that point toward x. Then the tree $\mathbf{T}$ gotten from $T_0(SRW_x^G)$ by ignoring the root and the arrows is uniformly distributed over all ordinary spanning trees of G, which is what we wanted to prove.   □

2.3   Weighted graphs

It is time to remove the extra assumptions that G is D-regular and simple. It will make sense later to generalize from graphs to weighted graphs and, since the generalization of Theorem 2.1 is as easy for weighted graphs as for unweighted graphs, we may as well introduce weights now.

A weighted graph is just a graph to each edge e of which is assigned a positive real number called its weight and written $w(e)$. Edge weights are not allowed to be zero, though one may conceptually identify a graph with an edge of weight zero with the same graph minus the edge in question. An unweighted graph may be thought of as a graph with all edge weights equal to one, as will be clear from the way random trees and random walks generalize. Write $d(v)$ for the sum of the weights of all edges incident to v. Corresponding to the old notion of a uniform random spanning tree is the *weight-selected* random spanning tree (WST). A WST, **T**, is defined to have

$$\mathbf{P}(\mathbf{T} = t) = \frac{\prod_{e \in t} w(e)}{\sum_u \prod_{e \in u} w(e)}$$

so that the probability of any individual tree is proportional to its weight which is by definition the product of the weights of its edges.

Corresponding to a simple random walk from a vertex x is the weighted random walk from x, WRW_x^G which is a Markov Chain in which the transition probabilities from a vertex v are proportional to the weights of the edges incident to v (among which the walk must choose). Thus if v has two neighbors w and x, and there are four edges incident to v with respective weights $1, 2, 3$ and 4 that connect v respectively to itself, w, x and x, then the probabilities of choosing these four edges are respectively $1/10, 2/10, 3/10$ and $4/10$. Formally, the probability of walking along an edge e incident to the current position v is given by $w(e)/d(v)$. The bookkeeping is a little unwieldly since knowing the set of vertices $WRW(0), WRW(1), \ldots$ visited by the WRW does not necessarily determine which edges were travelled now that the graph is not required to be simple. Rather than invent some clumsy *ad hoc* notation to include the edges, it is easier just to think that a WRW includes this information (so it is *not* simply given by its positions $WRW(j) : j \geq 0$), but that we will refer to this information in words when necessary. If G is a connected weighted graph then WRW^G has a unique stationary distribution denoted by positive numbers $\pi^G(v)$ summing to one. This will not in general be uniform, but its existence is enough to guarantee the existence of a stationary Markov chain with the same transition probabilities. We call this stationary Markov chain $SWRW$ the few times the need arises. The new and improved theorem then reads:

Theorem 2.2 *Let G be any finite, connected weighted graph and let x be any vertex of G. Run a weighted random walk WRW_x^G and let $\mathbf{T}$ be the random spanning tree gotten by ignoring the arrows and root of the random directed spanning tree $\tau(WRW_x^G)$. Then $\mathbf{T}$ has the distribution of WST.*

The proof of Theorem 2.1 serves for Theorem 2.2 with a few alterations. These will now be described, thought not much would be lost by taking these details on faith and skipping to the next section.

The groundskeeper's algorithm is unchanged with the provision that the WRW brings with it the information of which edge she should travel if more than one edge connects $WRW(i)$ to $WRW(i+1)$ for some i. The operation to get from the directed tree T_n to a candidate for T_{n-1} is basically the same, only, instead of there being D choices for how to do this, there is one choice for each edge incident to the root v of T_n: choose such an edge, add it to the tree oriented from v to its other endpoint x and remove the edge out of x. It is easy to see again that $\{T_n\}$ is a time-homogeneous Markov chain with transition probability from t to u zero unless u can be gotten from t by the above operation, and if so the probability is proportional to the weight of the edge that was added in the operation. (This is because if $T_n = t$ then $T_{n-1} = u$ if and only if u can be gotten from this operation and WRW, travelled along the edge added in this operation between times $n-1$ and n.)

The uniform distribution on vertices is no longer stationary for WRW since we no longer have D-regularity, but the distribution $\pi(v) = d(v)/\sum_x d(x)$ is easily seen to be stationary: Start a WRW with $WRW(0)$ having distribution π; then

$$\mathbf{P}(WRW(1) = v) = \sum_x \mathbf{P}(WRW(0) = x \text{ and } WRW(1) = v)$$

$$= \sum_x \frac{d(x)}{\sum_y d(y)} \left(\sum_{e \text{ connecting } x \text{ to } v} w(e)/d(x) \right)$$

$$= \frac{1}{\sum_y d(y)} \sum_{e \text{ incident to } v} w(e)$$

$$= \pi(v).$$

The stationary distribution π for the Markov chain $\{T_n\}$ gives a directed tree t rooted at v with probability

$$\pi(t) = K d(v) \prod_{e \in t} w(e),$$

where $K = (\sum_t d(\text{root}(t)) \prod_{e \in t} w(e))^{-1}$ is a normalizing constant. If t, rooted at v, can go to u, rooted at x, by adding an edge e and removing the edge

f, then $\pi(u)/\pi(t) = d(x)w(e)/d(v)w(f)$. To verify that π is a stationary distribution for T_n, write $\mathcal{C}(u)$ for the class of trees from which it is possible to get to u in one step. For each $t \in \mathcal{C}(u)$, write v_t, e_t, and f_t for the root of t, the edge added to t to get u, and the edge taken away from t to get u, respectively. If u is rooted at x, then

$$
\begin{aligned}
\mathbf{P}(T_{n-1} = u) &= \sum_t \mathbf{P}(T_n = t \text{ and } T_{n-1} = u) \\[2mm]
&= \sum_{t \in \mathcal{C}(u)} \pi(t)w(e_t)/d(v_t) \\[2mm]
&= \sum_{t \in \mathcal{C}(u)} [\pi(u)d(v_t)w(f_t)/d(x)w(e_t)]w(e_t)/d(v_t) \\[2mm]
&= \pi(u) \sum_{t \in \mathcal{C}(u)} w(f_t)/d(x) \\[2mm]
&= \pi(u),
\end{aligned}
$$

since, as t ranges over all trees that can get to u, f_t ranges over all edges incident to x.

Finally, we have again that $\tau(WRW_x^G(0))$ is distributed as $\tau(SWRW^G(0))$ conditioned on having root x and, since the unconditioned π is proportional to $d(x)$ times the weight of the tree (product of the edge weights), the factor of d is constant and $\mathbf{P}(\tau(WRW_x^G(0)) = t)$ is proportional to $\prod_{e \in t} w(e)$ for any t rooted at x. Thus $\tau(WRW_x^G(0))$ is distributed identically to WST. □

2.4   Applying the random walk construction to our model

Although the benefit is not yet clear, we have succeeded in translating the question of determining $\mathbf{P}(e \in \mathbf{T})$ from a question about uniform spanning trees to a question about simple random walks. To see how this works, suppose that e connects the vertices x and y, and then generate a uniform spanning tree by the random walk construction starting at x, i.e., let $\mathbf{T}$ be the tree gotten from $\tau(SRW_x^G)$ by ignoring the root and arrows. If $e \in \mathbf{T}$ then its orientation in $\tau(SRW_x^G)$ must be from y to x, and so $e \in \mathbf{T}$ if and only if $SRW(k-1) = x$ where k is the least k for which $SRW(k) = y$. In other words,

$$
\mathbf{P}(e \in \mathbf{T}) = \mathbf{P}(\text{first visit of } SRW_x^G \text{ to } y \text{ is along } e). \tag{1}
$$

The computation of this random walk probability turns out to be tractable.

More important is the fact that this may be iterated to get probabilities such as $\mathbf{P}(e, f \in \mathbf{T})$. This requires two more definitions. If G is a finite connected graph and e is an edge of G whose removal does not disconnect G, then the *deletion* of G by e is the graph $G \setminus e$ with the same vertex set and the same edges minus e. If e is any edge that connects distinct vertices x and y, then the *contraction* of G by e is the graph G/e whose vertices are the vertices of G with x and y replaced by a single vertex $x*y$. There is an edge $\rho(f)$ of G/e for every edge of f of G, where if one or both endpoints of f is x or y then that endpoint is replaced by $x*y$ in $\rho(f)$. We write $\rho(z)$ for the vertex corresponding to z in this correspondence, so $\rho(x) = \rho(y) = x*y$ and $\rho(z) = z$ for every $z \neq x, y$. The following example shows G_1 and G_1/e_4. The edge e_4 itself maps to a self-edge under ρ, e_5 becomes parallel to e_6 and D and E map to $D*E$.

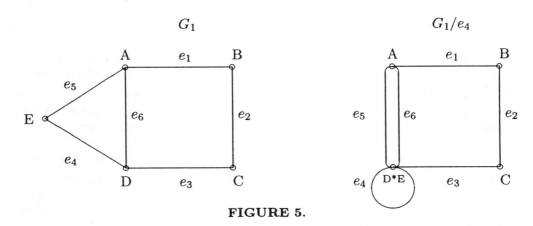

FIGURE 5.

It is easy to see that successive deletions and contractions may be performed in any order with the same result. If $e_1, \ldots, e_r$ are edges of G whose joint removal does not disconnect G then the successive deletion of these edges is permissible. Similarly, if $\{f_1, \ldots, f_s\}$ is a set of edges of G that contains no cycle, these edges may be successively contracted and the graph $G \setminus e_1, \ldots, e_r / f_1, \ldots, f_s$ is well-defined. It is obvious that the spanning trees of $G \setminus e$ are just those spanning trees of G that do not contain e. Almost as obvious is a one to one correspondence between spanning trees of G containing e and spanning trees of G/e: if t is a spanning tree of G containing e then there is a spanning tree of G/e consisting of $\{\rho(f) : f \neq e \in t\}$.

To translate $\mathbf{P}(e, f \in \mathbf{T})$ to the random walk setting, write this probability as $\mathbf{P}(e \in \mathbf{T})\mathbf{P}(f \in \mathbf{T} \mid e \in \mathbf{T})$. The first term has already been translated. The conditional distribution of a uniform random spanning tree given that it contains e is just uniform among those trees containing e, which is just $\mathbf{P}_{G/e}(\rho(f) \in \mathbf{T})$ where the subscript G/e refers to the fact that $\mathbf{T}$ is now taken to be a uniform random spanning tree of G/e. If f connects z and x then this is in turn equal to $\mathbf{P}(SRW_{\rho(x)}^{G/e}$ first hits $\rho(z)$ along $\rho(f))$. Both the terms have thus been translated; in general it should be clear how this may be iterated to translate the probability of any elementary event, $\mathbf{P}(e_1, \ldots, e_r \in \mathbf{T}$ and $f_1, \ldots, f_s \notin \mathbf{T})$ into a product of random walk probabilities. It remains to be seen how these probabilities may be calculated.

3. RANDOM WALKS AND ELECTRICAL NETWORKS

Sections 3.1 to 3.3 contain a development of the connection between random walks and electrical networks. The right place to read about this is in [8]; what you will see here is necessarily a bit rushed. Sections 3.5 and 3.6 contain similarly condensed material from other sources.

3.1   Resistor circuits

The electrical networks we discuss will have only two kinds of elements: resistors and voltage sources. Picture the resistors as straight pieces of wire. A resistor network will be built by soldering resistors together at their endpoints. That means that a diagram of a resistor network will just look like a finite graph with each edge bearing a number: the resistance. Associated with every resistor network H is a weighted graph G_H which looks exactly like the graph just mentioned except that the weight of an edge is not the resistance but the *conductance*, which is the reciprocal of the resistance. The distinction between H and G_H is only necessary while we are discussing precise definitions and will then be dropped. A voltage source may be a single battery that provides a specified voltage difference (explained below) across a specified pair of vertices or is may be a more complicated device to hold various voltages fixed at various vertices of the network. Here is an example of a resistor network on a familiar graph, with a one volt battery drawn as a dashed box. Resistances on the edges (made up arbitrarily) are given in ohms.

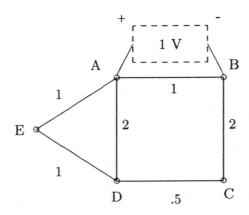

FIGURE 6.

The electrical properties of such a network are given by Kirchoff's laws. For the sake of exposition I will number the laws, although these do not correspond to the way Kirchoff actually stated the laws. The first law is that every vertex of the network has a voltage that is a real number. The second law gives every oriented edge (resistor) a current. Each edge has two possible orientations.

Say an edge connects x and y. Then the current through the edge is a real number whose sign depends on which orientation you choose for the edge. In other words, the current $I(\vec{xy})$ that flows from x to y is some real number and the current $I(\vec{yx})$ is its negative. [Note though that the weights $w(e)$ are always taken to positive; weights are functions of unoriented edges, whereas currents are functions of oriented edges.] If $I(e)$ denotes the current along an oriented edge $e = \vec{xy}$, $V(x)$ denotes the voltage at x and $R(e)$ denotes the resistance of e, then quantitatively, the second law says

$$I(\vec{xy}) = [V(x) - V(y)]R(e)^{-1}. \tag{2}$$

Kirchoff's third law is that the total current flowing into a vertex equals the total current flowing out, or in other words

$$\sum_{y \sim x} I(\vec{xy}) = 0. \tag{3}$$

This may be rewritten using (2). Recalling that in the weighted graph G_H, the weight $w(e)$ is just $R(e)^{-1}$ and that $d(v)$ denotes the sum of $w(e)$ over edges incident to v, we get at every vertex x an equation

$$0 = \sum_{y \sim x}[V(x) - V(y)]w(xy) = V(x)d(x) - \sum_{y \sim x}V(y)w(xy). \tag{4}$$

Since a voltage source may provide current, this may fail to hold at any vertex connected to a voltage source. The above laws are sufficient to specify the voltages of the network — and hence the currents — except that a constant may be added to all the voltages (in other words, it is the voltage differences that are determined, not the absolute voltages). In the above example the voltage difference across AB is required to be one. Setting the voltage at B to zero (since the voltages are determined only up to an additive constant), the reader may check that the voltages at A, C, D, and E are respectively $1, 4/7, 5/7$, and $6/7$ and the currents through AB, AE, ED, AD, DC, CB are respectively $1, 1/7, 1/7, 1/7, 2/7, 2/7$.

3.2 Harmonic functions

The voltages in a weighted graph G (which we are now identifying with the resistor network it represents) under application of a voltage source are calculated by finding a solution to Kirchoff's laws on G with specified boundary conditions. For each vertex x there is an unknown voltage $V(x)$. There is also a linear equation for every vertex not connected to a voltage source, and an equation given by the nature of each voltage source. Will these always be enough information so that Kirchoff's laws have a unique solution? The answer is yes and it is most easily seen in the context of harmonic functions.*

If f is a function on the vertices of a weighted graph G, define the *excess* of f at a vertex v, written $\Delta f(v)$ by

$$\Delta f(v) = \sum_{y \sim v} [f(v) - f(y)] w(vy).$$

You can think of Δ as an operator that maps functions f to other functions Δf, and is a discrete analog of the Laplacian operator. A function f from the vertices of a finite weighted graph G to the reals is said to be *harmonic* at a vertex v if and only if $\Delta f(v) = 0$. Note that for any function f, the sum of the excesses $\sum_{v \in G} \Delta f(v) = 0$, since each $[f(x) - f(y)] w(xy)$ cancels a $[f(y) - f(x)] w(yx)$ due to $w(xy) = w(yx)$. To see what harmonic functions are intuitively, consider the special case where G is unweighted, i.e., all of the edge weights are one. Then a function is harmonic if and only if its value at a vertex x is the average of the values at the neighbors of x. In the weighted case the same is true, but with a weighted average! Here is an easy but important lemma about harmonic functions.

Lemma 3.1 (maximum principle) *Let V be a function on the vertices of a finite connected weighted graph, harmonic everywhere except possibly at vertices of some set $X = \{x_1, \ldots, x_k\}$. Then V attains its maximum and minimum on X. If V is harmonic everywhere then it is constant.*

Proof: Let S be the set of vertices where V attains its maximum. Certainly S is nonempty, as it is finite. If $x \in S$ has a neighbor $y \notin S$ then V cannot be harmonic at x since $V(x)$ would then be a weighted average of values less than or equal to $V(x)$ with at least one strictly less. In the case where V is harmonic everywhere, this shows that no vertex in S has a neighbor not in S; hence, since the graph is connected, every vertex is in S and V is constant. Otherwise, suppose V attains its maximum at some $y \notin X$ and pick a path connecting y to some $x \in X$. The entire path must then be in S up until and including the first vertex along the path at which V is not harmonic. This is some $x' \in X$. The argument for the minimum is just the same. □

Kirchoff's third law (4) says that the voltage function is harmonic at every x not connected to a voltage source. Suppose we have a voltage source that provides a fixed voltage at some specified set of vertices. Say for concreteness that the vertices are $x_1, \ldots, x_k$ and the voltages produced at these vertices are $c_1, \ldots, c_k$. We now show that Kirchoff's laws determine the voltages everywhere else, i.e., there is at most one solution to them.

---

*There is also the question of whether any solution exists, but addressing that would take us too far afield. If you aren't convinced of its existence on physical grounds, wait until the next subsection where a probabilistic interpretation for the voltage is given, and then deduce existence of a solution from the fact that these probabilities obey Kirchoff's laws.

Theorem 3.2 *Let V and W be real-valued functions on the vertices of a finite weighted graph G. Suppose that $V(x_i) = W(x_i) = c_i$ for some set of vertices $x_1, \ldots, x_k$ and $1 \le i \le k$ and that V and W are harmonic at every vertex other than $x_1, \ldots, x_k$. Then $V = W$.*

Proof: Consider the function $V - W$. It is easy to check that being harmonic at x is a linear property, so $V - W$ is harmonic at every vertex at which both V and W are harmonic. Then by the maximum principle, $V - W$ attains its maximum and minimum at some x_i. But $V - W = 0$ at every x_i, so $V - W \equiv 0$. $\square$

Suppose that, instead of fixing the voltages at a number of points, the voltage source acts as a current source and supplies a fixed amount of current I_i to vertices x_i, $1 \le i \le k$. This is physically reasonable only if $\sum_{i=1}^{k} I_i = 0$. Then a net current of I_i will have to flow out of each x_i into the network. Using (2) gives

$$I_i = \sum_{y \sim x} w(x, y)(V(x) - V(y)) = \Delta V(x).$$

From this it is apparent that the assumption $\sum_i I_i = 0$ is algebraically as well as physically necessary since the excesses must sum to zero. Kirchoff's laws also determine the voltages (up to an additive constant) of a network with current sources, as we now show.

Theorem 3.3 *Let V and W be real-valued functions on the vertices of a finite weighted graph G. Suppose that V and W both have excess c_i at x_i for some set of vertices x_i and reals c_i, $1 \le i \le k$. Suppose also that V and W are harmonic elsewhere. Then $V = W$ up to an additive constant.*

Proof: Excess is linear, so the excess of $V - W$ is the excess of V minus the excess of W. This is zero everywhere, so $V - W$ is harmonic everywhere. By the maximum principle, $V - W$ is constant. $\square$

3.3 Harmonic random walk probabilities

Getting back to the problem of random walks, suppose G is a finite connected graph and x, a, b are vertices of G. Let's say that I want to calculate the probability that SRW_x reaches a before b. Call this probability $h_{ab}(x)$. It is not immediately obvious what this probability is, but we can get an equation by watching where the random walk takes its first step. Say the neighbors of x are $y_1, \ldots, y_d$. Then $\mathbf{P}(SRW_x(1) = y_i) = d^{-1}$ for each $i \le d$. If we condition on $\mathbf{P}(SRW_x(1) = y_i)$ then the probability of the walk reaching a before b is (by the Markov property) the same as if it had started out at y_i. This is just $h_{ab}(y_i)$. Thus

$$h_{ab}(x) \quad = \quad \sum_i \mathbf{P}(SRW_x(1) = y_i) h_{ab}(y_i)$$

$$= d^{-1} \sum_i h_{ab}(y_i).$$

In other words, h_{ab} is harmonic at x. Be careful though; if x is equal to a or b, it doesn't make sense to look one step ahead since $SRW_x(0)$ already determines whether the walk hit a or b first. In particular, $h_{ab}(a) = 1$ and $h_{ab}(b) = 0$, with h_{ab} being harmonic at every $x \neq a, b$.

Theorem 3.2 tells us that there is only one such function h_{ab}. This same function solves Kirchoff's laws for the unweighted graph G with voltages at a and b fixed at 1 and 0 respectively. In other words, the probability of SRW_x reaching a before b is just the voltage at x when a one volt battery is connected to a and b and the voltage at b is taken to be zero. If G is a weighted graph, we can use a similar argument — it is easy to check that the first-step transition probabilities $p(x, y) = w(\vec{xy})/\sum_z w(\vec{xz})$ show that $h_{ab}(x)$ is harmonic in the sense of weighted graphs. Summarizing this:

Theorem 3.4 *Let G be a finite connected weighted graph. Let a and b be vertices of G. For any vertex x, the probability of SRW_x^G reaching a before b is equal to the voltage at x in G when the voltages at a and b are fixed at one and zero volts respectively.*

Although more generality will not be needed, we remark that this same theorem holds when a and b are taken to be sets of vertices. The probability of SRW_x reaching a vertex in a before reaching a vertex in b is harmonic at vertices not in $a \cup b$ is zero on b and one on a. The voltage when vertices in b are held at zero volts and vertices in a are held at one volt also satisfies this, so the voltages and the probabilities must coincide.

Having given an interpretation of voltage in probabilistic terms, the next thing to find is a probabilistic interpretation of the current. The arguments are similar so they will be treated briefly; a more detailed treatment appears in [8]. First we will need to find an electrical analogue for the numbers $u_{ab}(x)$ which are defined probabilistically as the expected number of times a SRW_a hits x before the first time it hits b. This is defined to be zero for $x = b$. For any $x \neq a, b$, let $y_1, \ldots, y_r$ be the neighbors of x. Then the number of visits to x before hitting b is the sum over i of the number of times SRW_a hits y_i before b and goes to x on the next move (the walk had to be somewhere the move before it hit x). By the Markov property, this quantity is $u_{ab}(y_i)p(y_i, x) = u_{ab}(y_i)w(\vec{xy_i})/d(y_i)$. Letting $\phi_{ab}(z)$ denote $u_{ab}(z)/d(z)$ for any z, this yields

$$\phi_{ab}(x) = d(x)u_{ab}(x) = \sum_i u_{ab}(y_i)w(\vec{xy_i})/d(y_i) = \sum_i w(\vec{xy_i})\phi_{ab}(y_i).$$

In other words ϕ_{ab} is harmonic at every $x \neq a, b$. Writing K_{ab} for $\phi_{ab}(a)$ we then have that ϕ_{ab} is K_{ab} at a, zero at b and harmonic elsewhere; hence it is the same function as the the voltage induced by a battery of K_{ab} volts connected to a and b, with the voltage at b taken to be zero. Without yet knowing what

K_{ab} is, this determines ϕ_{ab} up to a constant multiple. This in turn determines u_{ab}, since $u_{ab}(x) = d(x)\phi_{ab}(x)$.

Now imagine that we watch SRW_a to see when it crosses over a particular edge $\vec{xy}$ and count plus one every time it crosses from x to y and minus one every time it crosses from y to x. Stop counting as soon as the walk hits b. Let $H_{ab}(\vec{xy})$ denote the expected number of signed crossings. (H now stands for harmonic, not for the name of a resistor network.) We can calculate H in terms of u_{ab} by counting the pluses and the minuses separately. The expected number of plus crossings is just the expected number of times the walk hits x, mulitplied by the probability on each of these occasions that the walk crosses to y on the next move. This is $u_{ab}(x)w(\vec{xy})/d(x)$. Similarly the expected number of minus crossings is $u_{ab}(y)w(\vec{xy})/d(y)$. Thus

$$H_{ab}(\vec{xy}) = u_{ab}(x)w(\vec{xy})/d(x) - u_{ab}(y)w(\vec{xy})/d(y)$$

$$= w(\vec{xy})(\phi_{ab}(x) - \phi_{ab}(y)).$$

But $\phi_{ab}(x) - \phi_{ab}(y)$ is just the voltage difference across $\vec{xy}$ induced by a K_{ab}-volt battery across a and b. Using (2) and $w(\vec{xy}) = R(\vec{xy})^{-1}$ shows that the expected number of signed crossings of $\vec{xy}$ is just the current induced in $\vec{xy}$ by a K_{ab}-volt battery connected to a and b. A moment's thought shows that the expected number of signed crossings of all edges leading out of a must be one, since the walk is guaranteed to leave a one more time than it returns to a. So the current supplied by the K_{ab}-volt battery must be one amp. Another way of saying this is that

$$\triangle \phi_{ab} = \delta_a - \delta_b. \tag{5}$$

Instead of worrying about what K_{ab} is, we may just as well say that the expected number of crossings of $\vec{xy}$ by SRW_a before hitting b is the current induced when one amp is supplied to a and drawn out at b.

3.4   Electricity applied to random walks
applied to spanning trees

Finally we can address the random walk question that relates to spanning trees. In particular, the claim that the probability in equation (1) is tractable will be borne out several different ways. First we will see how the probability may be "calculated" by an analog computing device, namely a resistor network. In the next subsection, the computation will be carried out algebraically and very neatly, but only for particularly nice symmetric graphs. At the end of the section, a universal method will be given for the computation which is a little messier. Finally in Section 4 the question of the individual probabilities in (1) will be avoided altogether and we will see instead how values for these probabilities (wherever they might come from) determine the probabilities for all contractions and deletions of the graph and therefore determine all the joint probabilities $\mathbf{P}(e_1, \ldots, e_k \in \mathbf{T})$ and hence the entire measure.

Let $e = x\vec{y}$ be any edge of a finite connected weighted graph G. Run SRW_x^G until it hits y. At this point either the walk just moved along e from x to y — necessarily for the first time — and e will be in the tree $\mathbf{T}$ given by $\tau(SRW_x^G)$ or else the walk arrived at y via a different edge in which case the walk never crossed e at all and $e \notin \mathbf{T}$. In either case the walk never crossed from y to x since it stops if it hits y. Then the expected number of signed crossings of $e = x\vec{y}$ by SRW_x up to the first time it hits y is equal to the probability of first reaching y along e which equals $\mathbf{P}(e \in \mathbf{T})$. Putting this together with the electrical interpretation of signed crossings give

Theorem 3.5 $\mathbf{P}(e \in \mathbf{T}) =$ *the fraction of the current that goes through edge e when a battery is hooked up to the two endpoints of e.* $\qquad\square$

This characterization leads to a proof of Theorem 1.1 provided we are willing to accept a proposition that is physically obvious but not so easy to prove, namely

Theorem 3.6 (Rayleigh's monotonicity law) *The effective resistance of a circuit cannot increase when a new resistor is added.*

The reason this is physically obvious is that adding a new resistor provides a new path for current to take while allowing the current still to flow through all the old paths. Theorem 1.1 says that the conditional probability of $e \in \mathbf{T}$ given $f \in \mathbf{T}$ must be less than or equal to the unconditional probability. Using Theorem 3.5 and the fact that the probabilities conditioned on $f \notin \mathbf{T}$ are just the probabilities for WST on $G \setminus f$, this boils down to showing that the fraction of current flowing directly across e is no greater on G than it is on $G \setminus f$. The battery across e meets two parallel resistances: e and the effective resistance of the rest of G. The fraction of current flowing through e is inversely proportional to the ratio of these two resistances. Rayleigh's theorem says that the effective resistance of the rest of G including f is at most the effective resistance of $G \setminus f$, so the fraction flowing through e on G is at most the fraction flowing through e on $G \setminus f$. In Section 4, a proof will be given that does not rely on Rayleigh.

3.5   Algebraic calculations for the square lattice

If G is a finite graph, then the functions from the vertices of G to the reals form a finite-dimensional real vector space. The operator $\triangle$ that maps a function V to its excess is a linear operator on this vector space. In this language, the voltages in a resistor network with one unit of current supplied to a and drawn out at b are the unique (up to additive constant) function V that solves $\triangle V = \delta_a - \delta_b$. Here δ_x is the function that is one at x and zero elsewhere. This means that V can be calculated simply by inverting $\triangle$ in the basis $\{\delta_x; x \in G\}$. Although $\triangle$ is technically not invertible, its nullspace has dimension one so it can be inverted on a set of codimension one. A classical determination of V for arbitrary graphs is carried out in the next subsection. The point of this subsection is to show how the inverse can be obtained in a simpler way for nice graphs.

The most general "nice" graphs to which the method will apply are the infinite $\mathbf{Z}^d$-periodic lattices. Since in this article I am restricting attention to finite graphs, I will not attempt to be general but will instead show a single example. The reader may look in [6] for further generality. The example considered here is the square lattice. This is just the graph you see on a piece of graph paper, with vertices at each pair of integer coordinates and four edges connecting each point to its nearest neighbors. The exposition will be easiest if we consider a finite square piece of this and impose wrap-around boundary conditions. Formally, let T_n (T for torus) be the graph whose vertices are pairs of integers $\{(i,j) : 0 \le i, j \le n-1\}$ and for which two points are connected if and only if they agree in one component and differ by one mod n in the other component. Figure 7 shows this with $n = 3$ and the broken edges denoting edges that wrap around to the other side of the graph. The graph is unweighted (all edge weights are one.)

Let $\zeta = e^{2\pi i/n}$ denote the first n^{th} root of unity. To invert $\triangle$ we exhibit its eigenvectors. Since the vector space is a space of functions, the eigenvectors are called eigenfunctions. For each pair of integers $0 \le k, l \le n-1$ let f_{kl} be the function on the vertices of T_n defined by

$$f_{kl}(i,j) = \zeta^{ki+lj}.$$

If you have studied group representations, you will recognize f_{kl} as the representations of the group $T_n = (\mathbf{Z}/n\mathbf{Z})^2$ and in fact the rest of this section may be restated more compactly in terms of characters of this abelian group.

It is easy to calculate

$$
\begin{aligned}
\triangle f_{kl}(i,j) &= 4\zeta^{ki+lj} - \zeta^{ki+l(j+1)} - \zeta^{ki+l(j-1)} - \zeta^{k(i+1)+lj} - \zeta^{k(i-1)+lj} \\
&= \zeta^{ki+lj}(4 - \zeta^k - \zeta^{-k} - \zeta^l - \zeta^{-l}) \\
&= \zeta^{ki+lj}(4 - 2\cos(2\pi k/n) - 2\cos(2\pi l/n)).
\end{aligned}
$$

Since the multiplicative factor $(4 - 2\cos(2\pi k/n) - 2\cos(2\pi l/n))$ does not depend on i or j, this shows that f_{kl} is indeed an eigenfunction for $\triangle$ with eigenvalue $\lambda_{kl} = 4 - 2\cos(2\pi k/n) - 2\cos(2\pi l/n)$.

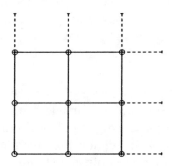

FIGURE 7.

Now if $\{v_k\}$ are eigenvectors for some linear operator A with eigenvalues $\{\lambda_k\}$, then for any constants $\{c_k\}$,

$$A^{-1}\left(\sum_k c_k v_k\right) = \sum_k \lambda_k^{-1} c_k v_k. \tag{6}$$

If some λ_k is equal to zero, then the range of A does not include vectors w with $c_k \neq 0$, so $A^{-1}w$ does not exist for such w and indeed the formula blows up due to the λ_k^{-1}. In our case $\lambda_{kl} = 4 - 2\cos(2\pi k/n) - 2\cos(2\pi l/n) = 0$ only when $k = l = 0$. Thus to calculate $\triangle^{-1}(\delta_a - \delta_b)$ we need to figure out coefficients c_{kl} for which $\delta_a - \delta_b = \sum_{kl} c_{kl} f_{kl}$ and verify that $c_{00} = 0$. For this puropose, it is fortunate that the eigenfunctions $\{f_{kl}\}$ are actually a unitary basis in the inner product $< f, g > = \sum_{ij} f(i,j)\overline{g(i,j)}$. You can check this by writing

$$< f_{kl}, f_{k'l'} > = \sum_{ij} \zeta^{ki+lj}\overline{\zeta^{k'i+l'j}}.$$

Elementary algebra show this to be one if $k = k'$ and $l = l'$ and zero otherwise, which is what it means to be unitary. Unitary bases are great for calculation because the coefficients $\{c_{kl}\}$ of any V in a unitary eigenbasis $\{f_{kl}\}$ are given by $c_{kl} = < V, f_{kl} >$. In our case, this means $c_{kl} = \sum_{ij} V(i,j)\overline{f_{kl}(i,j)}$. Letting a be the vertex $(0,0)$, b be the vertex $(1,0)$, and $V = \delta_a - \delta_b$, this gives $c_{kl} = 1 - \overline{\zeta}^k$ and hence

$$\delta_a - \delta_b = \sum_{k,l}(1 - \overline{\zeta}^k)f_{kl}.$$

We can now plug this into equation (6), since clearly $c_{00} = 0$. This gives

$$\triangle V \;=\; \delta_a - \delta_b$$

$$\Leftrightarrow \quad V(i,j) \;=\; cf_{00}(i,j) + \sum_{(k,l)\neq(0,0)} (1-\zeta^k)\lambda_{kl}^{-1} f_{kl}(i,j)$$

$$= \; c + \sum_{(k,l)\neq(0,0)} \frac{1-\zeta^k}{4 - 2\cos(2\pi k/n) - 2\cos(2\pi l/n)} \zeta^{ki+lj}. \quad (7)$$

This sum is easy to compute exactly and to approximate efficiently when n is large. In particular as $n \to \infty$ the sum may be replaced by an integral which by a small miracle admits an exact computation. Details of this may be found in [16, page 148]. You may check your arithmetic against mine by using (7) to derive the voltages for a one volt battery placed across the bottom left edge e of T_3 and across the bottom left edge e' of T_4:

5/8	3/8	1/2		56/90	34/90	40/90	50/90
5/8	3/8	1/2		50/90	40/90	42/90	48/90
1	0	1/2		56/90	34/90	40/90	50/90
				1	0	34/90	56/90

Section 5 shows how to put these numbers to good use, but we can already make one calculation based on Theorem 3.5. The four currents flowing out of the bottom left vertex under the voltages shown are given by the voltage differences: $1, 3/8, 1/2$, and $3/8$. The fraction of the current flowing directly through the bottom left edge e is $8/18$, and according to Theorem 3.5, this is $\mathbf{P}(e \in \mathbf{T})$. An easy way to see this is by the symmetry of the graph T_3. Each of the 18 edges should be equally likely to be in $\mathbf{T}$ and, since every spanning tree has 8 edges, the probability of any given edge being in the tree must be $8/18$.

3.6   Electrical networks and spanning trees

The order in which topics have been presented so far makes sense from an expository viewpoint but is historically backward. The first interest in enumerating spanning trees came from problems in electrical network theory. To set the record straight and also to close the circle of ideas:

> spanning trees → random walks → electrical networks → spanning trees.

I will spend a couple of paragraphs on this remaining connection.

Let G be a finite weighted graph. Assume there are no voltage sources and the quantity of interest is the *effective resistance* between two vertices a and b. This is defined to be the voltage it is necessary to place across a and b to induce a unit current flow. A classical theorem known to Kirchoff is:

Theorem 3.7 *Say s is an a, b-spanning bitree if s is a spanning forest with two components, one containing a and the other containing b. The effective resistance between a and b may be computed from the weighted graph G by taking the quotient N/D where*

$$D = \sum_{\text{spanning trees } t} \left(\prod_{e \in t} w(e) \right)$$

is the sum of the weights of all spanning trees of G and

$$N = \sum_{a, b\text{-spanning bitrees } s} \left(\prod_{e \in s} w(e) \right)$$

is the analogous sum over a, b-spanning bitrees.

To see how this is implied by Theorem 3.5 and equation (1), imagine adding an extra one ohm resistor from a to b. The probability of this edge being chosen in a WST on the new graph is by definition given by summing the weights of trees containing the new edge and dividing by the total sum of the weights of all spanning trees. Clearly D is the sum of the weights of trees not containing the extra edge. But the trees containing the extra edge are in one-to-one correspondence with a, b-spanning bitrees (the correspondence being to remove the extra edge). The extra edge has weight one, so the sum of the weights of trees that do contain the extra edge is N and the probability of a WST containing the extra edge is $N/(N + D)$. By equation (1) and Theorem 3.5, this must then be the fraction of current flowing directly through the extra edge when a battery is placed across a and b. Thinking of the new circuit as consisting of the extra edge in parallel with G, the fractions of the current passing through the two components are proportional to the inverses of their resistances, so the ratio of the resistance of the extra edge to the rest of the circuit must be $D : N$. Since the extra edge has resistance one, the effective resistance of the rest of the circuit is N/D.   □

The next problem of course is to efficiently evaluate the sum of the weights of all spanning trees of a weighted graph. The solution to this problem is almost as well known and can be found among other places in [7].

Theorem 3.8 *(matrix-tree theorem) Let G be a finite, simple, connected, weighted graph and define a matrix indexed by the vertices of G by letting $M(x, x) = d(x)$, $M(x, y) = -w(x\bar{y})$ if x and y are connected by an edge, and $M(x, y) = 0$ otherwise. Then for any vertex x the sum of the weights of all spanning trees of G is equal to the determinant of the matrix gotten from M by deleting by the row and column corresponding to x.*

The matrix M is nothing but a representation of $\triangle$ with respect to the basis $\{\delta_x\}$. Recalling that the problem essentially boils down to inverting $\triangle$, the only other ingredient in this theorem is the trick of inverting the action of a singular matrix on an element of its range by inverting the largest invertible principal minor of the matrix. Details can be found in [7]. □

4. TRANSFER-IMPEDANCES

In the last section we saw how to calculate $\mathbf{P}(e \in \mathbf{T})$ in several ways: by Theorems 3.5 or 3.7 in general and by equations such as (7) in particularly symmetric cases. By repeating the calculations in Theorem 3.5 and 3.7 for contractions and deletions of a graph (see Section 2.4), we could then find enough conditional probabilities to determine the probability of any elementary event $\mathbf{P}(e_1, \dots, e_r \in \mathbf{T}$ and $f_1, \dots, f_s \notin \mathbf{T})$. Not only is this inefficient, but it fails to apply to the symmetric case of equation (7) since contracting or deleting the graph breaks the symmetry. The task at hand is to alleviate this problem by showing how the data we already know how to get — current flows on G — determine the current flows on contractions and deletions of G and thereby determine all the elementary probabilities for WST on G. This will culminate in a proof of Theorem 1.2, which encapsulates all of the necessary computation into a single determinant.

4.1   An electrical argument

To keep notation to a minimum, this subsection will only deal with unweighted, D-regular graphs. Begin by stating explicitly the data that will be used to determine all other probabilities. For oriented edges $e = \vec{xy}$ and $f = \vec{zw}$ in a finite connected graph G, define the *transfer-impedance* $H(e, f) = \phi_{xy}(z) - \phi_{xy}(w)$ which is equal to the voltage difference across f, $V(z) - V(w)$, when one amp of current is supplied to x and drawn out at y. We will assume knowledge of $H(e, f)$ for every pair of edges in G [presumably via some analog calculation, or in a symmetric case by equation (7) or something similar] and show how to derive all other probabilities from these transfer-impedances.

Note first that $H(e, e)$ is the voltage across e for a unit current flow supplied to one end of e and drawn out of the other. This is equal to the current flowing directly along e under a unit current flow and is thus $\mathbf{P}(e \in \mathbf{T})$. The next step is to try a computation involving a single contraction. For notation, recall the map ρ which projects vertices and edges of G to vertices and edges of G/f. Fix edges $e = \vec{xy}$ and $f = \vec{zw}$ and let $\{V(v) : v \in G/f\}$ be the voltages we need to solve for — voltages at vertices of G/f when a unit current is supplied to $\rho(x)$ and drawn out at $\rho(y)$. As we have seen, this means $\triangle V(v) = +1, -1$, or 0 according to whether $v = x, y$, or neither. Suppose we lift this to a function $\overline{V}$ on the vertices of G by letting $\overline{V}(x) = V(\rho(x))$. Let's calculate the excess $\triangle\overline{V}$ of $\overline{V}$. Each edge of G corresponds to an edge in G/f, so for any $v \neq z, w$ in G, $\triangle\overline{V}(v) = \triangle V(\rho(v))$; this is equal to $+1$ if $v = x$, -1 if $v = y$, and zero otherwise. Since ρ maps both z and w onto the same vertex $v * w$, we can't tell what $\triangle \overline{V}$ is at z or w individually, but $\triangle \overline{V}(z) + \triangle \overline{V}(w)$ will equal $\triangle V(z*w)$ which will equal $+1$ if z or w coincides with x, -1 if z or w coincides with y, and zero otherwise (or if both coincide!). The last piece of information we have is that $\overline{V}(z) = \overline{V}(w)$. Summarizing,

> (i) $\triangle \overline{V} = \delta_x - \delta_y + c(\delta_z - \delta_w)$,

> (ii) $\overline{V}(z) = \overline{V}(w)$,

where c is some unknown constant. To see that this uniquely defines $\overline{V}$ up to an additive constant, note that the difference between any two such functions has excess $c(\delta_z - \delta_w)$ for some c, hence by the maximum principle reaches its maximum and minimum on $\{z, w\}$; on the other hand the values at z and w are equal, so the difference is constant.

Now it is easy to find $\overline{V}$. Recall from equation (5) that ϕ satisfies $\triangle \phi_{ab} = \delta_a - \delta_b$. The function $\overline{V}$ we are looking for is then $\phi_{xy} + c\phi_{zw}$ where c is chosen so that

$$\phi_{xy}(z) + c\phi_{zw}(z) = \phi_{xy}(w) + c\phi_{zw}(w).$$

In words, $\overline{V}$ gives the voltages for a battery supplying unit current in at x and out at y plus another battery across z and w just strong enough to equalize the voltages at z and w. How strong is that? The battery supplying unit current to x and y induces by definition a voltage $H(\vec{xy}, \vec{zw})$ across z and w. To counteract that, we need a $-H(\vec{xy}, \vec{zw})$ volt battery across z and w. Since supplying one unit of current in at z and out at w produces a voltage across z and w of $H(\vec{zw}, \vec{zw})$, the current supplied by the counterbattery must be $c = -H(\vec{xy}, \vec{zw})/H(\vec{zw}, \vec{zw})$. [We do not need to worry about $H(\vec{zw}, \vec{zw})$ being zero since this means that $\mathbf{P}(f \in \mathbf{T}) = 0$, so we shouldn't be conditioning on $f \in \mathbf{T}$.] Going back to the original problem,

$$
\begin{aligned}
\mathbf{P}(e \in \mathbf{T} \mid f \in \mathbf{T}) &= V(\rho(x)) - V(\rho(y)) \\[1em]
&= \overline{V}(x) - \overline{V}(y) \\[1em]
&= H(\vec{xy}, \vec{xy}) + H(\vec{zw}, \vec{xy}) \frac{-H(\vec{xy}, \vec{zw})}{H(\vec{zw}, \vec{zw})} \\[1em]
&= \frac{H(\vec{xy}, \vec{xy})H(\vec{zw}, \vec{zw}) - H(\vec{xy}, \vec{zw})H(\vec{zw}, \vec{xy})}{H(\vec{zw}, \vec{zw})}.
\end{aligned}
$$

Multiplying this conditional probability by the unconditional probability $\mathbf{P}(f \in \mathbf{T})$ gives the probability of both e and f being in $\mathbf{T}$ which may be written as

$$\mathbf{P}(e, f \in \mathbf{T}) = \begin{vmatrix} H(\vec{xy}, \vec{xy}) & H(\vec{xy}, \vec{zw}) \\ H(\vec{zw}, \vec{xy}) & H(\vec{zw}, \vec{zw}) \end{vmatrix}.$$

Thus $\mathbf{P}(e, f \in \mathbf{T}) = \det M(e, f)$ where M is the matrix of values of H as in Theorem 1.2.

Theorem 1.2 has in fact now been proved for $r = 1, 2$. The procedure for general r will be similar. Write $\mathbf{P}(e_1, \ldots e_r \in \mathbf{T})$ as a product of conditional probabilities $\mathbf{P}(e_i \in \mathbf{T} \mid e_{i+1}, \ldots, e_r \in \mathbf{T})$. Then evaluate this conditional probability by solving for voltages on $G/e_{i+1} \cdots e_r$. This is done by placing batteries across $e_1, \ldots, e_r$ so as to equalize voltages across all $e_{i+1}, \ldots, e_r$ simultaneously. Although in the $r = 2$ case it was not necessary to worry about dividing by zero, this problem does come up in the general case which causes an extra step in the proof. We will now summarily generalize the above discussion on how to solve for voltages on contractions of a graph and then forget about electricity altogether.

Lemma 4.1 *Let G be a finite D-regular connected graph and let $f_1, \ldots, f_r$ and $e = \vec{xy}$ be edges of G that form no cycle. Let ρ be the map from G to $G/f_1 \ldots f_r$ that maps edges to corresponding edges and maps vertices of G to their equivalence classes under the relation of being connected by edges in $\{f_1, \ldots, f_r\}$. Let $\overline{V}$ be a function on the vertices of G such that*

(i) *If $\vec{zw} = f_i$ for some i then $\overline{V}(z) = \overline{V}(w)$;*

(ii) *$\sum_{z \in \rho^{-1}(v)} \Delta \overline{V}(z) = +1$ if $\rho(x) = v$, -1 if $\rho(y) = v$, and zero otherwise.*

If $\mathbf{T}$ is a uniform spanning tree for G then $\mathbf{P}(e \in \mathbf{T} \mid f_1, \ldots, f_r \in \mathbf{T}) = \overline{V}(x) - \overline{V}(y)$.

Proof: As before, we know that $\mathbf{P}(e \in \mathbf{T} \mid f_1, \ldots, f_r \in \mathbf{T})$ is given by $V(\rho(x)) - V(\rho(y))$ where V is the voltage function on $G/f_1 \cdots f_r$ for a unit current supplied in at x and out at y. Defining $\overline{V}(v)$ to be $V(\rho(v))$, the lemma will be proved if we can show that $\overline{V}$ is the unique function on the vertices of G satisfying (i) and (ii). Seeing that $\overline{V}$ satisfies (i) and (ii) is the same as before. Since ρ provides a one to one correspondence between edges of G and edges of $G/f_1, \ldots, f_r$, the excess of $\overline{V}$ at vertices of $\rho^{-1}(v)$ is the sum, over edges leading out of vertices in $\rho^{-1}(v)$, of the difference of $\overline{V}$ across that edge, which is the sum over edges leading out of $\rho(v)$ of the difference of V across that edge. This is the excess of V at $\rho(v)$, which is $= 1, -1$, or 0 according to whether x or y or neither is in $\rho^{-1}(v)$.

Uniqueness is also easy. If $\overline{W}$ is any function satisfying (i), define a function W on the vertices of $G/f_1 \cdots f_r$ by $W(\rho(v)) = \overline{W}(v)$. If $\overline{W}$ satisfies (ii) as well then it is easy to check that W satisfies $\Delta W = \delta_{\rho(x)} - \delta_{\rho(y)}$ so that $W = V$ and $\overline{W} = \overline{V}$. $\qquad \square$

4.2   Proof of the transfer-impedance theorem

First of all, though is is true that the function H in the previous subsection and the statement of the theorem is symmetric, I'm not going to include a proof — nothing else we talk about relies on symmetry of H and a proof may be found in any standard treatment of the Green's function, such as [16]. Secondly, it is easiest to reduce the problem to the case of D-regular graphs immediately so as to be able to use the previous lemma. Suppose G is any finite connected graph. Let D be the maximum degree of any vertex in G and to any vertex of lesser degree k, add $D - k$ self-edges. The resulting graph is D-regular (though not simple) and furthermore it has the same spanning trees as G. To prove Theorem 1.2 for finite connected graphs, it therefore suffices to prove the theorem for finite, connected, D-regular graphs. Restating what is to be proved:

Theorem 4.2 *Let G be any finite, connected, D-regular graph and let $\mathbf{T}$ be a uniform random spanning tree of G. Let $H(\vec{xy}, \vec{zw})$ be the voltage induced across $\vec{zw}$ when one amp is supplied from x to y. Then for any $e_1, \ldots, e_r \in G$,*

$$\mathbf{P}(e_1, \ldots, e_r \in \mathbf{T}) = \det M(e_1, \ldots, e_r)$$

where $M(e_1, \ldots, e_r)$ is the r by r matrix whose i, j-entry is $H(e_i, e_j)$.

The proof is by induction on r. We have already proved it for $r = 1, 2$, so now we assume it for $r - 1$ and try to prove it for r. There are two cases. The first possibility is that $\mathbf{P}(e_1, \ldots, e_{r-1} \in \mathbf{T}) = 0$. This means that no spanning tree of G contains $e_1, \ldots, e_r$ which means that these edges contain some cycle. Say the cycle is $e_{n(0)}, \ldots, e_{n(k-1)}$ where there are vertices $v(i)$ for which $e_{n(i)}$ connects $v(i)$ to $v(i + 1 \bmod k)$. For any vertices x, y, ϕ_{xy} is the unique solution up to an additive constant of $\Delta \phi_{xy} = \delta_x - \delta_y$. Thus $\Delta \left(\sum_{i=0}^{k-1} \phi_{v(i) \, v(i+1 \bmod k)} \right) = 0$ which means that $\sum_{i=0}^{k-1} \phi_{v(i) \, v(i+1 \bmod k)}$ is constant. Then for any $\vec{xy}$,

$$\sum_{i=0}^{k-1} H(e_{n(i)}, \vec{xy}) = \sum_{i=0}^{k-1} \phi_{v(i) \, v(i+1 \bmod k)}(x) - \sum_{i=0}^{k-1} \phi_{v(i) \, v(i+1 \bmod k)}(y)$$

$$= 0.$$

This says that, in the matrix $M(e_1, \ldots, e_r)$, the rows $n(1), \ldots, n(k)$ are linearly dependent, summing to zero. Then $\det M(e_1, \ldots, e_r) = 0$ which is certainly the probability of $e_1, \ldots, e_r \in \mathbf{T}$.

The second possibility is that $\mathbf{P}(e_1, \ldots, e_{r-1} \in \mathbf{T}) \neq 0$. We can then write

$$\mathbf{P}(e_1, \ldots, e_r \in \mathbf{T}) = \mathbf{P}(e_1, \ldots e_{r-1} \in \mathbf{T}) \mathbf{P}(e_r \in \mathbf{T} \mid e_1, \ldots e_{r-1} \in \mathbf{T})$$

$$= \det M(e_1, \ldots, e_{r-1}) \mathbf{P}(e_r \in \mathbf{T} \mid e_1, \ldots e_{r-1} \in \mathbf{T})$$

by the induction hypothesis. To evaluate the last term we look for a function $\overline{V}$ satisfying the conditions of Lemma 4.1 with e_r instead of e and $e_1, \ldots, e_{r-1}$ instead of $f_1, \ldots, f_r$. For $i \leq r-1$, let x_i and y_i denote the vertices connected by e_i. For any $v \in G/e_1 \cdots e_{r-1}$ and any $i \leq r-1$, $\sum_{z \in \rho^{-1}(v)} \triangle \phi_{x_i y_i}(z) = \sum_{z \in \rho^{-1}(v)} \delta_{x_i}(z) - \delta_{y_i}(z)$ which is zero since the class $\rho^{-1}(v)$ contains both x_i and y_i or else contains neither. The excess of $\phi_{x_r y_r}$ summed over $\rho^{-1}(v)$ is just 1 if $\rho(x_r) = v$, -1 if $\rho(y_r) = v$, and zero otherwise. By linearity of excess, this implies that the sum of $\phi_{x_r y_r}$ with any linear combination of $\{\phi_{x_i y_i} : i \leq r-1\}$ satisfies (ii) of the lemma.

Satisfying part (i) is then a matter of choosing the right linear combination, but the lovely thing is that we don't have to actually compute it! We do need to know it exists and here's the argument for that. The ith row of $M(e_1, \ldots, e_r)$ lists the values of $\phi_{x_i y_i}(x_j) - \phi_{x_i y_i}(y_j)$ as j runs from 1 to r. Looking for $c_1, \ldots, c_{r-1}$ such that $\phi_{x_r y_r} + \sum_{i=1}^{r-1} \phi_{x_i y_i}$ is the same on x_j as on y_j for $j \leq r-1$ is the same as looking for c_i for which the rth row of M plus the sum of C_i times the ith row of M has zeros for every entry except the rth. In other words we want to row-reduce, using the first $r-1$ rows to clear $r-1$ zeros in the last row. There is a unique way to do this precisely when the determinant of the upper $r-1$ by $r-1$ submatrix is nonzero, which is what we have assumed. So these $c_1, \ldots, c_{r-1}$ exist and $\overline{V}(v) = \phi_{x_r y_r}(v) + \sum_{i=1}^{r-1} \phi_{x_i y_i}(v)$.

The lemma tells us that $\mathbf{P}(e_r \in \mathbf{T} \,|\, e_1, \ldots, e_{r-1} \in \mathbf{T})$ is $\overline{V}(x_r) - \overline{V}(y_r)$. This is just the r, r-entry of the row-reduced matrix. Now calculate the determinant of the row-reduced matrix in two ways. Firstly, since row-reduction does not change the determinant of a matrix, the determinant must still be $\det M(e_1, \ldots, e_r)$. On the other hand, since the last row is all zeros except the last entry, expanding along the last row gives that the determinant is the r, r-entry times the determinant of the upper $r-1$ by $r-1$ submatrix, which is just $\mathbf{P}(e_r \in \mathbf{T} \,|\, e_1, \ldots, e_{r-1} \in \mathbf{T}) \det M(e_1, \ldots, e_{r-1})$. Setting these two equal gives

$$\mathbf{P}(e_r \in \mathbf{T} \,|\, e_1, \ldots, e_{r-1} \in \mathbf{T}) = \det M(e_1, \ldots, e_r)/\det M(e_1, \ldots, e_{r-1}).$$

The induction hypothesis says that

$$\mathbf{P}(e_1, \ldots, e_{r-1} \in \mathbf{T}) = \det M(e_1, \ldots e_{r-1})$$

and multiplying the conditional and unconditional probabilities proves the theorem. $\square$

4.3 A few computational examples

It's time to take a break from theorem-proving to see how well the machinery we've built actually works. A good place to test it is the graph T_3, since the calculations have essentially been done, and since even T_3 is large enough to prohibit enumeration of the spanning trees directly by hand (you can use the

matrix-tree theorem with all weights on to check that there are 11664 of them).
Say we want to know the probability that the middle vertex A is connected to
B, C, and D in a uniform random spanning tree $\mathbf{T}$ of T_3.

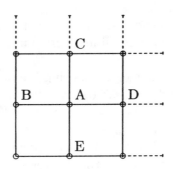

FIGURE 8.

We need then to calculate the transfer-impedance matrix for the edges
AB, AC and AD. Let's say we orient them all toward A. The symmetry
of T_3 under translation and 90° rotation allows us to rely completely on the
voltages calculated at the end of 3.5. Sliding the picture upward one square
and multiplying the given voltages by 4/9 to produce a unit current flow from
B to A gives voltages

$$
\begin{array}{ccc}
5/18 & 3/18 & 4/18 \\
8/18 & 0 & 4/18 \\
5/18 & 3/18 & 4/18
\end{array}
$$

which gives transfer-impedances $H(BA, BA) = 8/18$, $H(BA, CA) = 3/18$,
and $H(BA, DA) = 4/18$. The rest of the values follow by symmetry, giving

$$
M(BA, CA, DA) = \frac{1}{18} \begin{pmatrix} 8 & 3 & 4 \\ 3 & 8 & 3 \\ 4 & 3 & 8 \end{pmatrix}.
$$

Applying Theorem 4.2 gives $\mathbf{P}(BA, CA, DA \in \mathbf{T}) = \det M(BA, CA, DA) = \frac{312}{5832}$, or in other words just 624 of the 11664 spanning trees of T_3 contain all
these edges. Compare this to using the matrix-tree theorem to calculate the
same probability. That does not require the preliminary calculation of the
voltages, but it does require an eight by eight determinant.

Suppose we want now to calculate the probability that A is a *leaf* of $\mathbf{T}$, that
is to say, there is only one edge in $\mathbf{T}$ incident to A. By symmetry this edge will
be AB 1/4 of the time, so we need to calculate $\mathbf{P}(BA \in \mathbf{T} \text{ and } CA, DA, EA \notin$

T) and then multiply by four. As remarked earlier, we can use inclusion-exclusion to get the answer. This would entail writing

$$\mathbf{P}(BA \in \mathbf{T} \text{ and } CA, DA, EA \notin \mathbf{T})$$

$$= \mathbf{P}(BA \in \mathbf{T}) - \mathbf{P}(BA, CA \in \mathbf{T}) - \mathbf{P}(BA, DA \in \mathbf{T}) - \mathbf{P}(BA, EA \in \mathbf{T})$$
$$+ \mathbf{P}(BA, CA, DA \in \mathbf{T}) + \mathbf{P}(BA, CA, EA \in \mathbf{T}) + \mathbf{P}(BA, DA, EA \in \mathbf{T})$$
$$- \mathbf{P}(BA, CA, DA, EA \in \mathbf{T}).$$

This is barely manageable for four edges, and gets exponentially messier as we want to know about probabilities involving more edges. Here is an easy but useful theorem telling how to calculate the probability of a general *cylinder* event, namely the event that $e_1, \ldots, e_r$ are in the tree, while $f_1, \ldots, f_s$ are not in the tree.

Theorem 4.3 *Let $M(e_1, \ldots, e_k)$ be a k by k transfer-impedance matrix. Let $M^{(r)}$ be the matrix for which $M^{(r)}(i,j) = M(i,j)$ if $i \leq r$ and $M^{(r)}(i,j) = 1 - M(i,j)$ if $r + 1 \leq i \leq k$. Then $\mathbf{P}(e_1, \ldots, e_r \in \mathbf{T} \text{ and } e_{r+1}, \ldots, e_k \notin \mathbf{T}) = \det M^{(r)}$.*

Proof: The proof is by induction on $k - r$. The initial step is when $r = k$; then $M^{(r)} = M$ so the theorem reduces to Theorem 4.2. Now suppose the theorem to be true for $k - r = s$ and let $k - r = s + 1$. Write

$$\mathbf{P}(e_1, \ldots, e_r \in \mathbf{T} \text{ and } e_{r+1}, \ldots, e_k \notin \mathbf{T})$$

$$= \mathbf{P}(e_1, \ldots, e_r \in \mathbf{T} \text{ and } e_{r+2}, \ldots, e_k \notin \mathbf{T})$$
$$- \mathbf{P}(e_1, \ldots, e_{r+1} \in \mathbf{T} \text{ and } e_{r+2}, \ldots, e_k \notin \mathbf{T})$$

$$= \det M(e_1, \ldots, e_r, e_{r+2}, \ldots e_k) - \det M(e_1, \ldots, e_{r+1}, e_{r+2}, \ldots e_k),$$

since the induction hypothesis applies to both of the last two probabilities. Call these last two matrices M_1 and M_2. The trick now is to stick an extra row and column into M_1: let M' be $M(e_1, \ldots, e + k)$ with the $r + 1^{st}$ row replaced by zeros except for a one in the $r + 1^{st}$ position. Then M' is M_1 with an extra row and column inserted. Expanding along the extra row gives $\det M' = \det M_1$. But M' and M_2 differ only in the $r + 1^{st}$ row, so by multilinearity of the determinant,

$$\det M_1 - \det M_2 = \det M' - det M_2 = \det M''$$

where M'' agrees with M' and M_2 except that the $r + 1^{st}$ row is the difference of the $r + 1^{st}$ rows of M' and M_2. The induction is done as soon as you realize that M'' is just $M^{(r)}$.   □

Applying this to the probability of A being a leaf of T_3, we write

$$\mathbf{P}(BA \in \mathbf{T} \text{ and } CA, DA, EA \notin \mathbf{T})$$

$$= \det M^{(3)}(BA, CA, DA, EA)$$

$$= \begin{vmatrix} 8/18 & 3/18 & 4/18 & 3/18 \\ -3/18 & 10/18 & -3/18 & -4/18 \\ -4/18 & -3/18 & 10/18 & -3/18 \\ -3/18 & -4/18 & -3/18 & 10/18 \end{vmatrix} = \frac{10584}{18^4} = \frac{1176}{11664}$$

so A is a leaf of $4 \cdot 1176 = 4704$ of the 11664 spanning trees of T_3. This time, the matrix-tree theorem would have required evaluation of several different eight by eight determinants. If T_3 were replaced by T_n, the transfer-impedance calculation would not be significantly harder, but the matrix-tree theorem would require several n^2 by n^2 determinants. If n goes to ∞, as it might when calculating some sort of limit behavior, these large determinants would not be tractable.

5. POISSON LIMITS

As mentioned in the introduction, the random degree of a vertex in a uniform spanning tree of G converges in distribution to one plus a Poisson(1) random variable as G gets larger and more highly connected. This section investigates some such limits, beginning with an example symmetric enough to compute explicitly. The reason for this limit may seem clearer at the end of the section when we discuss a stronger limit theorem. Proofs in this section are mostly sketched since the details occupy many pages in [6].

5.1   The degree of a vertex in K_n

The simplest situation in which to look for a Poisson limit is on the complete graph K_n. This is pictured here for $n = 8$.

Calculating the voltages for a complete graph is particularly easy because of all the symmetry. Say the vertices of K_n are called $v_1, \ldots, v_n$, and put a one volt battery across v_1 and v_2, so $V(v_1) = 1$ and $V(v_2) = 0$. By Theorem 3.4, the voltage at any other vertex v_j is equal to the probability that $SRW_{v_j}^{K_n}$ hits v_1 before v_2. This is clearly equal to $1/2$. The total current flow out of v_1 with these voltages is $n/2$, since one amp flows along the edge to v_2 and $1/2$ amp flows along each of the $n - 2$ other edges out of v_1. Multiplying by $2/n$ to get a unit current flow gives voltages

$$V(v_i) = \begin{cases} 2/n & if & i = 1, \\ 0 & if & i = 2, \\ 1/n & & \text{otherwise.} \end{cases}$$

The calculations will of course come out similarly for a unit current flow supplied across any other edge of K_n.

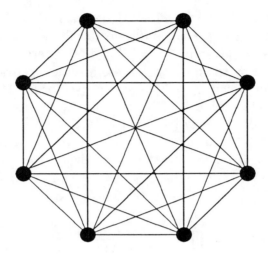

FIGURE 9.

The first distribution we are going to examine is of the degree in $\mathbf{T}$ of a vertex, say v_1. Since we are interested in which of the edges incident to v_1 are in $\mathbf{T}$, we need to calculate $H(\overline{v_1v_i}, \overline{v_1v_j})$ for every $i, j \neq 1$. Orienting all of these edges away from v_1 and using the voltages we just worked out gives

$$H(\overline{v_1v_i}, \overline{v_1v_j}) = \begin{cases} 2/n & if \quad i = j, \\ 1/n & \text{otherwise.} \end{cases}$$

Denoting the edge from v_1 to v_i by e_i, we have the $n-1$ by $n-1$ matrices

$$M(e_2, \ldots, e_n) = \begin{pmatrix} \frac{2}{n} & \frac{1}{n} & \cdots & \frac{1}{n} \\ \frac{1}{n} & \frac{2}{n} & \cdots & \frac{1}{n} \\ & & \vdots & \\ \frac{1}{n} & \frac{1}{n} & \cdots & \frac{2}{n} \end{pmatrix} \qquad M^{(n-1)}(e_2, \ldots, e_n) = \begin{pmatrix} \frac{n-2}{n} & \frac{-1}{n} & \cdots & \frac{-1}{n} \\ \frac{-1}{n} & \frac{n-2}{n} & \cdots & \frac{-1}{n} \\ & & \vdots & \\ \frac{-1}{n} & \frac{-1}{n} & \cdots & \frac{n-2}{n} \end{pmatrix}.$$

There must be at least one edge in $\mathbf{T}$ incident to v_1, so Theorem 4.3 says $\det M^{(n-1)} = \mathbf{P}(e_2, \ldots, e_n \notin \mathbf{T}) = 0$. This is easy to verify: the rows sum to zero. We can use $M^{(n-1)}$ to calculate the probability that e_2 is the only edge in $\mathbf{T}$ incident to v_1 by noting that this happens if and only if $e_3, \ldots, e_n \notin \mathbf{T}$.

This is the determinant of $M^{(n-2)}(e_3, \ldots, e_n)$ which is a matrix smaller by one than $M^{(n-1)}(e_2, \ldots, e_n)$, but which still has $(n-2)/n$'s down the diagonal and $-1/n$'s elsewhere. This is a special case of a *circulant* matrix, which is a type of matrix whose determinant is fairly easy to calculate.

A k by k circulant matrix is an M for which $M(i, j)$ is some number $a(i-j)$ depending only on $i - j \mod k$. Thus M has a_0's all down the diagonal for some a_0, a_1's on the next diagonal, and so forth. The eigenvalues of a circulant matrix $\lambda_0, \ldots, \lambda_{k-1}$ are given by $\lambda_j = \sum_{t=0}^{k-1} a_t \zeta^{jt}$ where $\zeta = e^{2\pi i/n}$ is the nth root of unity. It is easy to verify that these are the eigenvalues, by checking that the vector $\vec{w}$ for which $w_t = \zeta^{tj}$ is an eigenvector for M (no matter what the a_i are) and has eigenvalue λ_j. The determinant is then the product of the eigenvalues. Details of this may be found in [17].

In the case of $M^{(n-2)}$, $a_0 = (n-2)/n$ and $a_j = -1/n$ for $j \neq 0$. Then $\lambda_0 = \sum_j a_j = 1/n$. To calculate the other eigenvalues, note that, for any $j \neq 0 \mod n - 2$, $\sum_{t=1}^{n-3} \zeta^{jt} = 0$. Then $\lambda_j = (n-2)/n \sum_{t=1}^{n-3}(-1/n)\zeta^{jt} = (n-1)/n - (1/n) \sum_{t=0}^{n-3} \zeta^{tj} = (n-1)/n$. This gives

$$\det M^{(n-2)} = \prod_{j=0}^{n-3} \lambda_j = \frac{1}{n}\left(\frac{n-1}{n}\right)^{n-3} = \frac{1+o(1)}{ne}$$

as $n \to \infty$.[†] Part of the Poisson limit has emerged: the probability that v_1 has degree one in **T** is (by symmetry) $n-1$ times the probability that the particular edge e_2 is the only edge in **T** incident to v_1; this is $(n-1)(1+o(1))/en$, so it converges to e^{-1} as $n \to \infty$. This is $\mathbf{P}(X = 1)$ where X is one plus a Poisson(1), i.e., a Poisson of mean one.

Each further part of the Poisson limit requires a more careful evaluation of the limit. To illustrate, we carry out the second step. Use one more degree of precision in the Taylor series for $\ln(x)$ and $\exp(x)$ to get

$$n^{-1}\left(\frac{n-1}{n}\right)^{n-3} = n^{-1}\exp[(n-3)(-n^{-1} - n^{-2}(1/2 + o(1)))]$$

$$= n^{-1}\exp[-1 + (5/2 + o(1))n^{-1}]$$

$$= n^{-1}e^{-1}[1 + (5/2 + o(1))n^{-1}].$$

The reason we need this precision is that we are going to calculate the probability of v_1 having degree 2 by summing the $\mathbf{P}(e, f$ are the only edges incident to v_1 in **T**) over all pairs of edges e, f coming out of v_1. By symmetry this is just $(n-1)(n-2)/2$ times the probability that the particular edges e_2 and e_3 are the only edges in **T** incident to v_1. This probability is the determinant of a matrix that is not a circulant and to avoid calculating a difficult determinant it

---

[†]Here, $o(1)$ signifies a quantity going to zero as $n \to \infty$. This is a convenient and standard notation that allows manipulation such as $(2 + o(1))(3 + o(1)) = 6 + o(1)$.

is better to write this probability as the following difference — the probability that no edges other that e_2 and e_3 are incident to v_1 minus the probability that e_2 is the only edge incident to v_1 minus the probability that e_2 is the only edge incident to v_3. Since the final probability is this difference multiplied by $(n-1)(n-2)/2$, the difference should be of order n^{-2}, which explains why this degree of precision is required for the latter two probabilities.

The probability of $\mathbf{T}$ containing no edges incident to v_1 other than e_2 and e_3 is the determinant of $M^{(n-3)}(e_4, \ldots, e_n)$, which is an $n-3$ by $n-3$ circulant again having $(n-2)/n$ on the diagonal and $-1/n$ elsewhere. Then $\lambda_0 = \sum_{j=0}^{n-4} a_j = 2/n$ and $\lambda_j = (n-1)/n$ for $j \neq 0$ mod $n-3$, yielding

$$\det M^{(n-3)} = 2n^{-1}\left(\frac{n-1}{n}\right)^{n-4} = 2n^{-1}e^{-1}[1 + (7/2 + o(1))n^{-1}]$$

in the same manner as before. Subtracting off the probabilities of e_2 or e_3 being the only edge in $\mathbf{T}$ incident to v_1 gives

$$\mathbf{P}(e_2, e_3 \in \mathbf{T}, e_4, \ldots, e_n \notin \mathbf{T})$$

$$= 2n^{-1}e^{-1}[1 + (7/2 + o(1))n^{-1}] - 2n^{-1}e^{-1}[1 + (5/2 + o(1))n^{-1}]$$

$$= (2 + o(1))n^{-2}e^{-1}.$$

Multiplying by $(n-1)(n-2)/2$ gives

$$\mathbf{P}(v_1 \text{ has degree 2 in } \mathbf{T}) \to e^{-1}$$

as $n \to \infty$, which is $\mathbf{P}(X = 2)$ where X is one plus a Poisson(1).

5.2 Another point of view

The calculations of the last section may be continued *ad infinitum*, but each step requires a more careful estimate, so it pays to look for a way to do all the steps at once. The right alternative method will be more readily apparent if we generalize to graphs other than K_n which do not admit such a precise calculation (if a tool that is difficult to use breaks, you may discover a better one).

The important feature about K_n was that the voltages were easy to calculate. There is a large class of graphs for which the voltages are just as easy to calculate approximately. The term "approximately" can be made more rigorous by considering sequences of graphs G_n and stating approximations in terms of limits as $n \to \infty$. Since I've always wanted to name a technical term after my dog, call a sequence of graphs G_n *Gino-regular* if there is a sequence D_n such that

(i) The maximum and minimum degree of a vertex in G_n are $(1 + o(1))D_n$ as $n \to \infty$ and

(ii) The maximum and minimum over vertices $x \neq y, z$ of G_n of the probability that $SRW_x^{G_n}$ hits y before z are $1/2 + o(1)$ as $n \to \infty$.

Condition (ii) implies that $D_n \to \infty$, so the graphs G_n are growing locally. It is not hard to see that the voltage $V(z)$ in a unit current flow across any edge $e = \vec{xy}$ of a graph G_n in a Gino-regular sequence is $(1 + o(1))D_n^{-1}(\delta_x - \delta_y)(z)$ uniformly over all choices of $x, y, z \in G_n$ as $n \to \infty$. The complete graphs K_n are Gino-regular. So are the n-cubes, B_n, whose vertex sets are all the n-long sequences of zeros and ones and whose edges connect sequences differing in only one place.

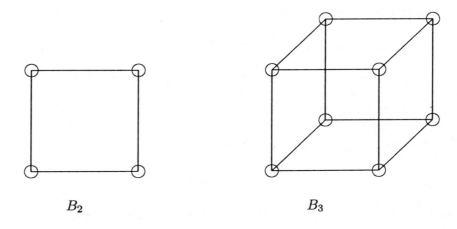

$$B_2 \qquad\qquad\qquad B_3$$

FIGURE 10.

To see why $\{B_n\}$ is Gino-regular, consider the "worst case" when x is a neighbor of y. There is a small probability that $SRW_x(1)$ will equal y, small because this is degree$(x)^{-1} = (1 + o(1))D_n^{-1}$ which is going to zero. There are even smaller probabilities of reaching y in the next few steps; in general, unless SRW_x hits y in one step, it tends to get "lost" and by the time it comes near y or z again it is thoroughly random and is equally likely to hit y or z first. In fact Gino-regular sequences may be thought of as graphs that are nearly degree-regular, where SRW gets lost quickly.

The approximate voltages give approximate transfer-impedances $H(e, f) = (2 + o(1))/n$ if $e = f$, $(1 + o(1))/n$ if e and f meet at a single vertex (choose orientations away from the vertex), and $o(1)/n$ if e and f do not meet. The determinant of a matrix is continuous in its entries, so it may seem that we have everything necessary to calculate limiting probabilities as limits of determinants of transfer-impedance matrices. If v is a vertex in G_k and $e_1, \ldots, e_n$

are the edges incident to v in G_k (so $n \approx D_k$), then the probability of e_2 being the only edge in $\mathbf{T}$ incident to v is the determinant of

$$M^{(n-1)}(e_2, \ldots, e_n) =$$
$$\begin{pmatrix} (n-2+o(1))/n & (-1+o(1))/n & \cdots & (-1+o(1))/n \\ (-1+o(1))/n & (n-2+o(1))/n & \cdots & (-1+o(1))/n \\ & & \vdots & \\ (-1+o(1))/n & (-1+o(1))/n & \cdots & (n-2+o(1))/n \end{pmatrix}.$$

Unfortunately, the matrix is changing size as $n \to \infty$, so convergence of each entry to a known limit does not give us the limit of the determinant.

If the matrix were staying the same size, the problem would disappear. This means we can successfully take the limit of probabilities of events as long as they involve a bounded number of edges. Thus, for any fixed edge e_1, $\mathbf{P}(e_1 \in \mathbf{T}) = \det M(e_1) = (1 + o(1))(2/n)$. For any fixed pair of edges e_1 and e_2 incident to the same vertex,

$$\begin{aligned} \mathbf{P}(e_1, e_2 \in \mathbf{T}) &= \det M(e_1, e_2) \\ &= \begin{vmatrix} (2+o(1))/n & (1+o(1))/n \\ (1+o(1))/n & (2+o(1))/n \end{vmatrix} \\ &= (3+o(1))n^{-2}. \end{aligned}$$

In general if $e_1, \ldots, e_r$ are all incident to v then the transfer-impedance matrix is n^{-1} times an r by r matrix converging to the matrix with 2s down the diagonal and 1s elsewhere. The eigenvalues of this circulant are $\lambda_0 = r + 1$ and $\lambda_j = 1$ for $j \neq 0$, yielding

$$\mathbf{P}(e_1, \ldots, e_r \in \mathbf{T}) = (r + 1 + o(1))n^{-r}.$$

What can we do with these probabilities? Inclusion-exclusion fails for the same reason as the large determinants fail — the $o(1)$ errors pile up. On the other hand, these probabilities determine certain expectations. Write $e_1, \ldots, e_n$ again for the edges adjacent to v and I_i for the indicator function which is one when $e_i \in \mathbf{T}$ and zero otherwise; then

$$\sum_i \mathbf{P}(e_i \in \mathbf{T}) = \sum_i \mathbf{E} I_i = \mathbf{E} \sum_i I_i = \mathbf{E} \deg(v).$$

This tells us that $\mathbf{E} \deg(v) = n(2+o(1))n^{-1} = 2+o(1)$. If try this with ordered pairs of edges, we get

$$\sum_{i \neq j} \mathbf{P}(e_i, e_j \in \mathbf{T}) = \sum_{i \neq j} \mathbf{E} I_i I_j = \mathbf{E} \sum_{i \neq j} I_i I_j.$$

This last quantity is the sum over all distinct ordered pairs of edges incident to v of the quantity: 1 if they are both in the tree and 0 otherwise. If $\deg(v) = r$ then a one occurs in this sum $r(r-1)$ times, so the sum is $\deg(v)(\deg(v) - 1)$. The determinant calculation gave $\mathbf{P}(e_i, e_j \in \mathbf{T}) = (3 + o(1))n^{-2}$ for each i, j, so

$$\mathbf{E}(\deg(v)(\deg(v) - 1)) = n(n-1)(3 + o(1))n^{-2} = 3 + o(1).$$

In general, using ordered r-tuples of distinct edges gives

$$\begin{aligned} &\mathbf{E}(\deg(v)(\deg(v) - 1) \cdots (\deg(v) - r + 1)) \\ = \quad &n(n-1) \cdots (n - r + 1)(r + 1 + o(1))n^{-r} \\ = \quad &r + 1 + o(1). \end{aligned}$$

Use the notation $(A)_r$ to denote $A(A - 1) \cdots (A - r + 1)$ which is called the rth lower factorial of A. If Y_n is the random variable $\deg(v)$ then we have succinctly

$$\mathbf{E}(Y_n)_r = r + 1 + o(1). \tag{8}$$

$\mathbf{E}(Y_n)_r$ is called the rth factorial moment of Y_n.

If you remember why we are doing these calculations, you have probably guessed that $\mathbf{E}(X)_r = r + 1$ when X is one plus a Poisson(1). This is indeed true and can be seen easily enough from the logarithmic moment generating function $\mathbf{E}t^X$ via the identity

$$\mathbf{E}(X)_r = \left(\frac{d}{dt}\right)^r \bigg|_{t=1} \mathbf{E}t^X,$$

using $\mathbf{E}t^X = \mathbf{E}e^{X \ln(t)} = \phi(\ln(t)) = te^{t-1}$; consult [14, page 301] for details. All that we need now for a Poisson limit result is a theorem saying that, if the factorial moments of Y_n are each converging to the factorial moments of X, then Y_n is actually converging in distribution to X. This is worth spending a short subsection on because it is algebraically very neat.

5.3   The method of moments

A standard piece of real analysis shows that, if all the factorial moments of a sequence of random variables converging to a limit are finite, then for each r the limit of the rth factorial moments is the rth factorial moment of the limit. (This is essentially the Lebesgue-dominated convergence theorem.)   Another standard result is that, if the moments of a sequence of random variables converge, then the sequence, or at least some subsequence, is converging in distribution to some other random variable whose moments are the limits of the moments in the sequence.   Piecing together these straight-forward facts leaves a serious gap in our prospective proof: What if there is some random variable Z distributed differently from X with the same factorial moments? If this could happen, then there would be no reason to think that Y_n converged in distribution to X rather than Z. This scenario can actually happen — there really are differently distributed random variables with the same moments! (See the discussion of the lognormal distribution in [9].) Luckily this only happens when X is badly behaved, and a Poisson plus one is not badly behaved. Here then is a proof of the fact that the distribution of X is the only one with r^{th} factorial moment $r + 1$ for all r. I will leave it to you to piece together, look up in [9], or take on faith how this fact plus the results from real analysis imply $Y \xrightarrow{D} X$.

Theorem 5.1 *Let X be a random variable with $\mathbf{E}(X)_r \leq e^{kr}$ for some k. Then no random variable distributed differently from X has the same factorial moments.*

Proof: The factorial moments $\mathbf{E}(X)_r$ determine the regular moments $\mu_r = \mathbf{E}X^r$ and *vice versa* by the linear relations $(X)_1 = X^1; (X)_2 = X^2 - X^1$, etc. From these linear relations it also follows that factorial moments are bounded by some e^{kr} if and only if regular moments are bounded by some e^{kr}, thus it suffices to prove the theorem for regular moments. Not only do the moments determine the distribution, it is even possible to calculate $\mathbf{P}(X = j)$ directly from the moments of X in the following manner.

The *characteristic function* of X is the function $\phi(t) = \mathbf{E}e^{itX}$ where $i = \sqrt{-1}$. This is determined by the moments since $\mathbf{E}e^{itX} = \mathbf{E}(1 + (itX) + (itX)^2/2! + \cdots) = 1 + it\mu_1 + (it)^2\mu_2/2! + \cdots$. We use the exponential bound on the growth of μ_r to deduce that this is absolutely convergent for all t (though a somewhat weaker condition would do). The growth condition also shows that $\mathbf{E}e^{itX}$ is bounded and absolutely convergent for $y \in [0, 2\pi]$. Now $\mathbf{P}(X = j)$ can be determined by Fourier inversion:

$$\frac{1}{2\pi} \int_0^{2\pi} \mathbf{E}e^{itX} e^{-ijt} dt \quad = \quad \frac{1}{2\pi} \int_0^{2\pi} [\sum_{r \geq 0} e^{itr} \mathbf{P}(X = r)] e^{-ijt} dt$$

$$= \quad \frac{1}{2\pi} \sum_{r \geq 0} \mathbf{P}(X = r) \int_0^{2\pi} e^{itr} e^{-ijt} dt$$

(switching the sum and integral is OK for bounded, absolutely convergent integrals)

$$= \quad \frac{1}{2\pi} \sum_{r \geq 0} \mathbf{P}(X = r) \delta_0(r - j)$$

$$= \quad \mathbf{P}(X = j).$$

$\square$

5.4   A branching process

I promised in the last half of section 1.4 to explain how convergence in distribution of $\deg(v)$ was a special case of convergence of **T** near v to a distribution called $\mathcal{P}_1$. (You might want to go back and reread that section before continuing.) The infinite tree $\mathcal{P}_1$ is interesting in its own right and I'll start making good on the promise by describing $\mathcal{P}_1$.

This begins with a short description of *Galton-Watson branching processes*. You can think of a Galton-Watson process as a family tree for some fictional amoebas. These fictional amoebas reproduce by splitting into any number of smaller amoebas (unlike real amoebas that can only split into two parts at a time). At time $t = 0$ there is just a single amoeba, and at each time $t = 1, 2, 3, \ldots$, each living amoeba $\mathcal{A}$ splits into a random number $N = N_t(\mathcal{A})$ of amoebas, where the random numbers are independent and all have the same distribution $\mathbf{P}(N_t(\mathcal{A}) = j) = p_j$. Allow the possibility that $N = 0$ (the amoeba died) or that $N = 1$ (the amoeba didn't do anything). Let $\mu = \sum_j j p_j$ be the mean number of amoebas produced in a split. A standard result from the theory of branching processes [4] is that if $\mu > 1$ then there is a positive probability that the family tree will survive forever, the population exploding exponentially as in the usual Malthusian forecasts for human population in the twenty-first century. Conversely when $\mu < 1$ the amoeba population dies out with probability 1 and in fact the chance of it surviving n generations decreases exponentially with n. When $\mu = 1$ the branching process is said to be *critical*. It must still die out, but the probability of it surviving n generations decays more slowly, like a constant times $1/n$. The theory of branching processes is quite large and you can find more details in [4] or [10].

Specialize now to the case where the random number of offspring has a Poisson(1) distribution, i.e., $p_j = e^{-1}/j!$. Here's the motivation for considering this case. Imagine a graph G in which each vertex has N neighbors and N is so large it is virtually infinite. Choose a subgraph U by letting each edge be included independently with probability N^{-1}. Fix a vertex $v \in G$ and look at the vertices connected to v in U. The number of neighbors of v in U has a Poisson(1) distribution by the standard characterization of a Poisson as the limit of number of occurrences of rare events. For each neighbor y of v in U, there are $N - 1$ edges out of y other than the one to v, and the number of those in U will again be Poisson(1) (since $N \approx \infty$, subtracting one does not matter) and continuing this way shows that the connected component of v in U is distributed as a Galton-Watson process with Poisson(1) offspring.

Of course U is not distributed like a uniform spanning tree **T**. For one thing, U may with probability e^{-1} fail to have any edges out of v. Even if this doesn't happen, the chance of U having more than n vertices goes to zero as $n \to \infty$ (a critical Galton-Watson process dies out) whereas **T**, being a spanning tree of an almost infinite graph, goes on as far as the eye can see. The next hope is that **T** looks like U conditioned not to die out. This should in fact seem plausible: you can check that U has no cycles near v since virtually all of the N edges out of each neighbor of v lead further away from v; then a uniform spanning tree should be a random cycle-free graph U that treats each edge as equally likely, conditioned on being connected.

The conditioning must be done carefully, since the probability of U living forever is zero, but it turns out fine if you condition on U living for at least n generations and take the limit as $n \rightarrow \infty$. The random infinite tree $\mathcal{P}_1$ that results is called the *incipient infinite cluster* at v, so named by percolation theorists (people who study connectivity properties of random graphs). It turns out there is an alternate description for the incipient infinite cluster. Let $v = v_0, v_1, v_2, \ldots$ be a single line of vertices with edges $\overline{vv_1}, \overline{v_1v_2}, \ldots$. For each of the vertices v_i independently, make a separate independent copy U_i of the critical Poisson(1) branching process U with v_i as the root and paste it onto the line already there. Then this collage has the same distribution as $\mathcal{P}_1$. This fact is the "whole tree" version of the fact that a Poisson(1) conditioned to be nonzero is distributed as one plus a Poisson(1) (you can recover this fact from the fact about $\mathcal{P}_1$ by looking just at the neighbors of v).

5.5   Tree moments

To prove that a uniform spanning tree $\mathbf{T}_n$ of G_n converges in distribution to $\mathcal{P}_1$ when G_n is Gino-regular, we generalize factorial moments to trees. Let t be a finite tree rooted at some vertex x and let W be a tree rooted at v. W is allowed to be infinite but it must be locally finite — only finitely many edges incident to any vertex. Say that a map f from the vertices of t to the vertices of W is a *tree-map* if f is one to one, maps x to v, and neighbors to neighbors. Let $N(W; t)$ count the number of tree-maps from t into W. For example, in Figure 11, $N(W; t) = 4$, since C and D can map to H and I in either order with A mapping to E, and B can map to F or G.

Define the tth *tree-moment* of a random tree Z rooted at v to be $\mathbf{E}N(Z; t)$. If t is an n-star, meaning a tree consisting of n edges all emanating from x, then a tree-map from t to W is just a choice of n distinct neighbors of v in order, so $N(W; t) = (\deg(v))_n$. Thus $\mathbf{E}N(Z; t) = \mathbf{E}(\deg(v))_n$, the nth factorial moment of $\deg(v)$. This is to show you that tree-moments generalize factorial moments. Now let's see what the tree-moments of $\mathcal{P}_1$ are. Let t be any finite tree and let $|t|$ denote the number of vertices in t.

Lemma 5.2 *Let U be a Galton-Watson process rooted at v with Poisson(1) offspring. Then $\mathbf{E}N(U; t) = 1$ for all finite trees t.*

Proof: Use induction on t, the lemma being clear when t is a single vertex. The way the induction step works for trees is to show that, if a fact is true for a collection of trees $t_1, \ldots, t_n$, then it is true for the tree t_* consisting of a root x with n neighbors $x_1, \ldots, x_n$ having subtrees $t_1, \ldots t_n$, respectively, as in Figure 12.

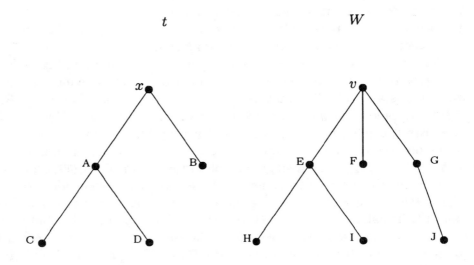

FIGURE 11.

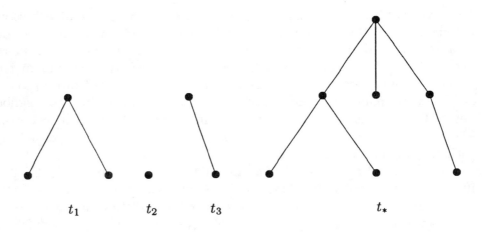

FIGURE 12.

So let $t_1, \ldots, t_n$ and t_* be as above. Any tree-map $f : t_* \to U$ must map the n neighbors of v into distinct neighbors of U and the expected number of ways to do this is $\mathbf{E}(\deg(v))_n$ which is one for all n since $\deg(v)$ is a Poisson(1) [9]. Now for any such assignment of f on the neighbors of v, the number of ways of completing the assignment to a tree-map is the product over $i = 1, \ldots, n$ of the number of ways of mapping each t_i into the subtree of U below $f(x_i)$. After conditioning on what the first generation of U looks like, the subtrees below any neighbors of v are independent and themselves Galton-Watsons with Poisson(1) offspring. (This is what it means to be Galton-Watson.) By induction then, the expected number of ways of completing the assignment of f is the product of a bunch of ones and is therefore one. Thus $EN(U;t) = E(\deg(v))_n \prod_{i=1}^n 1 = 1$. □

Back to calculating $EN(\mathcal{P}_1;t)$. Recall that $\mathcal{P}_1$ is a line $v_0, v_1, \ldots$ with Poisson(1) branching processes U_i stapled on. Each tree-map $f : t \to \mathcal{P}_1$ hits some initial segment $v_0, \ldots v_k$ of the original line, so there is some vertex $y_f \in t$ such that $f(y_f) = v_k$ for some k but v_{k+1} is not in the image of f. For each $y \in t$, we count the expected number of tree-maps f for which $y_f = y$. There is a path $x = f^{-1}(v_0), \ldots, f^{-1}(v_k) = y$ in t going from the root x to y. The remaining vertices of t can be separated into $k+1$ subtrees below each of the $f^{-1}(v_i)$. These subtrees must then get mapped respectively into the U_i. By the lemma, the expected number of ways of mapping anything into a U_i is one, so the expected number of f for which $y_f = y$ is $\prod_{i=1}^k 1 = 1$. Summing over y then gives

$$EN(\mathcal{P}_1;t) = |t|. \tag{9}$$

The last thing we are going to do to in proving the stronger Poisson convergence theorem is to show

Lemma 5.3 *Let G_n be a Gino-regular sequence of graphs, and let T_n be a uniform spanning tree of G_n rooted at some v_n. Then for any finite rooted tree t, $EN(T_n;t) \to |t|$ as $n \to \infty$.*

It is not trivial from here to establish that $T \wedge r$ converges in distribution to $\mathcal{P}_1 \wedge r$ for every r. The standard real analysis facts I quoted in section 5.3 about moments need to be replace by some not-so-standard (but not too hard) facts about tree-moments. Suffice it to say that the previous two lemmas do in the end prove (see [6] for details)

Theorem 5.4 *Let G_n be a Gino-regular sequence of graphs, and let T_n be a uniform spanning tree of G_n rooted at some v_n. Then for any r, $T_n \wedge r$ converges in distribution to $\mathcal{P}_1 \wedge r$ as $n \to \infty$.*

Sketch of proof of Lemma 5.3 Fix a finite t rooted at x. To calculate the expected number of tree-maps from t into T_n we will sum over every possible image of a tree-map the probability that all of those edges are actually present in T_n. By an image of a tree-map, I mean two things: (1) a collection $\{v_x : x \in t\}$ of vertices of G_n indexed by the vertices of t for which $v_x \sim v_y$ in G whenever $x \sim y$ in t; (2) a collection of edges e_ϵ connecting v_x and v_y for every edge $\epsilon \in t$ connecting some x and y. Fix such an image.

The transfer-impedance theorem tells us that the probability of finding all the edges v_e in T is the determinant of $M(e_\epsilon : \epsilon \in t)$. Now for edges $e, e' \in G$, Gino-regularity gives that $H(e, e') = D_n^{-1}(o(1) + \kappa)$ uniformly over edges of G_n, where κ is 2, 1, or 0 according to whether $e = e'$, they share an endpoint, or they are disjoint. The determinant is then well approximated by the corresponding determinant without the $o(1)$ terms, which can be worked out as exactly $|t| D_n^{1-|t|}$.

This must now be summed over all possible images, which amounts to multiplying $|t| D_n^{1-|t|}$ by the number of possible images. I claim the number of possible images is approximately $D_n^{|t|-1}$. To see this, imagine starting at the root x, which must get mapped to v_n, and choosing successively where to map each next vertex of t. Since there are approximately D_n edges coming out of each vertex of G_n, there are always about D_n choices for the image of the next vertex (the fact that you are not allowed to choose any vertex already chosen is insignificant as D_n gets large). There are $|t| - 1$ choices, so the number of maps is about $D_n^{|t|-1}$. This proves the claim. The claim implies that the expected number of tree-maps from t to $\mathbf{T}_n$ is $|t| D_n^{1-|t|} D_n^{|t|-1} = |t|$, proving the lemma. □

6. INFINITE LATTICES, DIMERS, AND ENTROPY

There is, believe it or not, another model that ends up being equivalent to the uniform spanning tree model under a correspondence at least as surprising as the correspondence between spanning trees and random walks. This is the so-called *dimer* or *domino tiling* model, which was studied by statistical physicists quite independently of the uniform spanning tree model. The present section is intended to show how one of the fundamental questions of this model, namely calculating its entropy, can be solved using what we know about spanning trees. Since it's getting late, there will be pictures but no detailed proofs.

6.1 Dimers

A dimer is a substance that on the molecular level is made up of two smaller groups of atoms (imagine two spheres of matter) adhering to each other via a covalent bond; consequently it is shaped like a dumbbell. If a bunch of dimer molecules are packed together in a cold room and a few of the less significant laws of physics are ignored, the molecules should array themselves into some sort of regular lattice, fitting together as snugly as dumbbells can. To model this, let r be some positive real number representing the length of one of the dumbbells. Let L be a lattice, i.e., a regular array of points in three-space, for which each point in L has some neighbors at distance r. For example r could be 1 and L could be the standard integer lattice $\{(x, y, z) : x, y, z \in \mathbb{Z}\}$, so r is the minimum distance between any two points of L (see the picture below). Alternatively r could be $\sqrt{2}$ or $\sqrt{3}$ for the same L. Make a graph G whose vertices are the points of L, with an edge between any pair of points at distance r from each other. Then the possible packings of dimers in the lattice are just the ways of partitioning the lattice into pairs of vertices, each pair (representing one molecule) being the two endpoints of some edge. The following picture shows part of a packing of the integer lattice with nearest-neighbor edges:

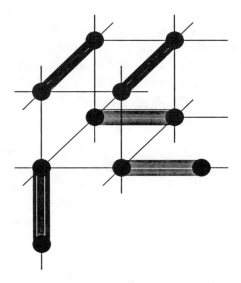

FIGURE 13.

Take a large finite box inside the lattice, containing N vertices. If N is even and the box is not an awkward shape, there will be not only one but many ways to pack it with dimers. There will be several edges incident to each vertex v, representing a choice to be made as to which other vertex will be covered by the molecule with one atom covering v. These choices obviously cannot be made independently, but it should be plausible from this that the total number of configurations is approximately γ^N for some $\gamma > 1$ as N goes to infinity. This number can be written alternatively as e^{hN} where $h = \ln(\gamma)$ is called the entropy of the packing problem. The thermodynamics of the resulting substance depend on, among other things, the entropy h.

The case that has been studied the most is where L is the two-dimensional integer lattice with $r = 1$. The graph G is then the usual nearest-neighbor square lattice. Physically this corresponds to packing the dimers between two slides. You can get the same packing problem by attempting to tile the plane with dominos — vertical and horizontal 1 by 2 rectangles — which is why the model also goes by the name of domino tiling.

6.2   Dominos and spanning trees

We have not yet talked about spanning trees of an infinite graph, but the definition remains the same: a connected subgraph touching each vertex and contaning no cycles. If the subgraph need not be connected, it is a spanning forest. Define an *essential spanning forest* or ESF to be a spanning forest that has no finite components. Informally, an ESF is a subgraph that you can't distinguish from a spanning tree by only looking at a finite part of it (since it has no cycles or *islands*).

Let G_2 denote the nearest-neighbor graph on the two dimensional integer lattice. Since G_2 is a planar graph, it has a *dual* graph G_2^*, which has a vertex in each cell of G_2 and an edge e^* crossing each edge e of G_2. In the following picture, filled circles and heavy lines denote G_2 and open circles and dotted lines denote G_2^*. Note that G_2, together with G_2^* and the points where edges cross dual edges, forms another graph $\tilde{G}_2$ that is just G_2 scaled down by a factor of two.

Each subgraph H of G has a dual subgraph H^* consisting of all edges e^* of G^* dual to edges e *not* in H. If H has a cycle, then the duals of all edges in the cycle are absent from H^*, which separates H^* into two components: the interior and exterior of the cycle. Similarly, an island in H corresponds to a cycle in H^* as in figure 15.

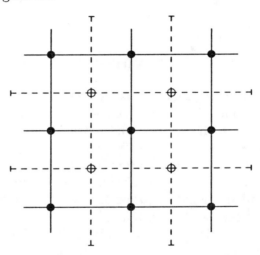

FIGURE 14.

From this description, it is clear that T is an essential spanning forest of G_2 if and only if T^* is an essential spanning forest of $G_2^* \cong G_2$.

Let T now be an infinite tree. We define *directed* a little differently from in the finite case: say T is directed if the edges are oriented so that every vertex has precisely one edge leading out of it. Following the arrows from any vertex gives an infinite path and it is not hard to check that any two such paths from different vertices eventually merge. Thus directedness for infinite trees is like directedness for finite trees, toward a vertex at infinity.

Say an essential spanning forest of G_2 is directed if a direction has been chosen for each of its components and each of the components of its dual. Here then is the connection between dominos and essential spanning forests.

Let T be a directed essential spanning forest of G_2, with dual T^*. Construct a domino tiling of $\tilde{G}_2$ as follows. Each vertex $v \in V(G_2) \subseteq V(\tilde{G}_2)$ is covered by a domino that also covers the vertex of $\tilde{G}_2$ in the middle of the edge of T that leads out of v. Similarly, each vertex $v^* \in V(G_2^*)$ is covered by a domino also covering the middle of the edge of T^* leading out of v. It is easy to check that this gives a legitimate domino tiling: every domino covers two neighboring vertices, and each vertex is covered by precisely one domino.

Conversely, for any domino tiling of $\tilde{G}_2$, directed essential spanning forests T and T^* for G_2 and G_2^* can be constructed as follows. For each $v \in V(G_2)$, the oriented edge leading out of v in T is the one along which the domino covering v lies (i.e., the one whose midpoint is the other vertex of $\tilde{G}_2$ covered by the domino covering v). Construct T^* analogously. To show that T and T^* are directed ESFs, amounts to showing there are no cycles, since clearly T and T^* will have one edge coming out of each vertex. This is true because, if you set up dominos in such a way as to create a cycle, they will always enclose an odd number of vertices (check it yourself!). Then there is no way to extend this configuration to a legitimate domino tiling of $\tilde{G}_2$.

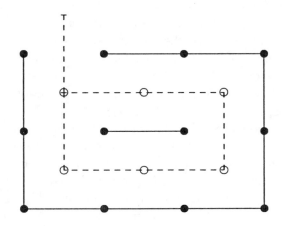

FIGURE 15.

It is easy to see that the two operations above invert each other, giving a one to one correspondence between domino tilings of $\tilde{G}_2$ and directed essential spanning forests of G_2. To bring this back into the realm of finite graphs requires ironing out some technicalities that I am instead going to ignore. The basic idea is that domino tilings of the $2n$-torus T_{2n} correspond to spanning trees of T_n almost as well as domino tilings of $\tilde{G}_2$ correspond to spanning trees of G_2. Going from directed essential spanning forests to spanning trees is one of the details glossed over here, but explained somewhat in the next subsection. The entropy for domino tilings is then one quarter the entropy for spanning trees, since T_{2n} has four times as many vertices as T_n. Entropy for spanning trees just means the number h for which T_n has approximately e^{hn^2} spanning trees. To calculate this, we use the matrix-tree theorem.

The number of spanning trees of T_n according to this theorem is the determinant of a minor of the matrix indexed by vertices of T_n whose v, w-entry is 4 if $v = w$, -1 if $v \sim w$, and 0 otherwise. If T_n were replaced by n edges in a circle, then this would be a circulant matrix. As is, it is a generalized circulant, with symmetry group $T_n = (\mathbb{Z}/n\mathbb{Z})^2$ instead of $Z/n\mathbb{Z}$. The eigenvalues can be gotten via group representations of T_n, resulting in eigenvalues $4 - 2\cos(2\pi k/n) - 2\cos(2\pi l/n)$ as k and l range from 0 to $n - 1$. The determinant we want is the product of all of these except for the zero eigenvalue at $k = l = 0$. The log of the determinant divided by n^2 is the average of these as k and l vary, and the entropy is the limit of this as $n \to \infty$, which is given by

$$\int_0^1 \int_0^1 \ln(4 - 2\cos(2\pi x) - 2\cos(2\pi y)) \; dx \, dy.$$

6.3   Miscellany

The limit theorems in Section 5 involved letting G_n tend to infinity locally, in the sense that each vertex in G_n had higher degree as n grew larger. Instead, one may consider a sequence such as $G_n = T_n$; clearly the n-torus converges in some sense to G_2 as $n \to \infty$, so there ought to be some limit theorem. Let $\mathbf{T}_n$ be a uniform spanning tree of G_n. Since G_n is not Gino-regular, the limit may not be $\mathcal{P}_1$ and in fact cannot be since the limit has degree bounded by four. It turns out that $\mathbf{T}_n$ converges in distribution to a random tree $\mathbf{T}$, called the uniform random spanning tree for the integer lattice. This works also for any sequence of graphs converging to the three or four dimensional integer lattices [13]. Unfortunately the process breaks down in dimensions five and higher. There the uniform spanning spanning trees on G_n do converge to a limiting distribution, but, instead of a spanning tree of the lattice, you get an essential spanning forest that has infinitely many components. If you can't see how the limit of spanning trees could be a spanning forest, remember that an essential spanning forest is so similar to a spanning tree that you can't tell them apart with any finite amount of information.

Another result from this study is that in dimensions 2, 3, and 4 the uniform random spanning tree $\mathbf{T}$ has only one path to infinity. What this really means is that any two infinite paths must eventually join up. Not only that, but $\mathbf{T}^*$ has the same property. That means there is only one way to direct $\mathbf{T}$, so that each choice of $\mathbf{T}$ uniquely determines a domino tiling of $\tilde{G}_2$. In this way it makes sense to speak of a uniform random domino tiling of the plane:

just choose a uniform random spanning tree and see what domino tiling it corresponds to.

That takes care of one of the details glossed over in the previous subsection. It also just about wraps up what I wanted to talk about in this article. As a parting note, let me mention an open problem. Let G be the infinite nearest neighbor graph on the integer lattice in d dimensions and let $\mathbf{T}$ be the uniform spanning tree on G gotten by taking a distributional limit of uniform spanning trees on d-dimensional n-tori as $n \to \infty$, as explained above.

Conjecture 2 *Suppose $d \geq 5$. Then with probability one, each component of the essential spanning forest has only one path to infinity, in the sense that any two infinite paths must eventually merge.*

References

[1] **Aldous, D.,** Asymptotic fringe distributions for general families of random trees, *Ann. Appl. Prob.*, 1, 228–266, 1990.

[2] **Aldous, D.,** The random walk construction of uniform spanning trees and uniform labelled trees, *SIAM J. Disc. Math.*, 3, 450–465, 1990.

[3] **Anatharam, V. and Tsoucas, P.,** A proof of the Markov chain tree theorem, *Stat. Prob. Let.*, 8, 189–192, 1989.

[4] **Athreya, K. and Ney, P.,** *Branching processes*, Springer-Verlag, New York, 1972.

[5] **Broder, A.,** Generating random spanning trees, in *Symp. foundations of computer sci.*, Institute for Electrical and Electronic Engineers, New York, 1988, 442–447.

[6] **Burton, R. and Pemantle, R.,** Local characteristics, entropy and limit theorems for spanning trees and domino tilings via transfer impedances, *Ann. Probab.*, 21, 1329–1371, 1993.

[7] **Cvetkovic, D., Doob, M. and Sachs, H.,** *Spectra of graphs: theory and application*, Academic Press, New York, 1980.

[8] **Doyle, P. and Snell, J. L.,** *Random walks and electric networks*, Carus Mathematical Monograph number 22, Mathematical Association of America, 1984.

[9] **Feller, W.,** *An introduction to probability theory and its applications*, volumes I and II, John Wiley & Sons, New York, 1967.

[10] **Harris, T.,** *The theory of branching processes*, Grundlehren #119, Springer-Verlag, New York, 1963.

[11] **Isaacson, D. and Madsen, R.,** *Markov chains: theory and applications*, John Wiley & Sons, New York, 1976.

[12] **Kastelyn, P.,** The statistics of dimers on a lattice, *Physica*, **27**, 1209–1225, 1961.

[13] **Pemantle, R.,** Choosing a spanning tree for the integer lattice uniformly, *Ann. Probab.*, **19**, 1559–1574, 1991.

[14] **Ross, S.,** *A first course in probability*, 3rd edition, Macmillan, New York, 1988.

[15] **Seymour, P. and Welsh, D.,** Combinatorial applications of an inequality from statistical mechanics, *Math. Proc. Camb. Phil. Soc.*, **77**, 485–495, 1975.

[16] **Spitzer, F.,** *Principles of random walk*, Van Nostrand, Princeton, 1964.

[17] **Stanley, R.,** *Enumerative combinatorics*, Wadsworth, Monterey, 1986.

[18] **Weinberg, L.,** *Network analysis and synthesis*, McGraw-Hill, New York, 1962.

[19] **White, N.,** Theory of Matroids. *Encyclopedia of mathematics and its applications*, volume 26, Cambridge University Press, New York, 1986.

Chapter 2

RANDOM WALKS: SIMPLE AND SELF-AVOIDING

Gregory F. Lawler
Department of Mathematics
Duke University

ABSTRACT

Two models for random motion on the integer lattice are considered, simple random walk and self-avoiding random walk. In both models the walkers take nearest neighbor steps. The major difference lies in the restriction that the self-avoiding walk cannot visit any site that has already been visited. This restriction makes for a significant difference both in the overall behavior of the walks and in the relative difficulty in trying to analyze the models rigorously.

1. INTRODUCTION

Consider a particle performing random motion along the integer lattice

$$Z^d = \{(z_1, \ldots, z_d) : z_i \text{ integer}\}.$$

One model of random motion is *simple random walk* where the particle at each time unit moves to one of its $2d$ nearest neighbors, each neighbor having the same probability. If we let $\omega(n)$ denote the position of the particle after n steps, then $\omega(0) = 0$ and for $n > 0$

$$\omega(n) = X_1 + \cdots + X_n,$$

where $X_j = \omega(j) - \omega(j-1)$ is the jump taken on the jth step. The X_j are independent, identically distributed random variables with $P\{X_j = e\} = 1/2d$ for each unit vector $e \in Z^d$. Another way of looking at this is to consider the set of random walk *paths* of length n,

$$\Lambda_n = \{\omega = [\omega(0), \ldots, \omega(n)] : \omega(i) \in Z^d; \omega(0) = 0; |\omega(i) - \omega(i-1)| = 1\},$$

and give each $\omega \in \Lambda_n$ probability $(2d)^{-n}$.

A somewhat similar model arises in chemical physics as a model of long polymer chains. Loosely speaking, a polymer is composed of a large number of monomers connected in a long chain. The chain tends to form itself "randomly" with the one restriction that the monomers cannot cross over each other. This leads to consideration of random walk paths on the integer lattice which do not have any self-intersections. We call a random walk $\omega \in \Lambda_n$ *self-avoiding* if it does not visit any point more than once, i.e., if $\omega(i) \neq \omega(j)$ for $0 \leq i < j \leq n$. Let Γ_n be the set of self-avoiding walks (SAWs) of length n. Polymer chains are then modeled by SAWs or, more precisely, by the uniform distribution on Γ_n.

While the simple random walk and the self-avoiding walk are both simple models to define, there is a big difference in the difficulty of analyzing these models. Because the simple random walk can be considered as the sum of independent, identically distributed random variables, the tools of probability theory are well suited for analyzing its behavior. The self-avoiding walk, however, which is defined in terms of a conditional probability for each n, cannot be analyzed using the standard tools of probability theory. Most of the interesting questions about SAWs remain open problems today, and the proofs of the few rigorous results have used either combinatorial arguments or techniques derived from mathematical physics. One of the reasons that it is hard to study SAWs from a probabilistic viewpoint is that a SAW is not a stochastic process that grows in time — any particle doing a "random self-avoiding walk" eventually gets trapped and cannot continue. A two-dimensional example is pictured in Figure 1.

FIGURE 1. A self-avoiding walk of length 8 that cannot be extended to a longer SAW.

Despite the fact that the SAW cannot be analyzed rigorously, there is enough heuristic and numerical evidence to see that the the model displays interesting phenomena. These include a number of dimension dependent "critical exponents". In this paper we will describe a number of these exponents, giving the conjectures on their values. We will start by defining the corresponding exponents for simple random walk (where most, but not all, of the exponents can be established rigorously).

2. THE NUMBER OF WALKS

The number of simple random walks of length n, i.e., the cardinality of Λ_n, is clearly $(2d)^n$, so the uniform probability measure $P = P_n$ on Λ_n gives measure $(2d)^{-n}$ to each walk. It is not so easy to compute the number of SAWs of length n. Let C_n be the cardinality of Γ_n. It is clear that a SAW cannot return to the point it has most recently visited. However, every simple random walk that always takes a step of $+1$ in whichever component it is moving is clearly self-avoiding. Since there are d positive directions,

$$d^n \leq C_n \leq 2d(2d-1)^{n-1}. \tag{1}$$

Any $(n+m)$-step SAW consists of an n-step SAW and an m-step walk joined together, although not every n-step SAW and m-step SAW can be combined to form an $(n+m)$-step SAW. This says that C_n is submultiplicative, i.e.,

$$C_{n+m} \leq C_n C_m,$$

and that $\ln C_n$ is subadditive,

$$\ln C_{n+m} \leq \ln C_n + \ln C_m.$$

If a_n is any subadditive sequence, then

$$\lim_{n \to \infty} \frac{a_n}{n} = \inf_{n > 0} \frac{a_n}{n}.$$

(To see this, it clearly suffices to show that for every m,

$$\limsup_{n \to \infty} \frac{a_n}{n} \leq \frac{a_m}{m}.$$

But if we let $\alpha = \sup\{a_j : 1 \leq j \leq m\}$ and write any integer n as $n = jm + k$ with $k \in \{1, \ldots, m\}$, then subadditivity gives

$$a_n \leq j a_m + a_k \leq \frac{n a_m}{m} + \alpha,$$

and if we divide by n and take the limsup we get the result.) Therefore, there exists some a such that

$$\lim_{n \to \infty} \frac{\ln C_n}{n} = a,$$

or if we set $\mu = e^a$,

$$C_n \approx \mu^n.$$

(We write $a_n \sim b_n$ if the sequences are asymptotic, i.e., $a_n/b_n \to 1$, and we write $a_n \approx b_n$ if $\ln a_n$ and $\ln b_n$ are asymptotic.) The number $\mu = \mu_d$ is often called the *connective constant* for SAW. By (1), $\mu \in [d, 2d-1]$. The exact value of μ is not known, although μ_2 is thought to be about 2.64 and μ_3 about 4.68. When the values of μ were first being investigated, there was some speculation that μ_2 might be exactly equal to e. This has been ruled out rigorously, and today it is generally believed that μ is just some real number which is not necessarily related to any of the well known constants.

The number of SAWs of length n is therefore about μ^n; more precisely,

$$C_n = \mu^n r_n,$$

for some sequence satisfying $r_n^{1/n} \to 1$. The behavior of the sequence r_n is of interest. There is also a related sequence

$$\tilde{r}_n = \frac{C_{2n}}{C_n C_n} = \frac{r_{2n}}{r_n^2}.$$

We can interpret $\tilde{r}_n$ as the probability that two SAWs of length n can be combined to form a SAW of length $2n$, or in other words, the probability that the paths of two SAWs of length n have no points in common other than the origin. When we speak of probabilities here, we mean with respect to the uniform probability $Q = Q_n$ on Γ_n which assigns probability C_n^{-1} to each SAW.

3. PROPERTIES OF SIMPLE WALKS

Consider the position of a simple random walker after n steps, $\omega(n)$. Clearly $\langle \omega(n) \rangle = 0$ (here we will use the notation $\langle \cdot \rangle$ for expectation, which is common in physics). Also

$$
\begin{aligned}
\langle |\omega(n)|^2 \rangle &= \langle |X_1 + \cdots + X_n|^2 \rangle \\
&= \sum_{j=1}^{n} \langle |X_j|^2 \rangle + \sum_{i \neq j} \langle X_i \cdot X_j \rangle \\
&= \sum_{j=1}^{n} 1 + 0 \\
&= n,
\end{aligned}
$$

Thus the mean square displacement of $\omega(n)$ is equal to n and the typical distance from the origin of $\omega(n)$ is of order $n^{1/2}$. With the aid of the central limit theorem we can make this more precise. Since $\omega(n)$ is the sum of independent, identically distributed random vectors $X_i = (X_i^1, \ldots, X_i^d)$ with $\langle X_i^j X_i^k \rangle = (1/d)\delta_{jk}$, the central limit theorem states that if A is any open ball in R^d,

$$\lim_{n \to \infty} P\{\frac{\omega(n)}{\sqrt{n}} \in A\} = \int_A (\frac{d}{2\pi})^{d/2} \exp\{-\frac{d|x|^2}{2}\} dx_1 \cdots dx_d.$$

Suppose we want to know the probability that the random walk is at a particular $z = (z_1, \ldots z_d) \in Z^d$. We first note that if n is even then $\omega(n)$ must be of even parity, i.e., the sum of its components must be even. Suppose n is even and z is a point of even parity. Consider $B = B(n^{-1/2}z, \epsilon)$ the ball of radius ϵ about $n^{-1/2}z$. By the central limit theorem, $P\{n^{-1/2}\omega(n) \in B\}$ is approximately

$$\int_B (\frac{d}{2\pi})^{d/2} \exp\{-\frac{d|x|^2}{2}\} dx_1 \cdots dx_d.$$

If ϵ is small this can be approximated by

$$(\frac{d}{2\pi})^{d/2}\exp\{-\frac{d|z|^2}{2}\}v(B),$$

where $v(B)$ denotes the volume of B. Now B contains about $n^{d/2}v(B)$ points of the form $n^{-1/2}y$, $y \in Z^d$, and about half of these are of even parity. Therefore, if the simple random walk spreads itself relatively evenly among the possible points, one would guess that $P\{\omega(n) = z\}$ is approximately

$$2(\frac{d}{2\pi n})^{d/2}\exp\{-\frac{d|z|^2}{2n}\}.$$

This is the case and the result is called the *local central limit theorem* (see [3],[6] for precise formulations of this theorem).

One particular case of this theorem is when $z = 0$. Then we get if n is even

$$P\{\omega(n) = 0\} \sim 2(\frac{d}{2\pi})^{d/2}n^{-d/2}.$$

How often does the random walker return to 0? Let R be the number of visits to 0,

$$R = \sum_{n=0}^{\infty} I\{\omega(n) = 0\},$$

where I denotes the indicator function. It is then easy to see that

$$\langle R \rangle = \sum_{n=0}^{\infty} P\{\omega(n) = 0\}$$

is bounded above and below by constants times

$$1 + \sum_{n=1}^{\infty} 2(\frac{d}{2\pi})^{d/2}(2n)^{-d/2} \quad \begin{cases} = \infty, & d = 1, 2, \\ < \infty & d > 2. \end{cases}$$

Therefore, the expected number of visits is infinite for $d \leq 2$ and finite for $d > 2$. Let T be the time of the first return after $n = 0$,

$$T = \inf\{n > 0 : \omega(n) = 0\}.$$

Note that T is a random time and $T = \infty$ if and only if the random walk does not return to the origin, in which case $R = 1$. Suppose $T < \infty$. Then the motion of the random walker after time T is that of a random walker independent of what happened up through time T (this property is often called the *strong Markov property* of simple random walk). Therefore, the expected number of visits from time T on (if $T < \infty$) is the same as the expected number for the original walk, and hence $\langle R \mid T < \infty \rangle = 1 + \langle R \rangle$. Therefore,

$$\langle R \rangle = P\{T = \infty\} \cdot 1 + P\{T < \infty\}(\langle R \rangle + 1),$$

or
$$P\{T = \infty\} = \frac{1}{\langle R \rangle},$$

which is zero for $d \leq 2$ and positive for $d \geq 3$. The strong Markov property can be used again to show this implies

$$P\{R = \infty\} = \begin{cases} 1, & d \leq 2, \\ 0, & d \geq 3. \end{cases}$$

We say that simple random walk is *recurrent* in one or two dimensions and *transient* in three or more dimensions.

Let a_n be the probability that $\omega(j) \neq 0$ for $1 \leq j \leq n$. Then $a_n \to 0$ if $d = 1, 2$. We can get a good estimate of the rate at which it approaches 0. Let R_n be the number of visits up through time n,

$$R_n = \sum_{j=0}^{n} I\{\omega(j) = 0\}.$$

Then as $n \to \infty$

$$\langle R_n \rangle = \sum_{j=0}^{n} P\{\omega(j) = 0\} \sim 1 + \sum_{j=2}^{n/2} 2(\frac{d}{2\pi})^{d/2} (2j)^{-d/2} \sim \begin{cases} \sqrt{2/\pi}\, n^{1/2}, & d = 1, \\ (1/\pi) \ln n, & d = 2. \end{cases}$$

We will use a *last-exit decomposition*. Let σ_n be the time of the last visit to 0 before n,

$$\sigma_n = \sup\{j : j \leq n, \ \omega(j) = 0\}.$$

By the strong Markov property, $P\{\sigma_n = j\} = P\{\omega(j) = 0\}a_{n-j}$. Since every path has a unique last return to 0 before n,

$$1 = \sum_{j=0}^{n} P\{\sigma_n = j\} = \sum_{j=0}^{n} P\{\omega(j) = 0\}a_{n-j} \geq a_n \sum_{j=0}^{n} P\{\omega(j) = 0\} = a_n \langle R_n \rangle.$$

To get a bound in the opposite direction for a_n, consider τ_n the first return to 0 after time n,

$$\tau_n = \inf\{j : j > n, \ \omega(j) = 0\}.$$

Then,

$$\begin{aligned} P\{n < \sigma_{2n} \leq 2n\} &= P\{\omega(j) = 0 \text{ for some } n < j \leq 2n\} \\ &= P\{n < \tau_n \leq 2n\}. \end{aligned}$$

Note that $\sigma_{2n} \leq n$ if and only if $\tau_n > 2n$. Therefore,

$$\begin{aligned} 1 - P\{n < \tau_n \leq 2n\} &= \sum_{j=0}^{n} P\{\sigma_{2n} = j\} \\ &= \sum_{j=0}^{n} P\{\omega(j) = 0\}a_{2n-j} \\ &\leq a_n \langle R_n \rangle. \end{aligned}$$

Therefore,

$$[1 - P\{n < \tau_n \le 2n\}]\langle R_n \rangle^{-1} \le a_n \le \langle R_n \rangle^{-1}.$$

Again by the strong Markov property we can see that

$$\langle R_{3n} - R_n \mid n < \tau_n \le 2n \rangle \ge \langle R_n \rangle,$$

and since $\langle R_{3n} - R_n \rangle \ge P\{n < \tau_n \le 2n\}\langle R_{3n} - R_n \mid n < \tau_n \le 2n \rangle$, we get

$$P\{n < \tau_n \le 2n\} \le \frac{\langle R_{3n} - R_n \rangle}{\langle R_n \rangle} \sim \begin{cases} \sqrt{3} - 1, & d = 1, \\ (\ln 3)(\ln n)^{-1}, & d = 2. \end{cases}$$

Combining all the inequalities, we get that $a_n \sim \langle R_n \rangle^{-1}$ if $d = 2$. For $d = 1$, this argument only proves that $c\langle R_n \rangle^{-1} \le a_n \le \langle R_n \rangle^{-1}$ for some $c > 0$. (It can be shown for $d = 1$ that $a_{2n} = P\{\omega(2n) = 0\}$ and hence that $a_n \sim \sqrt{2/\pi}n^{-1/2}$. See [2] for a nice proof of this.) In both cases, the probability of no return in n steps is equal, up to a multiplicative constant, to the inverse of the expected number of returns.

The *Green's function* $G(z)$ for a simple random walk is just the expected number of times that the walk visits z,

$$G(z) = \langle \sum_{j=0}^{\infty} I\{\omega(j) = z\} \rangle = \sum_{j=0}^{\infty} P\{\omega(j) = z\}.$$

If $d \le 2$, $G(z) = \infty$ for each $z \in Z^d$. If $d \ge 3$, however, $G(z) < \infty$. The local central limit theorem can be used to give the asymptotics of $G(z)$ as $|z| \to \infty$,

$$G(z) \sim \sum_{n=0}^{\infty} (\frac{d}{2\pi})^{d/2} n^{-d/2} \exp\{-\frac{d|z|^2}{2n}\}.$$

By approximating the sum by an integral and doing the appropriate calculation we get

$$G(z) \sim c|z|^{2-d}, \quad d \ge 3,$$

where $c = c_d = (d/2)\Gamma(\frac{d}{2} - 1)\pi^{-d/2}$.

4. INTERSECTIONS OF SIMPLE RANDOM WALKS

How different is the behavior of self-avoiding walks from that of simple random walks? As a start toward understanding this, we should try to discover how likely it is for paths of simple random walks to intersect. Let ω_1 and ω_2 be independent simple random walks starting at the origin and let J be the number of times (j, k) at which $\omega_1(j) = \omega_2(k)$,

$$J = \sum_{j=0}^{\infty} \sum_{k=0}^{\infty} I\{\omega_1(j) = \omega_2(k)\}.$$

There is a one-to-one correspondence between walks ω_2 satisfying $\omega_2(0) = 0$, $\omega_2(k) = \omega_1(j)$ and walks $\tilde{\omega}_2$ satisfying $\tilde{\omega}_2(0) = \omega_1(j)$, $\tilde{\omega}_2(k) = 0$. From this "time reversibility" property we see that

$$P\{\omega_1(j) = \omega_2(k)\} = P\{\omega_1(j + k) = 0\}.$$

Then,

$$\langle J \rangle = \sum_{j=0}^{\infty}\sum_{k=0}^{\infty} P\{\omega_1(j) = \omega_2(k)\}$$

$$= \sum_{j=0}^{\infty}\sum_{k=0}^{\infty} P\{\omega_1(j + k) = 0\}.$$

For $j + k$ even, $P\{\omega_1(j + k) = 0\} \sim c(j + k)^{-d/2}$. It is then easy to check that this sum is infinite if and only if $d/2 \leq 2$, i.e.,

$$\langle J \rangle = \infty, \ \ d \leq 4,$$

$$\langle J \rangle < \infty, \ \ d \geq 5.$$

This clearly implies that $P\{J = \infty\} = 0$ if $d \geq 5$ and in analogy with the problem of returns to the origin we would expect that

$$P\{J = \infty\} = 1, \ \ d \leq 4,$$

$$P\{J = 1\} > 0, \ \ d \geq 5.$$

(Note that $J = 1$ means that there is no intersection of the paths other than the obvious one at $j = k = 0$.) While these facts are true, it is not quite so easy to prove them. The argument for R above, makes use of a time T which is the first return. Here we cannot talk of the "first" intersection of the two paths, because there are two time scales involved.

Four is the critical dimension for intersections of random walks. There is a simple intuitive way of seeing why this should be true. Assume $d \geq 2$. In n steps a random walker goes about distance $\sqrt{n}$ and hence visits about n points in the ball of radius $\sqrt{n}$ (actually it visits about cn for some $c > 0$ if $d \geq 3$ and about $cn/\ln n$ in $d = 2$). A set in Z^d which contains r^2 points in the ball of radius r can be thought of as a "two-dimensional" subset. Whether or not two random walk paths intersect can then be considered as the question "in which dimensions do two two-dimensional sets intersect?" By comparing to intersections of planes in R^d, one can see that they should intersect if $d < 4$, should not intersect if $d > 4$, and $d = 4$ is a marginal case.

Let

$$J_n = \sum_{j=0}^{n}\sum_{k=0}^{n} I\{\omega_1(j) = \omega_2(k)\},$$

be the number of intersections of two paths of length n, and let $b_n = P\{J_n = 1\}$ be the probability that the paths of length n have no intersection other than at $j = k = 0$. Note that b_n is the simple random walk analogue of the sequence $\tilde{r}_n$ defined previously for self-avoiding walks. As noted above, $\lim_{n\to\infty} b_n$ is positive for $d \geq 5$ and zero for $d \leq 4$. For $d \leq 4$, we can ask how quickly b_n goes to 0. As a start, we can show in a straightforward manner that

$$\langle J_n \rangle = \sum_{j=0}^{n} \sum_{k=0}^{n} P\{\omega_1(j) = \omega_2(k)\} \sim \begin{cases} cn^{(4-d)/2}, & d < 4, \\ c(\ln n), & d = 4. \end{cases}$$

In analogy to the problem of returns to the origin, one might guess that b_n decays like a constant times $\langle J_n \rangle^{-1}$. We can use a variation of the last-exit decomposition to get such a bound in one direction.

Call a pair of times (j, k) an n-last intersection time if $\omega_1(j) = \omega_2(k)$ and $\omega_1(\tilde{j}) \neq \omega_2(\tilde{k})$ for all $j \leq \tilde{j} \leq n, k \leq \tilde{k} \leq n, (\tilde{j}, \tilde{k}) \neq (j, k)$. Note that any pair of paths has at least one n-last intersection time, although it is possible that they have more than one. By the strong Markov property, if $0 \leq j, k \leq n$, $P\{(j, k) \text{ is } 2n\text{-last}\} \geq P\{\omega_1(j) = \omega_2(k)\}b_n$. Let W_n be the event

$$\{\omega_1(j) \neq \omega_2(k) : n < j \leq 2n, 0 \leq k \leq 2n; \text{ or } 0 \leq j \leq 2n, n < k \leq 2n\}.$$

Then, if W_n occurs, there must be a $2n$-last intersection time (j, k) with $0 \leq j, k \leq n$. It is not too difficult to show there is a constant $c > 0$ (independent of n) such that $P(W_n) \geq c$, and hence

$$c \leq \sum_{j=0}^{n} \sum_{k=0}^{n} P\{(j, k) \ 2n\text{-last}\} \leq b_n \sum_{j=0}^{n} \sum_{k=0}^{n} P\{\omega_1(j) = \omega_2(k)\} \leq b_n \langle J_n \rangle,$$

i.e., $b_n \geq c \langle J_n \rangle^{-1}$.

Does a similar bound hold in the opposite direction? Let L_n be the number of n-last times. Then by a similar argument we can see that

$$\langle L_n \rangle = \sum_{j=0}^{n} \sum_{k=0}^{n} P\{(j, k) \ n\text{-last}\} \geq b_n \sum_{j=0}^{n} \sum_{k=0}^{n} P\{\omega_1(j) = \omega_2(k)\} = b_n \langle J_n \rangle.$$

If it were true that most pairs of paths only had a few n-last points, then this argument would give a bound in the opposite direction. It turns out, however, that most pairs of paths have many n-last points and $\langle L_n \rangle \to \infty$. In fact, $\langle J_n \rangle^{-1}$ does *not* give the correct rate of decay. In four dimensions it has been shown that

$$b_n \approx (\ln n)^{-1/2}.$$

For $d < 4$, it has been shown that

$$b_n \approx n^{-\varsigma}, \quad d = 1, 2, 3$$

for some $\zeta = \zeta_d > 0$ called the *intersection exponent* for a simple random walk. Suppose $d = 1$. If two one-dimensional random walk paths are not going to intersect, one must stay above the x-axis and the other below it. Therefore, $b_n = 2(a_n/2)^2$, where a_n is as defined above. Since $a_n \sim cn^{-1/2}$, $\zeta_1 = 1$. The values of ζ_2 and ζ_3 are not known. Good (nonrigorous) estimates for the values can be obtained by Monte Carlo simulations, i.e., by generating random walks on the computer and seeing how many intersect. This has led to conjectures $\zeta_2 = 5/8$, $\zeta_3 \in [.28, .29]$. The best known rigorous bounds on the exponent are

$$\zeta_2 \in [\frac{1}{2} + \frac{1}{4\pi}, \frac{3}{4}), \quad \zeta_3 \in [\frac{1}{4}, \frac{1}{2}).$$

There is another problem about intersections for which the probability of no intersection does decay like the inverse of the expected number of intersections. Let $\omega_1, \omega_2, \omega_3$ be three independent random walks starting at 0 and let u_n be the probability that ω_1 intersects neither ω_2 nor ω_3 in n steps,

$$u_n = P\{\omega_1(j) \neq \omega_2(k), \omega_1(j) \neq \omega_3(k), or\ 0 \leq j, k \leq n, (j, k) \neq (0, 0)\}.$$

Let K_n be the total number of intersections

$$K_n = \sum_{j=0}^{n} \sum_{k=0}^{n} (I\{\omega_1(j) = \omega_2(k)\} + I\{\omega_1(j) = \omega_3(k)\}).$$

Then $\langle K_n \rangle = 2\langle J_n \rangle$, and it can be shown that u_n decays like a constant times $\langle K_n \rangle^{-1}$.

5. DISTRIBUTIONS OF SAW

What is the distribution of the endpoint of a SAW? Here we are considering $Q = Q_n$ the uniform measure on Γ_n. Since the number of SAWs is exponentially smaller than the number of simple random walk paths, one cannot expect that results about typical simple paths will hold for typical SAWs. The first question we ask is: what is the typical distance from the origin, or as is more convenient, what is the typical squared distance

$$\langle |\omega(n)|^2 \rangle_Q,$$

where we write $\langle \cdot \rangle_Q$ for expectations with respect to the SAW measure Q? The *root mean square displacement exponent* $\nu = \nu_d$ is defined by

$$\langle |\omega(n)|^2 \rangle_Q \approx n^{2\nu}.$$

It is a little cavalier to define ν in the above fashion without first showing that there exists a ν satisfying the above. Unfortunately, there is no proof that such an exponent exists and, moreover, there is no proof that any of the other exponents that we will define for SAWs exist. A chemist, Flory, first gave a heuristic argument which gives the conjecture

$$\nu = \begin{cases} 3/(2+d), & d \leq 4, \\ 1/2, & d \geq 4. \end{cases}$$

The argument has a number of flaws (see [1] for a discussion of the Flory argument), but, somewhat amazingly, seems to be quite accurate. It certainly gives the correct value for $d = 1$ (there are only 2 SAWs of length n for $d = 1$). Heuristic and numerical evidence suggest that it is also exactly correct for $d = 2$ and $d \geq 4$, although ν_3 is now believed to be about .59$\cdots$, a little smaller than the Flory estimate. In the critical dimension $d = 4$ there is also expected to be a logarithmic correction,

$$n^{-1}\langle|\omega(n)|^2\rangle_Q \approx (\ln n)^{1/4}.$$

The conjectures show that the SAW goes to infinity faster than the simple random walk (whose root mean square displacement exponent equals $1/2$ in all dimensions), especially in low dimensions. Above the critical dimension, however, the SAW path grows at a rate comparable to the simple walk. None of the results for dimensions $d = 2, 3, 4$ has been proved rigorously. In fact, no nontrivial bound on the growth rate for SAWs has been found in these dimensions. One recent breakthrough has been a proof the Flory conjecture is true for $d \geq 5$ and

$$\langle|\omega(n)|^2\rangle_Q \sim cn,$$

for some $c = c_d > 1$.

The central limit theorem says that the distribution of the endpoint of a simple random walk approaches a Gaussian distribution. The same result holds for the SAW in high dimensions-the random variables $n^{-1/2}\omega(n)$ under the measure Q converges is distribution to a Gaussian. It is expected that there will also be a normal limit in $d = 4$ if the walk is appropriately scaled, i.e.,

$$\frac{\omega(n)}{\langle|\omega(n)|\rangle_Q^{1/2}},$$

converges to a normal distribution. In low dimensions, however, where $\nu > 1/2$, a Gaussian limit is not expected.

6. THE NUMBER OF WALKS (CONTINUED)

We have already noted that the number of SAWs C_n satisfies

$$C_n = \mu^n r_n,$$

where $r_n^{1/n} \to \infty$. If

$$\tilde{r}_n = \frac{r_{2n}}{r_n^2},$$

then $\tilde{r}_n$ has the interpretation as the probability that two independent SAWs of length n have no point in common. In analogy with the corresponding quantity b_n for simple random walks one might conjecture that

$$\tilde{r}_n \geq c, \quad d \geq 5,$$

$$\tilde{r}_n \approx (\ln n)^{-a}, \quad d = 4,$$

$$\tilde{r}_n \approx n^{-a}, \quad d = 2, 3.$$

If $d \geq 5$, this has been proven (it is quite believable since one would expect SAWs to be "thinner" then simple random walks and hence less likely to intersect). For low dimensions this is only a conjecture. The conjecture for $d = 4$ is that $\tilde{r}_n \approx (\ln n)^{-1/4}$. It is standard to label the exponent a in $d = 2, 3$ as $\gamma - 1$. In terms of r_n the conjecture is

$$r_n \approx n^{\gamma - 1}, \quad d = 2, 3$$

for some $\gamma = \gamma_d$. The current conjectures are that $\gamma_2 - 1 = 11/32$ and $\gamma_3 - 1$ is about .16. (There is an interesting phenomenon in all these exponents, at least if the conjectures are true. In two dimensions the exponents take on rational values, while in three dimensions there is no belief that the value will be rational.) Note that $\gamma - 1 < \zeta$ which agrees with the intuition that SAWs should be less likely to intersect than simple walks. Again, it is not known rigorously that the exponent γ exists; in contrast, it *is* known that the simple random walk exponent ζ exists even though the exact value is not known.

7. SELF-AVOIDING POLYGONS

We have seen that the probability that a simple random walk is at the origin in n steps looks like $cn^{-d/2}$, assuming n is even. We now consider the corresponding question for SAWs. Of course, a SAW of length n cannot satisfy $\omega(n) = 0$ since this would violate the self-avoidance constraint. A *self-avoiding polygon (SAP)* of length n is a simple random walk path $\omega \in \Lambda_n$ satisfying $\omega(0) = \omega(n)$ and having no other self-intersections, i.e., $\omega(i) \neq \omega(j), 0 \leq i < j \leq n - 1$. Let A_n be the number of SAPs of length n. Then A_n/C_n can be viewed as the probability that a SAW is at the origin in n steps.

If $z \in Z^d$, let $C_n(z)$ be the number of SAW's of length n whose endpoint is z,

$$C_n(z) = \{\omega \in \Gamma_n : \omega(n) = z\}.$$

There is a one-to-one correspondence between SAPs of length n and SAWs of length $n - 1$ whose endpoint is a nearest neighbor of the origin. Hence,

$$A_n = \sum_{|e|=1} C_{n-1}(e) = (2d)C_{n-1}(e_1),$$

where e_1 is the unit vector whose first component is 1 (the second equality uses some symmetry properties of SAWs). In analogy with simple walks we might expect that $A_n/C_n \approx n^{-\delta}$ for some $\delta = \delta_d$. It has been shown that

$$A_n \approx \mu^n,$$

i.e., that SAPs have the same connective constant as SAWs, but the proof is not strong enough to conclude that the ratio A_n/C_n is larger than some power of n. The exponent $1 + \gamma - \delta$ is generally called α or α_{sing} so that

$$\mu^{-n} A_n \approx n^{\alpha-2}, \quad n \text{ even.}$$

(The exponents for SAWs have analogues in other models of mathematical physics and the somewhat unnatural labelling of exponents has to do with this.) If $z \in Z^d$ with even parity, it is also conjectured that

$$\mu^{-n} C_n(z) \approx n^{\alpha-2}, \quad n \text{ even,}$$

where the rate of convergence depends on z. This is analogous to the fact that for simple random walks, for fixed z, $P\{\omega(n) = z\} \sim cn^{-d/2}$.

There is a conjectured relationship between the exponents, ν and α, sometimes called hyperscaling,

$$\alpha - 2 = -d\nu.$$

We will sketch the intuition behind this conjecture here. A SAP of length $2n$ is composed of two SAWs of length n, ω_1 and ω_2 (here we consider ω_2 to be the second half of the polygon traversed "backwards"). These two SAWs must have the same endpoint, $\omega_1(n) = \omega_2(n)$, but otherwise must avoid each other. Therefore,

$$A_{2n} = \sum_{z \in Z^d} C_n(z)^2 s_n(z),$$

where $s_n(z)$ is the probability that two SAWs of length n, each with endpoint z, have no other intersections.

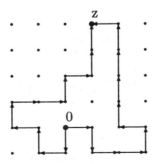

FIGURE 2. A SAP of length 22 consisting of two SAWs of length 11 ending at z.

We expect that the sum will be dominated by terms with $|z|$ equal to the typical distance of an n step SAW from the origin, n^ν (in this informal hand-waving argument we are ignoring all constants, which are irrelevant anyway for equivalence under $\approx$). Since there are on the order $n^{d\nu}$ such points we guess that A_{2n} is approximated by $n^{d\nu} C_n(z)^2 s_n(z)$ for z a typical point distance n^ν away. Again, since there are order $n^{d\nu}$ points distance ν away, we estimate $C_n(z)$ by $n^{-d\nu} C_n$. The factor $s_n(z)$ is trickier. The non-intersection constraint on ω_1 and ω_2 is strongest for points near 0 and points near z. One would expect the contribution at each of these places to be of order $\tilde{r}_n$ (or maybe $\tilde{r}_{n/2}$ which should be the same up to a constant). Combining this, we expect that a good approximation for A_{2n} should be $n^{d\nu}(n^{-d\nu} C_n)^2 \tilde{r}_n^2$. If we write this in terms of exponents, we get the hyperscaling conjecture.

If we combine the hyperscaling relation with the conjectures for ν, we get conjectures for α,

$$\alpha_2 = \frac{1}{2}; \quad \alpha_3 \doteq .23; \quad \alpha_d = 2 - \frac{d}{2}, \ d \geq 4.$$

One rigorous result is that, if the exponent α exists, then it must satisfy

$$\alpha_2 \leq \frac{5}{2}; \quad \alpha_3 \leq 2; \quad \alpha_d < 2, \ d \geq 4.$$

8. GREEN'S FUNCTION FOR SAW

Recall that the Green's function $G(z)$ for a simple random walk is the expected number of times that the walk visits z. Another way to define G is the following: let $B_n(z)$ be the number of simple random walk paths of length n ending at z, i.e., the cardinality of

$$\{\omega \in \Lambda_n : \omega(n) = z\}.$$

Then,

$$G(z) = \sum_{n=0}^{\infty} (2d)^{-n} B_n(z).$$

Using this as a guide, we define the *Green's function* or *two-point function for SAWs* $\overline{G}(z)$ by

$$\overline{G}(z) = \sum_{n=0}^{\infty} \mu^{-n} C_n(z).$$

Note that since

$$\mu^{-n} C_n(z) \approx n^{\alpha-2},$$

then the series converges for $d \geq 2$, assuming the conjectures for α are correct.

Recall that $G(z) \sim c|z|^{-(d-2)}$ for $d \geq 3$. The exponent $\eta = \eta_d$ for SAWs measures how much the asymptotic behavior of $\overline{G}$ differs from that of G,

$$\overline{G}(z) \approx |z|^{-(d-2+\eta)}, \quad d \geq 2.$$

There is a conjectured formula that relates η to ν and γ,

$$\gamma = \nu(2 - \eta).$$

For high dimensions, $\gamma = 1$ and $\nu = 1/2$ so this implies that $\eta = 0$, i.e., the exponent for the SAW Green's function is the same as the exponent for G (and the proofs about the SAW in high dimensions do also prove that $\eta = 0$). We will sketch here a heuristic argument why this relation should hold in low dimensions. Recall that we expect a SAW of length n to be distance n^ν from the origin. Therefore, if n is much smaller than $|z|^{1/\nu}$, its contribution to the above sum should be negligible. If n is comparable to $|z|^{1/\nu}$, then we would expect $\mu^{-n}C_n(z)$ to be on the order of $n^{-d}\mu^{-n}C_n$ or about $|z|^{(\gamma-1-d)/\nu}$. One can also check that values of n significantly larger than $|z|^{1/\nu}$ do not contribute significantly to the sum. Since there are on the order of $|z|^{1/\nu}$ values of n of order $|z|^{1/\nu}$ we get the estimate

$$G(z) \approx |z|^{1/\nu}|z|^{(\gamma-1-d)/\nu},$$

or $-(d - 2 + \eta) = (\gamma - d)/\nu$, and if we solve for γ we get the relationship.

9. UNIVERSALITY

One of the reasons the critical exponents are of interest is that the values of the exponents are not expected to depend on the exact form of the random walk. For example, suppose $S_n = X_1 + \cdots + X_n$ is any random walk taking values in Z^d whose increments X_i are independent, identically distributed with mean zero and finite non-zero variance. Then the relevant exponents for S_n are the same as for simple random walk, e.g.

$$\langle |S(n)|^2 \rangle \sim cn,$$

$$P\{S(n) = 0\} \sim cn^{-d/2}.$$

Similarly, the intersection exponent will be the same as for simple random walks. Probabilists refer to such results as *invariance principles* while a physicist would say that all such random walks belong to the same *universality class*.

For any such random walk one could also consider the corresponding self-avoiding walk. Then it is conjectured that all such self-avoiding walks are in the same universality class as the self-avoiding walk derived from nearest neighbor walks. It is this universality principle that makes the study of easily defined mathematical models potentially very relevant for describing complex physical phenomena, e.g., a polymer chain does not look like a SAW on the integer lattice, but the exponents that can be observed for such chains may well agree with the exponents of the SAW.

There are other models of random walks with self-repulsion interactions other than the uniform measure on SAWs. One example is the *weakly self-avoiding walk* or the *Domb-Joyce model*. Take a simple random walk of length n, $\omega_n \in \Lambda_n$ and let Y_n be the number of self-intersections,

$$Y_n(\omega) = \sum_{0 \le j < k \le n} I\{\omega(j) = \omega(k)\}.$$

For any $\beta > 0$, we can consider the measure $Q = Q_{n,\beta}$ on Λ_n given by

$$Q(\omega) = Z^{-1} \exp\{-\beta Y_n(\omega)\},$$

where $Z = \langle \exp\{-\beta Y_n\} \rangle$ is the normalization factor to make Q a probability measure. The case $\beta = 0$ corresponds to a simple random walk and the self-avoiding walk is the limit as $\beta \to \infty$. It is conjectured that for every $\beta > 0$ this model is in the same universality class as the usual SAW, i.e., that all the critical exponents, appropriately defined, are the same. (The case $\beta < 0$ corresponds to a self-attracting walk which is in a different universality class from either the simple random walk or the self-avoiding walk.)

10. KINETICALLY GROWING SAW

The SAW and the weakly self-avoiding walk are *configurational* models of random walks with a self-repulsion interaction. They are defined by considering all simple random walks of a given length and weighting the walks according to the number of self-intersections. (In the case of the uniform measure on self-avoiding walks, all walks of minimum energy are given the same weighting and all other walks are given zero probability.) Such models are natural from the viewpoint of statistical physics where it is generally postulated that walks try to form in a way that minimizes energy. Such models are not so natural from a probabilist's viewpoint, where one would like to think of a random walk growing with time. There are a number of models of *kinetically growing* self-avoiding walks; however, most of these models appear to be in a different universality class from the usual SAW model.

Probably the simplest way to try to define a kinetically growing self-avoiding walk is to consider a random walker which at each step chooses randomly among those nearest neighbor points it has not already visited. However, it is not too difficult to show that such a walker will eventually get trapped and find no new points to visit. If one does not require strict self-avoidance, this model can be modified so that returns are discouraged but not forbidden. One such model fixes a $\beta > 0$ and then has the walker choosing a new point among its nearest neighbors, with the probability of choosing point x being proportional to $\exp\{-\beta V(x)\}$, where $V(x)$ is the number of times the walk has already visited x. This walk has been called the *true* or *myopic* self-avoiding walk (the word myopic refers to the fact that the walker only looks at its nearest neighbors which choosing its next point). The behavior of this walk appears to be significantly different from that of the usual SAW; in fact, it is conjectured that the critical dimension for this model is two and that the root mean square displacement exponent is $1/2$ for $d \geq 2$. Clearly, if the conjectures are correct, this model is not in the same universality class as the usual SAW.

Another possible modification of this walk is to allow the random walker to choose a nearest neighbor at random, except that the walker may only choose a point which will not eventually trap the walker. This formulation allows for a kinetically growing strictly self-avoiding walk. However, it appears that this walk is in even another universality class. There are other possibilities as well — one has the walker choosing a new step with the probability of choosing x proportional to the probability that a simple random walk starting at x avoids the path up to that point. This last model has the advantage that some rigorous bounds can be given for the critical exponent. Its critical dimension is four, just as the usual SAW, but the values for the mean square displacement exponent below the critical dimension seem to differ. Therefore, this walk is in another universality class.

One may ask if it is possible to find a kinetically growing walk which is in the same universality class as the usual SAW. The answer is yes, at least in high dimensions; however, no one has come up with such a kinetically growing walk that has a transition function that is easy to calculate.

11. MONTE CARLO SIMULATIONS

Since it is extremely difficult to derive values of exponents rigorously, Monte Carlo simulations are often done to estimate exponents. The idea of such simulations is easy: in order to estimate the probability of an event, do as many independent trials as possible and see how often the event occurs. Of course, we can never *prove* facts using Monte Carlo simulations.

As an example of how to use Monte Carlo simulations to estimate an exponent, consider the intersection exponent ζ for simple random walks in two and three dimensions. In this case, the exponent is known to exist,

$$b_n \approx n^{-\zeta};$$

in fact, it is conjectured that $b_n \sim cn^{-\zeta}$ or, at least, $b_n \sim L(n)n^{-\zeta}$ for some slowly varying function L. (A slowly varying function is a function $L(n)$ such that for each $t > 0$, $L(tn) \sim L(n)$ as $n \to \infty$. Any power of a logarithm is slowly varying while n^α is not.) The first step in trying to estimate ζ is to produce a large number, say M, of pairs of independent random walks ω_1 and ω_2. The walks are run until the first time k such that $\omega_1(j) = \omega_2(k)$ or $\omega_2(k) = \omega_2(j)$ for some $0 \leq j \leq k$. Some maximum number of steps, N, is set and then for each $n \leq N$ we record $K(n)$, the number of pairs of walks that did not intersect up through time n. Our estimate of b_n is then $K(n)/M$.

How do we then estimate ζ? One way is to plot the estimates of b_n on log-log paper and find the slope of the best line. This method creates a number of problems in analyzing the possible statistical errors. We will give a simpler method. Suppose it were true that

$$b_n = cn^{-\zeta},$$

for some unknown c. Then

$$\zeta = \frac{\ln(b_n/b_{2n})}{\ln 2}.$$

For a given n, standard statistical techniques say that an approximate 95% confidence interval for b_{2n}/b_n is given by

$$\bar{p} \pm 2[\frac{\bar{p}(1-\bar{p})}{K(n)}]^{1/2},$$

where $\bar{p} = \bar{p}_n = K(2n)/K(n)$. This then gives a confidence interval for ζ. As an example, for $d = 3$, $M = 1,000,000$ pairs of walks produced $K(200) = 149,912; K(400) = 122,926; K(800) = 100,725$. Using $n = 200$ we get the 95% confidence interval $(.283, .290)$ for ζ, while using $n = 400$ we obtain the interval $(.284, .291)$. (An interesting fact to note is that these two intervals are "independent" measurements of ζ.)

This Monte Carlo simulation gives good support for the conjectured value of ζ_3. What possible errors can come in?

1. The random walks are being generated by a computer using a pseudo-random number generator. There is a possibility that the numbers are in some sense "not random enough" for the simulation.

2. In the analysis we assumed that $b_n = cn^{-\zeta}$. This is not exactly correct. If it is true that $b_n \sim cn^{-\zeta}$ (or $b_n \sim L(n)n^{-\zeta}$ for some slowly varying L) and we have chosen values of n sufficiently large so that the error is not very big, we can show that the error in the above analysis should not be very large.

3. There is also the possibility that our random sample of walks is a nonrepresentative sample, i.e., that we were unlucky. This is always a fear, but we can use statistical analysis to show how likely it is that the sample is extremely skewed.

12. SIMULATIONS OF SAW

Monte Carlo simulations of SAWs are more difficult than simulations of simple random walks. It is easy to give an efficient routine to produce simple walks of length n (at least if one has no worries about the random number generator). However, the naive routines for producing SAWs suffer from at least one of two difficulties: they are very inefficient (e.g., if one produces simple random walks of length n and then keeps only those walks which are self-avoiding, then the amount of time to produce a walk of length n grows exponentially in n) or they generate SAWs from some distribution other than the uniform distribution on Γ_n (e.g., if one tries to generate walks by giving a transition function that favors points that have not been visited, one can check that the distribution this produces on Γ_n is not the uniform distribution—in fact, the distribution may well lead to a model for random walks in a different universality class from the SAW). This leads to a question: how can one efficiently produce a random sample of walks from the uniform distribution on SAWs?

Some of the most efficient algorithms used today make use of Markov chains on the space of SAWs. Suppose Π is a stochastic matrix on Γ_n, i.e., a matrix $\Pi(\omega, \lambda)$, $\omega, \lambda \in \Gamma_n$, satisfying $0 \leq \Pi(\omega, \lambda)$ and

$$\sum_{\lambda \in \Gamma_n} \Pi(\omega, \lambda) = 1, \quad \omega \in \Gamma_n.$$

Suppose that Π is: *symmetric*, $\Pi(\omega, \lambda) = \Pi(\lambda, \omega)$; *ergodic*, for every ω, λ, there exists an m with $\Pi^m(\omega, \lambda) > 0$; and *aperiodic*, which will be guaranteed if $\Pi(\omega, \omega) > 0$. Then the uniform measure is the invariant measure for this irreducible Markov chain and it is standard that starting with any initial $\omega \in \Gamma_n$, the distribution of the chain approaches the uniform distribution of Γ_n.

Suppose then that such a Π can be found so that transitions of this Markov chain can be made efficiently. Then this suggests a way to get an approximately uniform sample from Γ_n: start with some $\omega \in \Gamma_n$; run the chain for M steps and use that walk as the first walk in the sample; do another M steps and use that walk for the second walk and continue. How large M has to be in order to make this sample nearly uniform depends on Π.

One example of a Π that has produced efficient algorithms is the *pivot algorithm*. Let $\mathcal{O}$ be the set of d-dimensional orthogonal transformations that leave Z^d invariant. For $d = 2$, this consists of rotations by integral multiples of $\pi/2$, reflections about the coordinate axes, and reflections about the diagonals. The pivot algorithm goes as follows: start with $\omega \in \Gamma_n$. Choose a $k, 0 \leq k < n$, randomly and also choose a $T \in \mathcal{O}$ at random. Consider the random walk path that fixes the first k steps of ω and performs the transformation T on the remaining walk (considering $\omega(k)$ to be the origin). This will produce a random walk path that may or may not be self-avoiding. If it is self-avoiding, we take this walk; otherwise, we stay with ω. It is easy to see that this corresponds to a reversible, aperiodic, Markov chain, and it can be shown that this chain is in fact ergodic. While at first this transformation may not appear to be very efficient (since many possible transformations will be rejected), it turns out that the pivot transformation allows for extremely accurate estimation of some exponents, e.g., the exponent ν.

13. NOTES

For more detailed treatments of simple random walk see [3], [6]. The first of these also discusses the problem of the intersection of random walks. The self-avoiding walk was first introduced by Hammersley and is still an area of active research. There is some discussion of self-avoiding walks in [3]; a more detailed treatment of SAWs can be found in [4]. A lot of work on polymer models has been done from a nonrigorous viewpoint, see e.g., [1]. No attempt has been made in this article to give credit for results. (See [3], [4] and their bibliographies for references to original articles.)

References

[1] **de Gennes, P-G.**, *Scaling Concepts in Polymer Physics*, Cornell University Press, 1979.

[2] **Feller, W.**, *An Introduction to Probability Theory and Its Applications*, Vol I, John Wiley & Sons, 1968.

[3] **Lawler, G.**, *Intersections of Random Walks*, Birkhäuser, Boston, 1991.

[4] **Madras, N. and Slade, G.**, *The Self-Avoiding Walk*, Birkhäuser, Boston, 1993.

[5] **Slade, G.**, The diffusion of self-avoiding walk in high dimensions, *Commun. Math. Phys.*, **110**, 661–683, 1987.

[6] **Spitzer, F.**, *Principles of Random Walk*, Springer-Verlag, 1976.

Chapter 3

SOME CONNECTIONS BETWEEN BROWNIAN MOTION AND ANALYSIS VIA STOCHASTIC CALCULUS[*]

R. J. Williams[†]
Department of Mathematics
University of California, San Diego

ABSTRACT

In this paper, some connections between the fundamental stochastic process Brownian motion and the mathematical subject of analysis are made using stochastic calculus. This calculus which was introduced by K. Itô enables one to compute with functions of Brownian motion. A distinctive feature of stochastic calculus is that the change of variables formula is different from that in ordinary Newton calculus because the sample paths of Brownian motion are of unbounded variation. In this note, some basic aspects of stochastic calculus are explained first. Then this calculus is used as a tool to make connections between Brownian motion and the following problems in analysis: the classical Dirichlet problem, the Schrödinger equation, and Laplace's equation with oblique derivative boundary conditions in a quadrant. The latter is relevant to the study of approximations to two station queueing systems in heavy traffic. References to some of the multitude of other properties and applications of Brownian motion are included at the end of this paper.

1. INTRODUCTION

Brownian motion in $\mathbb{R}^d$ $(d \geq 1)$ is a fundamental stochastic process because it lies at the intersection of many different topics in probability, analysis, and applied stochastics. In particular, if $B = \{B(t), \ t \geq 0\}$ is a Brownian motion in $\mathbb{R}^d$ that starts from the origin, then (i) B is a limit of renormalized simple symmetric random walks, (ii) B has the self-similarity property that it is equal in distribution to $\{\lambda^{-\frac{1}{2}}B(\lambda t), \ t \geq 0\}$ for any $\lambda > 0$, (iii) B has

---

[*]Part of the material discussed in this article was presented in an MAA Invited Lecture at the Annual Joint Mathematics Meeting held in Phoenix, Arizona, in January 1989.

[†]This article was written while the author was visiting the Technion, Israel, and was supported in part by an Alfred P. Sloan Fellowship.

independent components, stationary independent increments, and continuous sample paths, (iv) B is a Gaussian process, (v) B is a martingale with respect to its own filtration, (vi) any bounded harmonic function of B yields a martingale, (vii) B is a time-homogeneous Markov process with transition probability densities that satisfy the heat equation, and (viii) B plays the role of the key source of randomness in many models arising in applications in the physical, biological and social sciences.

This note focuses on some connections between Brownian motion and analysis, for which property (vi) above is a prototype. Some of the examples presented here are motivated by problems arising in applications. A key tool in the discussion that follows is a *stochastic calculus* that enables one to compute with functions of Brownian motion. A feature of this calculus is that the change of variables formula is different from the one in ordinary Newton calculus, because the sample paths of Brownian motion are of unbounded variation. Before the rudimentary aspects of stochastic calculus are described, the reader is reminded of some aspects of ordinary Riemann-Stieltjes integration. There is no need to consider Lebesgue-Stieltjes integrals because all of the processes considered here have continuous sample paths.

Let $f, g : \mathbb{R}_+ \to \mathbb{R}$ be continuous functions, and suppose that g is also locally of bounded variation, i.e., g is of bounded variation on each compact interval in $\mathbb{R}_+$. Under the rules of ordinary Riemann-Stieltjes integration, one can define

$$\int_0^t f(s)dg(s) = \lim_{n \to \infty} \sum_{t_i^n, t_{i+1}^n \in \pi_n} f(\tilde{t}_i^n)(g(t_{i+1}^n) - g(t_i^n)), \qquad (1)$$

where for each n, $\pi_n \equiv \{t_0^n, t_1^n, \ldots, t_n^n\}$ is a partition of $[0, t]$ such that $0 = t_0^n < t_1^n < \cdots < t_n^n = t$, $\tilde{t}_i^n \in [t_i^n, t_{i+1}^n]$ for each $i \in \{0, 1, \ldots, n-1\}$, and $|\pi_n| \equiv \max_{i=0}^{n-1} |t_{i+1}^n - t_i^n| \to 0$ as $n \to \infty$. Furthermore, by requiring that the following integration by parts formula hold,

$$f(t)g(t) - f(0)g(0) = \int_0^t f(s)dg(s) + \int_0^t g(s)df(s), \qquad (2)$$

one can define $\int_0^t g(s)df(s)$, because all of the other entities in (2) are well defined. Note that f need not be of bounded variation. However, if neither f nor g is locally of bounded variation, then in general one cannot make sense of the deterministic integral $\int_0^t f(s)dg(s)$. Finally, if g is locally of bounded variation and $F : \mathbb{R} \to \mathbb{R}$ is a continuously differentiable function, then one has the change of variables formula

$$F(g(t)) = F(g(0)) + \int_0^t F'(g(s))dg(s). \qquad (3)$$

Now, consider a Brownian motion B in $\mathbb{R}$. Since the sample paths of B are continuous, for a given sample path of B, one can take $f(s) = B(s)$ for

$s \geq 0$ in the above, i.e., for a given ω one can consider $f(s) = B(s, \omega)$. Then one can use the above to define $\int_0^t g(s)dB(s)$ for any continuous function g that is locally of bounded variation. Indeed, g can even be a sample path of a stochastic process having these properties sample path by sample path. However, one cannot use this procedure to define even such simple integrals as $\int_0^t B(s)dB(s)$, because the sample paths of Brownian motion are *not* locally of bounded variation. Indeed, they are only locally of finite *quadratic variation*, i.e., for each $t \in \mathbb{R}_+$ and sequence $\{\pi_n, n = 1, 2, \ldots\}$ of partitions of $[0, t]$ as described before,

$$[B]_t \equiv \lim_{n \to \infty} \sum_{t_i^n, t_{i+1}^n \in \pi_n} (B(t_{i+1}^n) - B(t_i^n))^2 \tag{4}$$

exists as a non-trivial limit in probability. Indeed, by a bare-hands calculation one can show that $[B]_t = t$. For many practical purposes one would like to be able to define integrals of the form $\int_0^t f(B(s))dB(s)$ for continuous functions $f :$ $\mathbb{R} \to \mathbb{R}$. In particular, such integrals play an essential role in the development of a change of variables formula for sufficiently differentiable functions of B. With regard to this, note that by use of a telescoping series,

$$(B(t))^2 = (B(0))^2 + \sum_{t_i^n, t_{i+1}^n \in \pi_n} 2B(t_i^n)(B(t_{i+1}^n) - B(t_i^n))$$

$$+ \sum_{t_i^n, t_{i+1}^n \in \pi_n} (B(t_{i+1}^n) - B(t_i^n))^2 \tag{5}$$

where by (4) the last sum on the right tends to t in probability as $n \to \infty$. Thus, if one defines $\int_0^t B(s)dB(s)$ to equal the limit in probability of $\sum_{t_i^n, t_{i+1}^n \in \pi_n} B(t_i^n)(B(t_{i+1}^n) - B(t_i^n))$ as $n \to \infty$, then one obtains from (5) that

$$(B(t))^2 = (B(0))^2 + 2 \int_0^t B(s)dB(s) + t. \tag{6}$$

This suggests that a change of variables formula for B does not have the same form as (3) in general, due to an extra contribution from the quadratic variation of B. In fact, one can make sense of integrals such as $\int_0^t B(s)dB(s)$ as limits in probability of approximating sums of the form indicated above, and there is a change of variables formula for *twice* continuously differentiable functions of B. Since the examples that follow involve d-dimensional Brownian motions with various starting points and sometimes additional processes with sample paths that are locally of bounded variation, stochastic integrals and the associated change of variables formula are described below in sufficient generality to accommodate these examples. First, a definition of d-dimensional Brownian motion (or equivalently, Brownian motion in $\mathbb{R}^d$) is given.

A d-dimensional Brownian motion that starts from the origin is a stochastic process $B \equiv \{B(t) : t \geq 0\}$ with continuous sample paths and independent components $B_1, \ldots, B_d$, such that for each $j \in \{1, \ldots, d\}$, B_j is a one-dimensional Brownian motion characterized by (i)–(iii) below:

(i) for any $0 = t_0 < t_1 < \ldots < t_\ell < \infty$, $\{B_j(t_k) - B_j(t_{k-1}), \ k = 1, \ldots, \ell\}$ are independent random variables,

(ii) for any $0 \leq s < t < \infty$, $B_j(t) - B_j(s)$ is a normally distributed random variable with mean zero and variance $t - s$,

(iii) $B_j(0) = 0$.

A d-dimensional Brownian motion that starts from $x \in \mathbb{R}^d$ is obtained by replacing (iii) by (iii'): $B_j(0) = x_j$, the jth component of x. In the sequel, B denotes a d-dimensional Brownian motion starting from some $x \in \mathbb{R}^d$, defined on a complete probability space $(\Omega, \mathcal{F}, P_x)$. Expectations with respect to P_x are denoted by E_x. For each $t \geq 0$, $\mathcal{F}_t \equiv \sigma\{B(s) : 0 \leq s \leq t\}$, the σ-field generated by B up to time t and augmented (denoted by the tilde) by the P_x-null sets in $\mathcal{F}$. Then, $\{B(t), \mathcal{F}_t, t \geq 0\}$ is a martingale [see (i)–(iii) below for a definition]. Let m be a non-negative integer and let $Y = \{Y(t), t \geq 0\}$ be an m-dimensional stochastic process defined on $(\Omega, \mathcal{F}, P_x)$, such that Y has continuous sample paths and for each t, $Y(t)$ is measurable with respect to $\mathcal{F}_t$, i.e., $Y(t)$ is measurable as a function from $(\Omega, \mathcal{F}_t)$ into $(\mathbb{R}^m, \mathcal{B}^m)$ where $\mathcal{B}^m$ denotes the family of Borel sets in $\mathbb{R}^m$. When referring to the latter property, one says that Y is *adapted* to $\{\mathcal{F}_t : t \geq 0\}$. It is further assumed that the sample paths of Y are locally of bounded variation. Let $f : \mathbb{R}^d \times \mathbb{R}^m \to \mathbb{R}^d$ be a continuous function. Then for each $t \in \mathbb{R}_+$,

$$M(t) \equiv \int_0^t f(B(s), Y(s)) \cdot dB(s)$$

$$= \lim_{n \to \infty} \sum_{j=1}^d \sum_{t_i^n, t_{i+1}^n \in \pi_n} f_j(B(t_i^n), Y(t_i^n))(B_j(t_{i+1}^n) - B_j(t_i^n)) \quad (7)$$

can be shown to exist as a limit in probability, where $\{\pi_n, n = 1, 2, \ldots\}$ is a sequence of partitions of $[0, t]$ as described before. Moreover, M can be taken to have continuous sample paths and if f is bounded, then $\{M(t), \mathcal{F}_t, t \geq 0\}$ is a martingale, i.e.,

(i) $M(t)$ is $\mathcal{F}_t$-measurable for each $t \geq 0$,

(ii) $E_x[\|M(t)\|] < \infty$ for each $t \geq 0$,

(iii) $E_x[M(t) \mid \mathcal{F}_s] = M(s)$ for all $0 \leq s < t < \infty$,

where $E_x[\ \cdot \mid \mathcal{F}_s]$ denotes conditional expectation given $\mathcal{F}_s$. In particular, $E_x[M(t)] = E_x[M(0)] = 0$ for all $t \geq 0$. In order to obtain the martingale property for M (which is inherited from that of B), it is important that one use $f_j(B, Y)$ evaluated at the *left* end-point of the intervals $[t_i^n, t_{i+1}^n]$ in the approximating sums of (7). Using other points in the interval can yield a different (non-martingale) value for the integral. This is in marked contrast to the situation for Riemann-Stieltjes integrals. Now, if $F : \mathbb{R}^d \times \mathbb{R}^m \to \mathbb{R}$ is such that $F = F(b, y)$ is twice continuously differentiable in $b \in \mathbb{R}^d$ and once continuously differentiable in $y \in \mathbb{R}^m$, then it can be shown that P_x-a.s. for all $t \geq 0$,

$$
\begin{aligned}
F(B(t), Y(t)) = {} & F(B(0), Y(0)) + \int_0^t \nabla_b F(B(s), Y(s)) \cdot dB(s) \\
& + \int_0^t \nabla_y F(B(s), Y(s)) \cdot dY(s) + \tfrac{1}{2} \int_0^t \triangle_b F(B(s), Y(s)) ds,
\end{aligned} \tag{8}
$$

where $\nabla_b F$ denotes the gradient of F with respect to its first d arguments, $\nabla_y F$ denotes the gradient of F with respect to its last m arguments, $\triangle_b F$ denotes the d-dimensional Laplacian of F with respect to its first d arguments, and the first integral in (8) is defined as a limit in probability as per (7), and the second integral is a sum of m Riemann-Stieltjes integrals defined path-by-path with respect to the locally bounded variation sample paths of $Y_1, \dots, Y_m$, i.e.,

$$
\int_0^t \nabla_y F(B(s), Y(s)) \cdot dY(s) = \sum_{k=1}^m \int_0^t \frac{\partial F}{\partial y_k}(B(s), Y(s)) dY_k(s).
$$

The last integral in (8) is defined path-by-path as an ordinary Riemann integral. Formula (8) is a version of Itô's change of variables formula in stochastic calculus. Note that this differs from the formula in ordinary Newton calculus by the addition of the last term in (8) that arises because the paths of Brownian motion are locally of finite quadratic variation rather than being locally of bounded variation. If $d = 1$ and $m = 0$ then (8) simplifies to

$$
F(B(t)) = F(B(0)) + \int_0^t F'(B(s)) dB(s) + \frac{1}{2} \int_0^t F''(B(s)) ds, \tag{9}
$$

and in particular, if $F(b) = b^2$ then one recovers (6). For a justification of (7) and (8) the reader is referred to [4]. Here the use of (8) will be illustrated with some examples. The first of these is a classical application to the Dirichlet problem.

2. DIRICHLET PROBLEM

Let D be a bounded domain in $\mathbb{R}^d$ with boundary ∂D. Let f be a continuous real-valued function defined on ∂D. Consider solutions $u \in C^2(D) \cap C(\bar{D})$ of the *Dirichlet problem:*

$$
\triangle u = 0 \quad \text{in } D, \tag{10}
$$

$$
u = f \quad \text{on } \partial D. \tag{11}
$$

Here $C^2(D)$ denotes the set of real-valued functions that are defined and twice continuously differentiable on D. The set of real-valued functions that are defined and continuous on the closure $\bar{D}$ of D is denoted by $C(\bar{D})$. Physically, a solution of this Dirichlet problem yields the equilibrium temperature distribution in the region D when the temperature at the boundary of the region has a fixed distribution determined by the function f. It is well known that in order to solve this Dirichlet problem in general, one must impose a regularity condition on the boundary. Here a probabilistic definition of regularity is given, which is equivalent to the usual analytic one.

Definition. Let $\tau_D = \inf\{t > 0 : B(t) \notin D\}$. A point $x \in \partial D$ is *regular* if

$$P_x(\tau_D = 0) = 1.$$

The boundary, ∂D, is said to be regular if every point $x \in \partial D$ is regular.

Thus, a point $x \in \partial D$ is regular if and only if Brownian motion started at x hits $D^c \equiv \mathbb{R}^d \backslash D$ immediately after time zero, with probability one. There are examples of domains with boundary points that are not regular, Lebesgue's thorn being a classical example in three dimensions. The reader is referred to [5, p. 248] or [9, Section 7.10] for more details on this and on necessary and sufficient conditions for regularity of boundary points.

Theorem. *Suppose ∂D is regular. The following are equivalent.*

(i) $u \in C^2(D) \cap C(\bar{D})$ *satisfies* (10) *and* (11).
(ii) $u(x) = E_x[f(B(\tau_D))]$ *for all* $x \in \bar{D}$.

Proof. Itô's formula (8) will be used to prove that (i) implies (ii). That is, it will be used to give a probabilistic representation for solutions of the Dirichlet problem. Indeed, such probabilistic representations provide a convenient means for establishing uniqueness of solutions of partial differential equations. Given a function u satisfying (i), the representation (ii) actually holds for $x \in D$ without the assumption that ∂D is regular, as can be seen from the proof below. The converse, (ii) implies (i), requires more knowledge of the behavior of Brownian motion than is assumed here. A proof can be found, for instance, in [1, Chapter 4].

Let $\{D_n\}_{n=0}^\infty$ be a sequence of subdomains of D such that $\bar{D}_n \subset D_{n+1} \subset D$ for all n and $\bigcup_n D_n = D$. Fix $x \in D$ and let n be sufficiently large that $x \in D_n$. Since $u \in C^2(D)$ and $\bar{D}_n$ is compact, u can be extended off $\bar{D}_n$ to a function $u_n \in C_b^2(\mathbb{R}^d)$, the space of twice continuously differentiable functions that together with their first and second partial derivatives are bounded on $\mathbb{R}^d$. Applying (8) with $m = 0$ and $F = u_n$, one obtains P_x-a.s. for all $t \geq 0$:

$$u_n(B(t)) = u_n(B(0)) + \int_0^t \nabla u_n(B(s)) \cdot dB(s) + \frac{1}{2}\int_0^t \triangle u_n(B(s))ds. \quad (12)$$

The stochastic integral with respect to dB defines a martingale since ∇u_n is continuous and bounded on $\mathbb{R}^d$. By truncating time at $\tau_{D_n} \equiv \inf\{s \geq 0 : B(s) \notin D_n\}$, t and u_n can be replaced by $t \wedge \tau_{D_n}$ and u respectively in (12). By Doob's optional stopping theorem [1, p. 30], $\{\int_0^{t \wedge \tau_{D_n}} \nabla u(B(s)) \cdot dB(s), \mathcal{F}_t, t \geq 0\}$ is a martingale and hence has zero expectation under P_x. Also, $\triangle u = 0$ on D_n, so the last integral in (12) with $t \wedge \tau_{D_n}$ in place of t is 0. Hence, after replacing t by $t \wedge \tau_{D_n}$ in (12) and taking expectations there, one obtains

$$E_x\left[u(B(t \wedge \tau_{D_n}))\right] = u(x). \tag{13}$$

Since D is bounded, by the properties of Brownian motion (cf. [4, Ex. 12, p. 116]),

$$P_x(\tau_D < \infty) = 1, \tag{14}$$

which implies that $P_x(\tau_{D_n} < \infty) = 1$ for each n. Thus, by bounded convergence and the continuity of u, one can let $t \to \infty$ in (13) to obtain

$$E_x\left[u(B(\tau_{D_n}))\right] = u(x). \tag{15}$$

Now since $D_n^c \downarrow D^c$ and $x \in D$, $\tau_{D_n} \uparrow \tau_D$ P_x-a.s., and so by bounded convergence and the continuity of u on $\bar{D}$, on letting $n \to \infty$ in (15) one obtains

$$E_x\left[u(B(\tau_D))\right] = u(x) \quad \text{for all} \quad x \in D. \tag{16}$$

Since $u = f$ on ∂D, it follows that (ii) holds for x in D.

If $x \in \partial D$, then by the regularity of ∂D, $\tau_D = 0$ P_x-a.s., and then

$$E_x\left[f(B(\tau_D))\right] = f(x) = u(x). \qquad \square$$

3. SCHRÖDINGER EQUATION

Let D be a bounded domain in $\mathbb{R}^d$ with regular boundary, let q be a bounded, continuous (if $d = 1$) or Hölder continuous (if $d \geq 2$), real-valued function defined on D, and let f be a continuous real-valued function defined on the boundary ∂D of D. Define τ_D as in the previous section. For notational convenience, extend q to be zero off D. For each $t \geq 0$, define

$$e_q(t) = \exp\left(\int_0^t q(B(s))ds\right). \tag{17}$$

This $e_q(\cdot)$ is called the *Feynman-Kac* functional associated with q. It is a continuous one-dimensional process adapted to $\{\mathcal{F}_t\}$ and its sample paths are locally of bounded variation. Consider solutions $u \in C^2(D) \cap C(\bar{D})$ of the reduced *Schrödinger equation*:

$$\frac{1}{2}\triangle u + qu = 0 \quad \text{in } D, \tag{18}$$

$$u = f \quad \text{on } \partial D. \tag{19}$$

If q is non-positive, there is a simple representation for such solutions. This extends to positive q's, or q's which change sign, provided a certain integral (or gauge) condition (20) is satisfied.

Theorem. *Suppose*

$$\varphi(x) \equiv E_x[e_q(\tau_D)] < \infty \quad \text{for some } x \in D. \tag{20}$$

The following are equivalent.

(i) $u \in C^2(D) \cap C(\bar{D})$ *satisfies* (18) *and* (19).
(ii) $u(x) = E_x\left[f(B(\tau_D))e_q(\tau_D)\right]$ *for all* $x \in \bar{D}$.

Proof. It is shown below that (i) implies (ii). The converse is more delicate, and in particular uses the assumption of continuity/Hölder continuity of q (see Chung [1, §4.7]).

Suppose (i) holds. Let $\{D_n\}_{n=0}^{\infty}$, $\{\tau_{D_n}\}_{n=0}^{\infty}$ be as in the previous section. Fix $x \in D$ and let n be sufficiently large that $x \in D_n$. As in Section 2, extend u off $\bar{D}_n$ to a function $u_n \in C_b^2(\mathbb{R}^d)$. Then by applying (8) with $m = 1$, $F(b, y) = u_n(b)y$ and $Y = e_q$, and then truncating time at τ_{D_n}, one obtains P_x-a.s. for all $t \geq 0$,

$$
\begin{aligned}
u(B(t \wedge \tau_{D_n}))\, e_q(t \wedge \tau_{D_n}) - u(B(0)) &= \int_0^{t \wedge \tau_{D_n}} e_q(s)\nabla u(B(s)) \cdot dB(s) \\
&+ \int_0^{t \wedge \tau_{D_n}} (qu)(B(s))e_q(s)ds \\
&+ \frac{1}{2}\int_0^{t \wedge \tau_{D_n}} \triangle u(B(s))e_q(s)ds. \tag{21}
\end{aligned}
$$

Since u satisfies (18) and $B(\cdot \wedge \tau_{D_n}) \in \bar{D}_n \subset D$ P_x-a.s., the sum of the last two integrals is zero. Now, $e_q \nabla u_n(B)$ is bounded on each compact time interval. It can be shown from this that $\{\int_0^t e_q(s)\nabla u_n(B(s)) \cdot dB(s), \mathcal{F}_t, t \geq 0\}$ is a martingale, and hence by Doob's optional stopping theorem, $\{\int_0^{t \wedge \tau_{D_n}} e_q(s)\nabla u(B(s)) \cdot dB(s), \mathcal{F}_t, t \geq 0\}$ is a martingale which has zero expectation. Thus taking expectations in (21) yields

$$E_x\left[u(B(t \wedge \tau_{D_n}))e_q(t \wedge \tau_{D_n})\right] = u(x). \tag{22}$$

Recall from Section 2 that $\tau_D < \infty$ P_x-a.s. It has been shown by Chung and Rao [2] (see also [4, §6.4]) that *under condition* (20), $\{u(B(t \wedge \tau_{D_n}))e_q(t \wedge \tau_{D_n}) :$

$t \geq 0$, $n \geq 0$} is uniformly integrable under P_x, and so, by combining this with the continuity of u on $\bar{D}$, one can pass to the limit as $t \to \infty$ and then $n \to \infty$ in (22) to obtain

$$E_x\left[u(B(\tau_D))e_q(\tau_D)\right] = u(x) \text{ for all } x \in D.$$

Since $u = f$ on ∂D, this again reduces to (ii) for $x \in D$. If $x \in \partial D$, the regularity of ∂D gives the representation there.   □

The function φ defined in (20) is known as the *Feynman-Kac gauge*. The following gives some equivalent conditions for finiteness of this gauge. For a proof, see [3]. (For $d = 1$, q is assumed to be Hölder continuous in [3]. However, scrutiny of the proof reveals that in this one-dimensional case the result still holds if q is simply bounded and continuous on D.)

Proposition. *The following conditions are equivalent.*

(i) *$\varphi(x) < \infty$ for some $x \in D$.*
(ii) *There is a solution $u \in C^2(D) \cap C(\bar{D})$ of (18)–(19) satisfying $u > 0$ on $\bar{D}$.*
(iii) *There is no $\lambda \geq 0$ such that the eigen-problem*

$$\begin{cases} \frac{1}{2}\Delta u + qu & = \lambda u \;\; \text{in } D \\ u & \equiv 0 \;\;\; \text{on } \partial D \end{cases}$$

has a non-trivial solution $u \in C^2(D) \cap C(\bar{D})$.

Remark. If $q \leq 0$, then (i) is easily seen to hold and hence the Proposition yields that (ii)–(iii) hold, which is a well known fact in analysis. The case $q \equiv 0$ corresponds to the Dirichlet problem treated in Section 2.

4. REFLECTED BROWNIAN MOTION IN A QUADRANT

In this section, a process which is a functional of Brownian motion and that arises as an approximation to the queue-length process in a simple queueing network model will be considered. The queueing model is described first.

Consider two single-server queues in parallel (see Figure 1). The two arrival processes for these queues are assumed to be renewal processes that are independent of one another. The service times for a given server are assumed to form a sequence of independent, identically distributed random variables. The two sequences for the two servers are assumed to be independent of one another and of the arrival processes. The arrival rates are assumed to be equal and so are the service rates. Each server has an infinite waiting room. If the first server is ever idle, customers are transferred from the queue of the second server to that of the first server.

As it stands, in this generality, the queueing model cannot be analyzed exactly. However, with very general assumptions on the interarrival time and service time distributions and service disciplines, under conditions of *heavy traffic* (mean interarrival times roughly equal to mean service times), one can approximate the two-dimensional queue length process for this pair of queues by a diffusion process that lives in the positive quadrant of $\mathbb{R}^2$ [11]. This diffusion behaves like Brownian motion in the interior of the quadrant, and it is confined to the quadrant by instantaneous "pushing" at the boundary, where the direction of push is constant on a given side and these directions are illustrated in Figure 2. For historical reasons the directions of push are called *directions of reflection,* although one should not think of the construction as by a mirror reflection, but rather by "deflection" or "pushing" at the boundary in the prescribed directions. The directions of reflection have a natural interpretation in terms of the original queueing model. Namely, the normal reflection on the horizontal boundary corresponds to the enforcement of the non-negativity constraint on the contents of queue 2, whereas the 45° downward reflection on the vertical boundary corresponds to the enforcement of the non-negativity constraint on queue 1 as well as the fact that, when server 1 has no customers from his queue to serve, he can serve customers from queue 2, causing a corresponding decrement in the contents of queue 2.

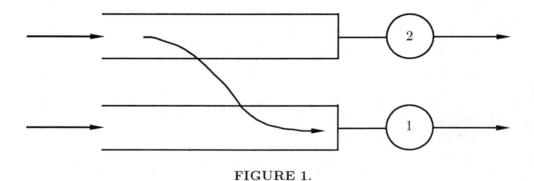

FIGURE 1.

In fact, there is an explicit representation for the diffusion process described above. Let B be a two-dimensional Brownian motion starting from some point x in the positive quadrant $\mathbb{R}^2_+$. Define for each $t \geq 0$,

$$
\begin{aligned}
Y_1(t) &= \left(-\min_{0 \leq s \leq t} B_1(s)\right)^+ \\
Z_1(t) &= B_1(t) + Y_1(t) \\
Y_2(t) &= \left(-\min_{0 \leq s \leq t} (B_2(s) - Y_1(s))\right)^+ \\
Z_2(t) &= B_2(t) - Y_1(t) + Y_2(t),
\end{aligned}
$$

where $y^+ = \max(y, 0)$ for $y \in \mathbb{R}$. Then $Z \equiv \begin{bmatrix} Z_1 \\ Z_2 \end{bmatrix}$ is a reflected Brownian motion in $\mathbb{R}^2_+$ that starts from x and has directions of reflection as indicated in

PLATE I. Nucleation of spiral pairs in a prototypical Greenberg–Hastings model.

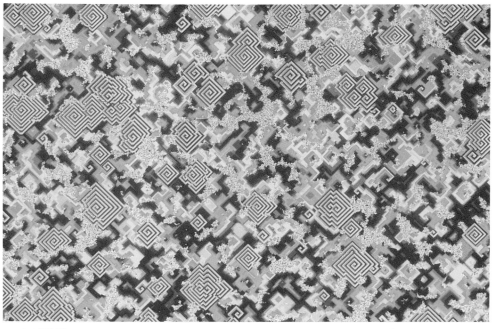

PLATE II. Clustering of wave fragments in a Greenberg–Hastings model with high threshold.

PLATE III. Self-organization of droplets and spirals in the basic cyclic cellular automaton.

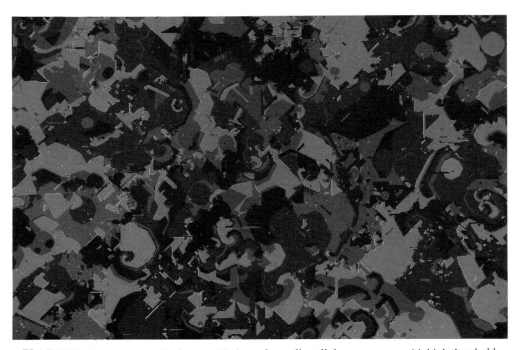

PLATE IV. Large-scale turbulent equilibrium of a cyclic cellular automaton with high threshold.

Figure 2. Indeed, Y_1, Y_2 may be characterized as the unique pair of continuous, non-decreasing processes such that

(i) $Z(t) \equiv B(t) + v_1 Y_1(t) + v_2 Y_2(t) \in \mathbb{R}_+^2$ for all $t \geq 0$, and

(ii) $Y_i(0) = 0$, Y_i can increase only when Z_i is zero, $i = 1, 2$,

where $v_1 = \left(\begin{smallmatrix}1\\-1\end{smallmatrix}\right)$, $v_2 = \left(\begin{smallmatrix}0\\1\end{smallmatrix}\right)$ are the directions of reflection (normalized to have inward normal component of length one) shown in Figure 2.

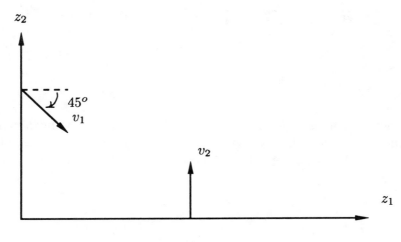

FIGURE 2.

One of the questions of interest for the process Z is whether it ever reaches the origin starting from $x \neq 0$. The following function, together with Itô's formula, allows us to answer this question in a precise manner. Let (r, θ) be polar coordinates in $\mathbb{R}_+^2$ with $r \geq 0$ and $\theta \in [0, \frac{\pi}{2}]$. Let

$$u(r, \theta) = r^{\frac{1}{2}} \cos\left(\frac{1}{2}\theta\right). \tag{23}$$

Then u is the real part of the complex function $z^{\frac{1}{2}}$ and consequently is harmonic in $\mathbb{R}_+^2 \backslash \{0\}$. Moreover, for $r > 0$,

$$\frac{1}{r}\frac{\partial u}{\partial \theta}\Big|_{\theta=0} = -\frac{1}{2}r^{-\frac{1}{2}}\sin\left(\frac{1}{2}\theta\right)\Big|_{\theta=0} = 0$$

and

$$\left(-\frac{1}{r}\frac{\partial u}{\partial \theta} - \frac{\partial u}{\partial r}\right)\Big|_{\theta=\frac{\pi}{2}} = \frac{1}{2}\left(r^{-\frac{1}{2}}\sin\left(\frac{1}{2}\theta\right) - r^{-\frac{1}{2}}\cos\left(\frac{1}{2}\theta\right)\right)\Big|_{\theta=\frac{\pi}{2}} = 0.$$

Interpreting these equations in Cartesian coordinates yields

$$v_2 \cdot \nabla u = 0 \quad \text{on} \quad \{z \in \mathbb{R}_+^2 \backslash \{0\} : z_2 = 0\}, \tag{24}$$

$$v_1 \cdot \nabla u = 0 \quad \text{on} \quad \{z \in \mathbb{R}_+^2 \backslash \{0\} : z_1 = 0\}. \tag{25}$$

Suppose $x \in \mathbb{R}_+^2 \backslash \{0\}$ and fix ϵ, R such that $0 < \epsilon < u(x) < R < \infty$. Let $\bar{D}_{\epsilon R} = \{z \in \mathbb{R}_+^2 : \epsilon \le u(z) \le R\}$. Then one can extend u off $\bar{D}_{\epsilon R}$ to a function $\tilde{u} \in C_b^2(\mathbb{R}^2)$. By applying Itô's formula (8) with $F : \mathbb{R}^2 \times \mathbb{R}^2 \to \mathbb{R}$ given by $F(b, y) = \tilde{u}(b + v_1 y_1 + v_2 y_2)$ for $b \in \mathbb{R}^2$, $y \in \mathbb{R}^2$, one obtains P_x-a.s. for all $t \ge 0$,

$$
\begin{aligned}
\tilde{u}(Z(t)) = \tilde{u}(Z(0)) \quad &+ \quad \int_0^t \nabla \tilde{u}(Z(s)) \cdot dB(s) \\
&+ \quad \sum_{i=1}^2 \int_0^t (v_i \cdot \nabla \tilde{u})(Z(s)) dY_i(s) \\
&+ \quad \frac{1}{2} \int_0^t \triangle \tilde{u}(Z(s)) ds.
\end{aligned} \tag{26}
$$

The stochastic integral with respect to dB defines a martingale since $\nabla \tilde{u}$ is continuous and bounded on $\mathbb{R}^2$. By truncating time at $\tau_{\epsilon R} \equiv \inf\{t \ge 0 : u(Z(t)) \le \epsilon$ or $u(Z(t)) \ge R\}$, and noting that $\triangle u = 0$ in $\bar{D}_{\epsilon R}$, one obtains P_x-a.s. for all $t \ge 0$,

$$
\begin{aligned}
u(Z(t \wedge \tau_{\epsilon R})) = u(x) \quad &+ \quad \int_0^{t \wedge \tau_{\epsilon R}} \nabla u(Z(s)) \cdot dB(s) \\
&+ \quad \sum_{i=1}^2 \int_0^{t \wedge \tau_{\epsilon R}} (v_i \cdot \nabla u)(Z(s)) dY_i(s),
\end{aligned} \tag{27}
$$

where by Doob's optional stopping theorem the first integral defines a martingale, which has zero expectation. Moreover, since Y_i can increase only when $Z_i = 0$, and $v_i \cdot \nabla u = 0$ on $\{z \in \mathbb{R}_+^2 \backslash \{0\} : z_i = 0\}$, the integrals in the last term of (27) are zero. Thus, taking expectations in (27) yields

$$E_x \left[u(Z(t \wedge \tau_{\epsilon R})) \right] = u(x). \tag{28}$$

Now, one would like to let $t \to \infty$ in (28). For this, the following is needed.

Proposition. *For each $K > 0$,*

$$P_x(\sigma_K < \infty) = 1,$$

where $\sigma_K = \inf\{t \ge 0 : |Z(t)| \ge K\}$.

Proof. From (i), for $v = \binom{1}{1}$,

$$v \cdot Z(t) = v \cdot B(t) + Y_2(t) \ge v \cdot B(t) \quad \text{for all } t \ge 0, \tag{29}$$

since $Y_2(t) \ge 0$. Now, $v \cdot B$ is equal in distribution to $\sqrt{2}$ times a one-dimensional Brownian motion and consequently

$$P_x(\limsup_{t \to \infty} v \cdot B(t) = +\infty) = 1,$$

and hence by (29),

$$P_x(\limsup_{t \to \infty} v \cdot Z(t) = +\infty) = 1.$$

Since $v \cdot Z \leq 2|Z|$, the Proposition follows. $\square$

Now, $u(z) \geq |z|^{\frac{1}{2}} c$ where $c = \inf_{\theta \in [0, \frac{\pi}{2}]} \cos(\frac{1}{2}\theta) > 0$, and so it follows from the above Proposition that

$$P_x(\tau_{\epsilon R} < \infty) = 1. \tag{30}$$

Hence, since u is bounded on $\bar{D}_{\epsilon R}$, one can let $t \to \infty$ in (28) to obtain by bounded convergence that

$$E_x[u(Z(\tau_{\epsilon R}))] = u(x). \tag{31}$$

By observing the values of u at $Z(\tau_{\epsilon R})$, one concludes from (31) that

$$\epsilon P_x(\tau_\epsilon < \tau_R) + R P_x(\tau_R < \tau_\epsilon) = u(x), \tag{32}$$

where $\tau_r = \inf\{t \geq 0 : u(Z(t)) = r\}$. Since $P_x(\tau_R < \tau_\epsilon) = 1 - P_x(\tau_\epsilon < \tau_R)$ by (30), one can rearrange (32) to obtain

$$P_x(\tau_\epsilon < \tau_R) = \frac{R - u(x)}{R - \epsilon}.$$

On letting $\epsilon \downarrow 0$ one obtains

$$P_x(\tau_0 < \tau_R) = 1 - \frac{u(x)}{R}.$$

Finally, letting $R \to \infty$ yields

$$P_x(\tau_0 < \infty) = 1.$$

Thus, the reflected Brownian motion Z hits the origin P_x-a.s. starting from any $x \in \mathbb{R}^2_+ \backslash \{0\}$.

Reflected Brownian motions in two-dimensional polygons and in three and higher dimensional polyhedrons, with constant oblique reflection on each boundary face, arise as approximations to other queueing network models under conditions of heavy traffic. For more discussion of such processes, the interested reader is referred to [6, 7, 8, 11, 12, 13].

Bibliographical note. The reader interested in pursuing more details and applications of stochastic calculus is referred to the books [4, 5], and for further details on many aspects of Brownian motion, see [1, 9, 10].

References

[1] **Chung, K. L.**, *Lectures from Markov Processes to Brownian Motion,* Springer-Verlag, New York, 1982.

[2] **Chung, K. L. and Rao, K. M.**, Feynman-Kac functional and the Schrödinger equation, *Seminar on Stochastic Processes 1981,* eds. E. Çinlar, K.L. Chung, R.K. Getoor, Birkhäuser, Boston, 1981, 1–29.

[3] **Chung, K. L., Li P. and Williams, R. J.**, Comparison of probability and classical methods for the Schrödinger equation, *Expositiones Mathematica,* **4**, 271–278, 1986.

[4] **Chung, K. L. and Williams, R. J.**, *Introduction to Stochastic Integration,* 2nd Edition, Birkhäuser, Boston, 1990.

[5] **Durrett, R.**, *Brownian Motion and Martingales in Analysis,* Wadsworth, Belmont, CA, 1984.

[6] **Harrison, J. M.**, *Brownian Motion and Stochastic Flow Systems,* John Wiley & Sons, New York, 1985.

[7] **Harrison, J. M. and Williams, R. J.**, Brownian models of open queueing networks with homogeneous customer populations, *Stochastics,* **22**, 77–115, 1987.

[8] **Harrison, J. M. and Williams, R. J.**, Brownian models of multiclass queueing networks, *Proc. 29th I.E.E.E. Conf. on Decision and Control,* 1990, 573-574.

[9] **Itô, K. and McKean, H. P. Jr.**, *Diffusion Processes and their Sample Paths,* Springer-Verlag, New York, 1974.

[10] **Karatzas, I. and Shreve, S. E.**, *Brownian Motion and Stochastic Calculus,* Springer-Verlag, 1988.

[11] **Reiman, M. I.**, Open queueing networks in heavy traffic, *Math. Oper. Res.,* **9**, 441–458, 1984.

[12] **Taylor, L. M. and Williams, R. J.**, Existence and uniqueness of semimartingale reflecting Brownian motions in an orthant, to appear in *Probability Theory and Related Fields,* 1993–1994.

[13] **Varadhan, S. R. S. and Williams, R. J.**, Brownian motion in a wedge with oblique reflection, *Comm. Pure Appl. Math.,* **38**, 405–443, 1985.

Chapter 4

CAN YOU FEEL THE SHAPE OF A MANIFOLD WITH BROWNIAN MOTION?*

Mark A. Pinsky
Department of Mathematics
Northwestern University

ABSTRACT

We review the formulation of Brownian motion on a Riemannian manifold, obtained as a limit of piecewise geodesic paths. The exit time and exit place from a geodesic ball can be studied by means of asymptotic expansions in the radius of the ball. In dimension less than six, the mean exit time characterizes the metric of the space in the sense of local isometry. Counter-examples are presented to show that, in dimension six or greater, distinct Riemannian metrics may give rise to the same exit time distribution, in particular the same mean exit time. Parallel results are obtained for the principal Dirichlet eigenvalue of a small geodesic ball and for the harmonic measure of a small sphere.

1. INTRODUCTION

Imagine an ensemble of particles moving at random, independently of one another on a surface or higher dimensional manifold. By performing the averages of suitable functionals of the motion, can we discover the intrinsic geometry of the manifold?

Admittedly imprecise as a mathematical problem, this question belongs to the increasingly large field of "inverse problems". These are perhaps more familiar in the case of inverse spectral problems in which one seeks to recover geometric information beginning with the eigenvalues of a differential operator [9,6,4,3].

Brownian motion on a surface or higher dimensional manifold is canonically defined in terms of the structure of the space. This means that, if we have a manifold M with Riemannian metric g, there is a canonical prescription for constructing a stochastic process $\{X(t), t > 0\}$ which is related to (M, g) in

---

*Research supported by the National Science Foundation. Reprinted with permission from *EXPOSITIONES MATHEMATICAE*, 2(1984).

such a way that specializes to the usual n-dimensional Brownian motion when we take $M = R^n$ with the standard Euclidean metric g_0, where $g_0(X, Y) = \sum_{i=1}^n X_i Y_i$.

For the expert in diffusion theory, Brownian motion may be defined as a diffusion process whose infinitesimal generator is a constant multiple of the Laplace-Beltrami operator. While this may be worked into a satisfactory theory, it lacks the intuitive appeal that one might hope for in a "geometric" theory of Brownian motion. To make the definition more natural, we propose to view Brownian motion as a limit of the "isotropic transport process" a Markov process on the tangent bundle of the Riemannian manifold. This can be done in a coordinate-free manner and gives the usual Brownian motion semigroup in the limit.

Having defined Brownian motion, we can state a precise conjecture as follows: *If the mean exit time of Brownian motion from every ball of radius $r > 0$ is the same as the mean exit time for the standard Brownian motion from a ball of radius r in the Euclidean space (R^n, g_0), prove that $(M, g) \cong (R^n, g_0)$ in the sense of local isometry.*

In the following sections we will describe the positive results obtained in dimension less than six and the counter examples that are found in higher dimensions. As a by-product of these techniques, we can describe results on the harmonic measure of a small geodesic sphere and the principal eigenvalue of a small geodesic ball.

2. BROWNIAN MOTION OF A RIEMANNIAN MANIFOLD

A Riemannian manifold is a pair (M, g) where M is a smooth manifold and g is a Riemannian metric, i.e., a positive-definite bilinear functional on each tangent space $M_m, m \epsilon M$ that varies smoothly: $m \to g(m; X, Y)$ is a smooth function for each pair of smooth vector fields X, Y.

A useful example of a Riemannian manifold is obtained by taking $M = U$, an open set in R^n and $g(m; X, Y) = \sum_{i,j} g_{i,j}(m) X_i(m) Y_j(m)$ where (g_{ij}) are smooth functions on M. If we think of $X = (dx_1, \cdots, dx_n) = Y$ as representing a small displacement, this is often written $ds^2 = \sum_{i,j} g_{ij} dx_i dx_j$. A general Riemannian manifold may be obtained by patching together open sets U to make an entire manifold.

A Riemannian manifold has the structure of a metric space, where the distance between two points is defined by

$$d(m_1, m_2) = \inf L(\gamma).$$

Here the infimum is taken over piecewise smooth curves $t \to \gamma(t), 0 \leq t \leq 1$ with $\gamma(0) = m_1, \gamma(1) = m_2$ and the length of the curve is defined by

$$L(\gamma) = \int_0^1 g(\dot{\gamma}(t), \dot{\gamma}(t))^{1/2} dt$$

where $\dot{\gamma}(t) \in M_{\gamma(t)}$ is the velocity vector of the curve γ at the time t. It is routine to verify that with this definition of distance (M, g) becomes a metric space.

A curve $t \to \gamma(t)$ is a *unit-speed geodesic* if $g(\dot{\gamma}(t), \dot{\gamma}(t)) = 1$ and for every t_0 there exists $\delta > 0$ such that

$$d(\gamma(t_0 - \delta), \gamma(t_0 + \delta)) = \int_{t_0 - \delta}^{t_0 + \delta} g(\dot{\gamma}(t), \dot{\gamma}(t))^{1/2} dt.$$

In other words a geodesic minimizes arc length locally. In the Euclidean space (R^n, g_0) every straight line is a geodesic. On the sphere (S^n, g_1) every great circle is a geodesic (of course the minimizing property only holds for arcs of length less than π in this case). On a general Riemannian manifold (M, g) we may construct geodesics by solving a system of second order ordinary differential equations that are precisely the Euler equations for the variational problem of minimizing the arc-length functional.

With the unit-speed geodesics we may define the *canonical horizontal vector field* Z by

$$Zf(m, \xi) = (d/dt)f(\gamma(t), \dot{\gamma}(t))|_{t=0} \qquad (m = \gamma(0), \xi = \dot{\gamma}(0))$$

where f is a smooth function on the unit sphere bundle $SM = \cup_{m \in M} S_m$ and $S_m = \{\xi \in M_m : g(m; \xi, \xi) = 1\}$. The Riemannian metric is used to define an *averaging operator* by

$$Pf(m) = \int_{S_m} f(m, \xi) \omega_m(d\xi)$$

where ω_m is the uniform probability measure on the unit sphere $S_m \subseteq M_m$. The operator Z maps $C^\infty(SM)$ into itself while P maps $C^\infty(SM)$ into $C^\infty(M)$ [16].

In case $(M, g) = (R^n, g_0)$, ω_m is the usual surface measure on the unit sphere; in a general Riemannian manifold one can easily see that there exists a unique probability measure that is invariant under the orthogonal group defined by the Riemannian metric at $m \in M$.

The operators Z and P have the further properties that $Pf = f$ and $PZf = 0$, for $f \in C^\infty(M)$ (i.e., f depends only on m — smoothly). In the Euclidean case $(M, g) = (R^n, g_0)$ this may be seen by writing

$$Pf(m) = \int_{S_m} f(m) \omega_m(d\xi) = f(m) \int_{S_m} \omega_m(d\xi) = f(m),$$

while

$$PZf(m) = \int_{S_m} \sum_i (\xi_i \partial f / \partial x_i) \omega_m(d\xi) = \sum_i \partial f / \partial x_i \int_{S_m} \xi_i \omega_m(d\xi) = 0$$

since the average of any cartesian coordinate over the unit sphere is zero. In the case of a general manifold the formula for Zf contains additional first-order terms, but these do not affect the validity of $PZf = 0$.

The *normalized Laplacian* of a Riemannian manifold (M, g) is defined by

$$\triangle f = PZ^2 f = P(Z(Zf)) \quad f \in C^\infty(M).$$

Indeed this is an operator on $C^\infty(M)$ since we have the mapping properties depicted below:

$$\text{in } SM: \quad Zf \rightarrow Z^2 f$$
$$\uparrow \qquad \downarrow$$
$$\text{in } M: \quad f \qquad PZ^2 f$$

In case $(M, g) = (R^n, g_0)$ we have

$$
\begin{aligned}
\triangle f &= \int_{S_m} Z^2 f(m, \xi) \omega_m(d\xi) \\
&= \int_{S_m} \sum_{i,j} [\xi_i \xi_j \partial^2 f / \partial x_i \partial x_j] \omega_m(d\xi) \quad \text{(since } \ddot{\gamma} = 0 \text{ here)} \\
&= \sum_{i,j} \partial^2 f / \partial x_i \partial x_j \int_{S_m} \xi_i \xi_j \omega_m(d\xi) \\
&= (1/n) \sum_i \partial^2 f / \partial x_i^2
\end{aligned}
$$

where we have used the fact that the average of the product of two different coordinates is zero, while the average of the square of the ith coordinate is independent of i and obtained by noting that $\sum_i \xi_i^2 = 1$.

On a general n-dimensional Riemannian manifold $\triangle$ is n^{-1} times the Laplace-Beltrami operator [16, Proposition 4.8], whose classical definition need not concern us here. We may simply use the definition PZ^2, which is coordinate-free. One may note that by introducing the unit sphere bundle we have succeeded in factoring the Laplacian in terms of first order differential operators, which is not possible on the base manifold M if $n > 1$. This representation of the Laplacian has the further advantage of providing probabilistic insight, which we now explain.

Imagine a particle that moves along unit-speed geodesics $t \rightarrow \gamma(t)$, except for certain random times $0 = t_0 < t_1 < t_2 < \ldots$ when suddenly it changes direction according to the uniform distribution $\omega_m(d\xi)$. This is a rough description of the *isotropic transport process*, $(m(t), \xi(t))$, a stochastic process which is well-defined on any complete Riemannian manifold. The infinitesimal generator of the isotropic transport process is defined by

$$Af(m, \xi) = (d/dt) E_{m,\xi} f(m(t), \xi(t))|_{t=0+}$$

where $E_{m,\xi}$ denotes the mathematical expectation for paths with $m(0) = m, \xi(0) = \xi$. We assume that the "inter-arrival times" $t_k - t_{k-1}$ are independent random variables and have the exponential distribution: Prob $(t_k - t_{k-1} > t)$

$= e^{-t}$, for $k \geq 1, t > 0$. This hypothesis assures that $(m(t), \xi(t))$ possesses the *Markov property*, which is commonly paraphrased by stating that, given the present state $(m(t_0), \xi(t_0))$, the past $(m(t), \xi(t))_{t<t_0}$ and the future $(m(t), \xi(t))_{t>t_0}$ are independent of one another (in fact the exponential probability law is known to be implied by the Markov property). It is important to note that the state space of this Markov process is the unit sphere bundle SM. The first component $m(t)$ does not possess the Markov property when viewed as a stochastic process taking values in M.

To compute the generator Af, one may first compute the Laplace transform of the Markov semigroup and use the Fubini theorem to get hold of the discrete random variables (t_n) [16, Theorem 3.1]. A more elementary approach is to note that, for small t, with probability $e^{-t} = 1 - t + o(t)$, the process moves along the original geodesic and, with probability $t + o(t)$, it chooses a new direction according the uniform measure ω_m. Thus

$$E_{m,\xi} f(m(t), \xi(t)) = (1-t)f(\gamma(t), \dot\gamma(t)) + t \int_{S_m} f(m, \nu)\omega_m(d\nu) + o(t) \qquad t \downarrow 0.$$

Substracting $f(m, \xi)$ and dividing by t yields

$$E_{m,\xi} f(m(t), \xi(t)) \quad - \quad f(m, \xi) =$$
$$f(\gamma(t), \dot\gamma(t)) \quad - \quad f(m, \xi) + t \int_{S_m} f(m, \nu)\omega_m(d\nu) - tf(\gamma(t), \dot\gamma(t)) + o(t).$$

Thus taking the limit $t \downarrow 0$ we have

$$Af(m, \xi) = Zf(m, \xi) + Pf(m) - f(m, \xi).$$

The infinitesimal generator of the isotropic transport process is a first order integro–differential operator where the derivative is taken along the "horizontal" vector field Z and the integration occurs in each "vertical" fiber S_m.

To obtain Brownian motion we insert a small parameter $\varepsilon > 0$ and consider a sequence of isotropic transport processes whose generators are of the form $A_\varepsilon f = (Zf)/\varepsilon + (Pf - f)/\varepsilon^2$. This can be effected by repeating the above construction with independent exponentially distributed inter-arrival times (t_k) with Prob $(t_k - t_{k-1}) > t) = e^{-t/\varepsilon^2}$ and geodesics with $|\dot\gamma(t)| = \varepsilon$. The state space of the resulting process is the ε-sphere bundle $\cup_{m \in M} \{\xi \in M_m : g(m; \xi, \xi) = \varepsilon\}$. The normalized Laplacian is obtained as the "extended limit" of A_ε in the following sense:

Lemma 2.1 *For each $f \in C^\infty(M)$ there exists a sequence $f_\varepsilon \in C^\infty(SM)$ such that*

$$\lim_{\varepsilon \downarrow 0} f_\varepsilon = f$$

$$\lim_{\varepsilon \downarrow 0} [\frac{Z}{\varepsilon} + \frac{(P-I)}{\varepsilon^2}] f_\varepsilon = PZ^2 f.$$

In fact one may choose $f_\varepsilon = f + \varepsilon(Zf) + \varepsilon^2(Z^2 f - PZ^2 f)$; nothing more complicated is required. The theory of convergence of semigroups, developed in general by Trotter and Kato [12] and adapted by Kurtz [13] for use in probability theory may now be applied. According to this theory we have

$$\lim_{\varepsilon \downarrow 0} \exp t [\frac{Z}{\varepsilon} + \frac{(P-I)}{\varepsilon^2}] f = \exp(t PZ^2) f, \quad f \in C^\infty(M)$$

provided that the limit semigroup $\exp[tPZ^2]$ satisfies some technical conditions; these are automatically satisfied in case M is compact but also hold for many non-compact manifolds [21]. This theorem of "weak convergence" is interpreted by the statement that Brownian motion (generated by PZ^2) is the weak limit of isotropic transport processes (generated by $Z/\varepsilon + (P-I)/\varepsilon^2$) when $\varepsilon \downarrow 0$. This is similar in spirit to the random walk approximation, a well-known method of approximating Brownian motion on Euclidean space. Because of the piecewise geodesic sample paths, we may be tempted to call the present construction a "random-ski approximation" to Brownian motion on a manifold.

It is remarkable that Brownian motion, which is a Markov process with state space M, is obtained as the limit of a sequence of isotropic transport processes that are Markov processes on the sphere bundle, but the first component $m(t)$ is not a Markov process with state space M. In this sense the first component $m(t)$ "becomes Markovian in the limit $\varepsilon \downarrow 0$". This phenomenon can be understood in the context of "homogenization" as was shown in [17].

3. EXIT TIMES, DYNKIN'S FORMULA AND SECTIONAL CURVATURE

We recall some methods from the theory of Markov processes. A Markov process $\{X(t), t > 0\}$ on a locally compact space M is called *Fellerian* if for each $t > 0$ the function $x \to E_x f(X(t))$ is continuous and vanishes at infinity whenever f is continuous and vanishes at infinity. For example Brownian motion on a compact manifold is Fellerian. A Fellerian process has the strong Markov property, i.e., given the present state $X(T)$, the past $\{X(t), t < T\}$ and the future $\{X(t), t > T\}$ are independent of one another for any "stopping time" T. For example, the first time T that $X(t)$ leaves an open subset of the locally compact space M is a typical example of a stopping time.

Dynkin's formula [5] states that

$$E_x(f(X(T)) - f(x) = E_x \int_0^T Af(X(s)) ds$$

where T is a stopping time with $E_x(T)$ finite, f is a smooth function, and A is the infinitesimal generator. For example if $X(s) = (\gamma(s), \dot{\gamma}(s))$ is the geodesic flow of a Riemannian manifold and $T = t$ a constant then $A = Z$ and Dynkin's formula reduces to $f(\gamma(t), \dot{\gamma}(t)) - f(m, \xi) = \int_0^t Z f(\gamma(s), \dot{\gamma}(s)) ds$, which is the integral form of the differential equation $df/dt = Zf$. In general we may think of Dynkin's formula as a sort of "fundamental theorem of calculus" for Markov processes.

Dynkin's formula provides a useful link between probability and partial differential equations. For example suppose that T is the first time that Brownian motion leaves an open ball B and f is the solution of $\Delta f = -1$ in the ball with $f = 0$ on the boundary. Substituting in Dynkin's formula gives $0 - f(x) = E_x(-T)$; hence, the mean exit time $E_x(T)$ is the solution of Poisson's equation $\Delta f = -1$ with zero boundary conditions. For example in the case of R^n the solution in a ball of radius r is $f(x) = \frac{1}{2}(r^2 - |x|^2)$, proving that $E_0(T) = \frac{1}{2}r^2$, a well known formula for Brownian motion.

Another useful tool is the "stochastic Taylor formula" [1] which is written in the form

$$E_x f(X(T)) - f(x) = E_x(TAf(X(T))) - E_x \int_0^T sA^2 f(X(s)) ds$$

valid whenever $E_x(T^2)$ is finite and f is smooth. This may be used to show that the second moment $E_x(T^2)$ is the solution of the "biharmonic" Poisson equation $\frac{1}{2}A^2 f = +1$ in the ball B with the boundary condition $f = Af = 0$. For example in the case of Brownian motion in R^n, f is a fourth degree polynomial and $E_0(T^2)$ is proportional to the fourth power of the radius of the ball.

These formulas may be applied to Brownian motion on a Riemannian manifold (M, g). Taking $f(x) = \frac{1}{2}(r^2 - d(x, m)^2)$, we have an approximate solution of the equation $\Delta f = -1$ with the boundary condition $f = 0$. Substituting in Dynkin's formula where T is the first time that Brownian motion leaves the ball, we have $E_m(T_r) = \frac{1}{2}r^2 + O(r^4), r \downarrow 0$. This reflects the locally Euclidean nature of the Riemannian geometry; the error term $O(r^4)$ contains much geometric information that will be exploited below to solve the inverse problem stated in the introduction. As a first application we define the *Riemannian sectional curvature* by

$$K_m(V) = \lim_{r \downarrow 0} \frac{E_m(vT_r) - \frac{1}{2}r^2}{r^4}$$

where vT_r is the time from a disc of radius r of a two-dimensional Brownian motion in the two-plane V. This process is defined by "pulling back" the metric g to the two-plane V by means of the exponential mapping and constructing a Brownian motion process on the resulting surface. This definition displays the stochastic meaning of sectional curvature: if $K_m > 0$, Brownian motion on V

takes longer to leave a disc than does its Euclidean cousin; while if $K_m < 0$, it takes less time than in the Euclidean case. Put otherwise, negative curvature causes repulsion from the origin while positive curvature causes attraction to the origin — in comparison with the flat Euclidean case.

It may be shown that our definition of $K_m(V)$ agrees to within a constant factor with the definition given in works on differential geometry [20]. In these works it is shown that $K_m \equiv 0$ iff (M, g) is locally isometric to (R^n, g_0). Therefore our conjecture is trivially true for surfaces; if the mean exit time from every disc is $\frac{1}{2}r^2$, then the surface is flat.

4. EXIT TIME FROM A GEODESIC BALL

To investigate higher-dimensional manifolds, we need to study in greater detail the mean exit time of Brownian motion from a ball of radius r when $r \downarrow 0$. It will be seen that in low dimensions ($n < 6$), we can retrieve the curvature of two-dimensional sections from the mean exit time of an n-dimensional ball.

The exit time from n-dimensional ball of radius r is the smallest time for which Brownian motion leaves the ball, more precisely

$$T_r^g = \inf\{t > 0 : d(X(t), m) = r\}.$$

We emphasize the dependence on the Riemannian metric g. This has the important scaling property that if $c > 0$,

$$T_r^{c^2 g} = c^2 T_{r/c}^g$$

where the equality sign is taken in the sense of probability law; $c^2 g$ is the metric on M obtained by multiplying the length of all tangent vectors by c, which multiplies all distances by c and induces the indicated scaling property. This has the following consequences:

(*i*) T_r^g / r^2 has a limiting probability distribution independent of (M, g) where $r \downarrow 0$,

(*ii*) $E_m(T_r^g)$ is a smooth function of r^2.

Property (i) is an interesting central limit theorem, but it does not give any further geometric information. Property (ii) suggests the possibility of an asymptotic expansion $E_m(T_r^g) \sim \sum_j \beta_j r^{2j} (r \downarrow 0)$ where (β_j) are geometric invariants. We have the following theorem [7]:

Theorem 4.1

$$E_m(T_r^g) = c_0 r^2 + c_1 \tau_m r^4 + r^6 [c_2 |R|^2 + c_3 |\varrho|^2 + c_4 \tau^2 + c_5 \triangle \tau] + O(r^8), r \downarrow 0$$

where $(c_i)_{0 \leq i \leq 5}$ *depend on* $n = \dim M$ *with* $c_0 > 0, c_1 > 0, 0 < c_2 = -c_3; \tau_m, |\varrho|_m^2,$ $|R|_m^2$ *are respectively the scalar curvature, the norm of the Ricci tensor and the norm of the full curvature tensor at* $m \in M$. *These are the orthogonal curvature invariants of H. Weyl* [2].

To solve our inverse problem in low dimensions we need the following technical fact [7]:

Lemma 4.2   *If* (M, g) *is a Riemannian manifold with* $\tau_m \equiv 0$, $|R|_m^2 - |\varrho|_m^2 \equiv 0$ *then either* $|R|_m^2 = |\varrho_m|^2 \equiv 0$ *or* $n = \dim M \geq 6$.

Combined with the above theorem, we have the following corollary.

Corollary 4.3   *If* (M, g) *is a Riemannian manifold with* $n < 6$ *and* $E_m(T_r^g) = \frac{1}{2}r^2 + O(r^8)$, $r \downarrow 0$, *then* (M, g) *is locally isometric to* (R^n, g_0).

This corollary settles affirmatively the conjecture stated in the Introduction, for manifolds of dimension less than 6.

One might obtain the impression that, in order to prove similar results in higher dimensions, it would suffice to find additional terms in the asymptotic expansion of the mean exit time. This was shown to be *false* by the following counter-example of H.R. Hughes [8], of the product manifold $M = \mathbf{S}^3 \times \mathbf{H}^3$ with the product metric; the first factor is the three-dimensional sphere of constant sectional curvature $= +k^2$ and the second factor is the three-dimensional hyperbolic space of constant sectional curvature $= -k^2$ where k is any positive number. The following result is obtained.

Proposition 4.4   *For any* $r < \frac{\pi}{k}$ *the probability law of the exit time* T_r *in* $\mathbf{S}^3 \times \mathbf{H}^3$ *coincides with the probability law of the exit time in six-dimensional Euclidean space* $\mathbf{R}^6$. *In particular the mean exit times agree to all orders of perturbation theory.*

Suitable counter-examples may be obtained in any dimension $n > 6$ by taking an additional flat factor in the form $M = \mathbf{S}^3 \times \mathbf{H}^3 \times \mathbf{R}^{n-6}$.

Hughes' original proof uses a suitable version of the Feynman-Kac-Girsanov formula. Subsequently we obtained an independent proof [19] using the properties of the *bi-radial Laplacian*, which is the generator of the joint six-dimensional process, when restricted to functions which depend on the pair of geodesic distances (r_1, r_2) in the product space.

5. EXIT PLACE FROM A GEODESIC SPHERE

Brownian motion of Euclidean space *cannot* be characterized by the law of its exit place, even in dimension two. To see this, think of a Brownian particle on the surface of the earth starting at the North Pole and hitting the Arctic Circle. More generally, on any space of *constant sectional curvature*, the exit place distribution from a geodesic sphere is the uniform distribution, just as on Euclidean space. This leads us to the revised goal of using the exit place distribution to study the *variation* of curvture. As a first approximation to this variation, one is led to the Ricci tensor, as we shall see below.

In order to study the distribution of the exit place from a sphere of radius r, it is convenient to work on the unit sphere of Euclidean space by means of the exponential mapping $\exp_m$. This is the differentiable mapping

$$\exp_m : \mathbf{R}^m \to M$$

which sends $0 \in \mathbf{R}^n$ to $m \in M$ and maps straight lines to geodesics emanating for $m \in M$. If Ψ is a continuous function on the sphere $\mathbf{S}^{n-1}$ we define the *harmonic measure operator* $\mu_m(r, d\theta)$ on the sphere by the operator identity

$$H_r \Psi(m) \equiv E_m \Psi(r^{-1} \exp_m^{-1}(X_{T_r})) = \int_{\mathbf{S}^{n-1}} \Psi(\theta)\mu_m(r, d\theta).$$

This family of measures indexed by $r > 0$ converges to the uniform measure on the sphere when $r \downarrow 0$. The following two-term correction formula was discoverd by Ming Liao [14].

Theorem 5.1 *For any n-dimensional Riemannian manifold, the harmonic measure operator has the asymptotic expansion*

$$H_r \Psi(m) = \int_{\mathbf{S}^{n-1}} [1 - \frac{1}{12}r^2 \rho_m^{\#}(\theta) - \frac{1}{24}r^3 \rho_m^{\#\#}(\theta)]\Psi(\theta)\mathrm{Leb}(d\theta) + O(r^4)$$

where $\rho^{\#}$ denotes the traceless Ricci tensor defined by $\rho_m^{\#}(\theta) = (\rho_{ij} - \frac{\delta_{ij}\tau_m}{n})\theta_i\theta_j$ and $\rho_m^{\#\#}(\theta) = \frac{\partial \rho_{ij}}{\partial x_k}\theta_i\theta_j\theta_k - \frac{\theta_k}{n+2}\frac{\partial \tau}{\partial x_k}$ where repeated indices imply a summation over $\{1, \ldots, n\}$.

For a two-dimensional manifold the traceless Ricci tensor is zero and we can infer that the uniform distribution of exit place implies constant curvature as follows.

Corollary 5.2 *Suppose that $n = 2$ and the exit place distribution satisfies $H_r\Psi(m) = \int_{\mathbf{S}^1} \Psi(\theta)\mathrm{Leb}(d\theta) + o(r^3), r \downarrow 0 \ \forall m \in M$. Then the manifold has constant sectional curvature.*

In the case of higher dimensions we obtain the following general result.

Corollary 5.3 *Suppose that for some $m \in M$ the exit place distribution satisfies $H_r\Psi(m) = \int_{\mathbf{S}^{n-1}} \Psi(\theta)\mathrm{Leb}(d\theta) + o(r^2), r \downarrow 0$. Then the traceless Ricci tensor $\rho_m^{\#} = 0$, In particular if the condition holds for all $m \in M$ then $\rho_m^{\#} \equiv 0$, i.e., the manifold is Einsteinian.*

When we combine this with the results of the previous section, we see that the combined hypotheses on the Euclidean mean exit time and the uniform exit place distribution suffice to characterize Euclidean space of *any* dimension.

Corollary 5.4 *Suppose that for all $m \in M$ the exit place distribution satisfies $H_r\Psi(m) = \int_{\mathbf{S}^{n-1}} \Psi(\theta)\mathrm{Leb}(d\theta) + o(r^2), r \downarrow 0$ and that the mean exit time satisfies $E_m(T_r) = \frac{r^2}{n} + o(r^2), r \downarrow 0$. Then the manifold is locally isometric to the Euclidean space $\mathbf{R}^n$.*

This is easily seen from the previous results when we note that the second hypothesis implies that the scalar curvature $\tau_m \equiv 0$; taken with the first hypothesis this shows that the entire Ricci tensor $\rho_{ij}(m) \equiv 0$. Using the second hypothesis again and the expansion for the mean exit time shows that the full curvature tensor $R_{ijkl}(m) \equiv 0$, which is well known to imply the local isometry with Euclidean space.

6. JOINT DISTRIBUTION OF EXIT TIME AND EXIT PLACE

On any Riemannian manifold of constant sectional curvature it is true that the exit time and exit place from a ball are independent random variables for Brownian paths starting at the center of the ball. This is easily seen by the separation-of-variables solution associated with the Laplace operator for such spaces. More generally, one may inquire to what extent independence characterizes a Riemannian manifold.

The random variables (T_r, X_{T_r}) are most conveniently studied through the Laplace transform which defines a family of measures $\mu_m^\alpha(r, \cdot)$ on the sphere $\mathbf{S}^{n-1}$ by

$$E_m(e^{\frac{-\alpha T_r}{r^2}} \Psi(r^{-1} \exp_m(X_{T_r})) = \int_{\mathbf{S}^{n-1}} \Psi(\theta)\mu_m^\alpha(d\theta)$$

When $\alpha = 0$ this reduces to the harmonic measure operator of the previous section. When $r \downarrow 0$ for fixed $\alpha > 0$, this factors into the product of the uniform measure with the Laplace transform of the Euclidean exit time distribution which is the statement of *asymptotic independence*. The following general necessary condition for independence has been obtained independently by three authors [10,14,18].

Proposition 6.1 *Suppose that $\forall m \in M$ the random variables T_r, X_{T_r} are independent, $\forall r < r(m)$. Then the scalar curvature is constant.*

One might conjecture that independence imples stronger curvature conditions, such as the Einstein condition which follows from the uniform exit distribution. This is *false* as was shown by H.R. Hughes[8], who obtained the following.

Proposition 6.2 *For any $r < \frac{\pi}{k}$ the exit time and exit place from the ball of radius r in the product manifold $\mathbf{S}^3 \times \mathbf{H}^3$ are independent random variables.*

7. RELATED RESULTS ON THE PRINCIPAL EIGENVALUE

As a curious by-product of the methods used to obtain the above probabilistic results, one can obtain corresponding asymptotic results for the *principal Dirichlet eigenvalue of a small geodesic ball*. This may be defined, for example as the solution of the following minimization problem:

$$\lambda_1(B_r) = \inf_{f: f \neq 0, f = 0 \text{ on } \partial B_r} \frac{\int_{B_r} |df|^2}{\int_{B_r} |f|^2}$$

Equivalently this is the negative of the first non-zero eigenvalue of the Laplacian, which is a non-positive essentially self-adjoint operator on a suitable Hilbert space.

The probabilistic interpretation of the principal eigenvalue is through the *large deviation asymptotics* of the exit time distribution from the ball of radius r. The acoustic interpretaion is that of *principal frequency*, if one can imagine a non-Euclidean drumhead that is stretched over the geodesic ball of radius r.

The following three-term asymptotic expansion for λ_1 has been obtained [11].

Theorem 7.1 *For any n-dimensional Riemannian manifold, the principal eigenvalue has the asymptotic expansion*

$$\lambda_1(B_r) = \frac{const}{r^2} - \frac{\tau_m}{6} + const.r^2[\|R\|^2 - |\rho|^2 + 5\Delta\tau_m] + O(r^4), \qquad r \downarrow 0$$

where the constants depend only on the dimension.

From this asymptotic expansion, one can obtain the following converse theorems that characterize Euclidean space in terms of the principal eigenvalue, in much the same fashion as we have done for the mean exit time in section 4.

Corollary 7.2 *Suppose that $n = 2$ and that $\forall m \in M$ the principal eigenvalue satisfies $\lambda_1(B_r) = \frac{const}{r^2} + o(1), r \downarrow 0$. Then M is locally isometric to the Euclidean plane.*

In higher dimensions we have

Corollary 7.3 Suppose that $n < 6$ and that $\forall m \in M$ the principal eigenvalue satisfies $\lambda_1(B_r) = \frac{const}{r^2} + o(r^2), r \downarrow 0$. Then M is locally isometric to the Euclidean space $\mathbf{R}^n$.

Just as with the mean exit time, we obtain the same counter-example in dimension six [19]:

Proposition 7.4 *For any $r < \frac{\pi}{k}$ the principal eigenvalue of the ball of radius r in $\mathbf{S}^3 \times \mathbf{H}^3$ satisfies $\lambda_1(B_r) \equiv \frac{const}{r^2}$, where the constant is the square of the first positive zero of the Bessel function J_2.*

ACKNOWLEDGMENTS

We would like to thank Prof. Laurie Snell and the staff of Dartmouth College for editorial assistance with the initial draft of this work.

References

[1] **Athreya, K. B. and Kurtz, T. G.,** A generalization of Dynkin's identity, *Annals of Probability,* 1, 570-579, 1973.

[2] **Berger, M., Gauduchon, P., Mazet, E.,** *Le Spectre d'une Variété Riemannienne*, Springer Verlag Lecture Notes in Mathematics, Vol. 194, 1971.

[3] **Bérard, P.,** Variétés Riemanniennes isospectrales non-isometriques, *Asterisque*, **177-178**, 127-154, 1989.

[4] **Brooks, R.,** Constructing isospectal manifolds, *American Mathematical Monthly*, **95**, 823-837, 1988.

[5] **Dynkin, E.,** *Markov Processes*, 2 vols., Springer-Verlag, 1965.

[6] **Gordon, C., Webb, D. and Wolpert, S.,** Isospectral plane domains and surfaces via Riemannian orbifolds, *Inventiones Mathematicae*, Springer-Verlag, **110**, 1-22, 1992.

[7] **Gray, A. and Pinsky, M.,** The mean exit time from a geodesic ball in a Riemannian manifold, *Bulletin des Sciences Mathématiques*, **107**, 345-370, 1983.

[8] **Hughes, H. R.,** Brownian exit distributions from normal balls in $S^3 \times H^3$, *Annals of Probability*, **20**, 655-659, 1992.

[9] **Kac, M.,** Can one hear the shape of a drum? *American Mathematical Monthly*, **73**, 1-23, 1966.

[10] **Kozaki, M. and Ogura, Y.,** On the independence of exit time and exit place from small geodesic balls on Riemannian manifolds, *Mathematische Zeitschrift*, **197**, 561-581, 1988.

[11] **Karp, L. and Pinsky, M.,** First eigenvalue of a small geodesic ball in a Riemannian manifold, *Bulletin des Sciences Mathématiques*, **111**, 222-239, 1987.

[12] **Kato, T.,** *Perturbation Theory of Linear Operators*, Springer-Verlag, 1966.

[13] **Ethier, S. T. and Kurtz, T. G.,** *Markov Processes: Characterization and Convergence* , John Wiley & Sons, 1986.

[14] **Liao, M.,** Hitting distributions of small geodesic spheres, *Annals of Probability*, **16**, 1029-1050, 1988 .

[15] **Pinsky, M.,** *Lectures on Random Evolutions*, World Scientific Press, 1991.

[16] **Pinsky, M.,** Isotropic transport process on a Riemannian manifold, *Transactions of the American Mathematical Society*, **218**, 353-360, 1976.

[17] **Pinsky, M.,** Homogenization in stochastic differential geometry, *Publ. Res. Inst. Math. Sci. Kyoto,* **17**, 235-244, 1981.

[18] **Pinsky, M.,** Local stochastic differential geometry, in *Geometry of Random Motion,* AMS Contemporary Mathematics, **73**, 263-272, 1988.

[19] **Pinsky, M.,** Feeling the shape of a manifold with Brownian motion — the last word in 1990, in *Stochastic Analysis,* Cambridge University Press **167**, 1991, 305-320, eds. M. T. Barlow and N. H. Bingham.

[20] **Spivak, M.,** *Differential Geometry,* Vol. 2, Publish or Perish, Inc., 1970.

[21] **Yau, S. T.,** The heat kernel of a complete Riemannian manifold, *Journal des Mathématiques Pures et Appliquées,* **57**, 191-201, 1978.

Chapter 5

SOME NEW GAMES FOR YOUR COMPUTER*

Rick Durrett
Department of Mathematics
Cornell University

ABSTRACT

This paper surveys recent research on the Greenberg-Hastings model and a related epidemic model.

GREENBERG-HASTINGS MODEL

In the 1970s, countless hours of CPU time were spent running Conway's game of life. In the 1980s, the Mandelbrot set was drawn so many times that its psychedelic potato shape could be seen on many computer screens even when they were turned off. This article describes some new games that are appropriate for the 1990s because they can be implemented on parallel machines (or enjoyed on personal computers). These games, called interacting particle systems, provide a flexible framework for modelling systems in which the spatially distributed "particles" interact locally, lead to challenging mathematical problems, and produce nice pictures. However, it is not our purpose here to promote particle systems as the replacement of self-organized criticality — i.e., as this year's theory that says nothing about everything — so we will climb down from our soapbox and concentrate on the recreational aspects of our subject.

In most interacting particle systems, the state at time t is described by a function ξ_t from $\mathbb{Z}^d$, the set of points in d-dimensional space with integer coordinates, to a finite set F of possible states. The points $x \in \mathbb{Z}^d$ are called sites and are thought of as spatial locations, whereas $\xi_t(x)$ gives the state of site x at time t. Typically, time is continuous, that is, $t \in [0, \infty)$, and the evolution is random, but, for simplicity, we will start with an example in which time is discrete, that is, $t \in \{0, 1, \ldots\}$, and the evolution is deterministic. In this special case, particle systems are usually referred to as cellular automata.

---

*Reprinted with permission from *NONLINEAR SCIENCE TODAY*, Vol. 1, No. 4, Springer-Verlag, New York, 1991.

In formulating the Greenberg-Hastings model, we are thinking of a network of neurons (e.g., those that cause cardiac tissue to contract) in which each site can be excited (i.e., be in state 1), rested (state κ), or in a sequence of recovery states $(2, \ldots, \kappa - 1)$. The playing field is two dimensional, and we formulate the evolution as follows.

(*i*) If $\xi_n(x) < \kappa$ then $\xi_{n+1}(x) = \xi_n(x) + 1$.

(*ii*) If $\xi_n(x) = \kappa$ and at least one neighbor is excited, then $\xi_{n+1}(x) = 1$;
     otherwise $\xi_{n+1}(x) = \kappa$.

Here the neighbors of x are the set of y with $y - x \in \mathcal{N}$, and $\mathcal{N}$ is the neighborhood set. We will consider a number of different choices for $\mathcal{N}$, but one common choice is $\mathcal{N} = \{(1,0), (0,1), (-1,0), (0,-1)\}$, the four nearest neighbors of (0,0).

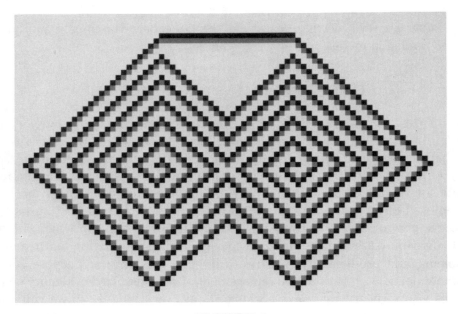

FIGURE 1.

In words, an excited site progresses through a sequence of recovery states until it is rested and then becomes excited again if at least one neighbor is excited. Clearly, if the human heart really worked like this, the whole species would have died out long ago. Nonetheless, this is a useful caricature: Generalizations of these game rules to incorporate nonuniformity in the duration of the refractory period (i.e., $\kappa - 2$) have been useful in investigating fibrillation (see [12], pp. 106-107). A second reason for interest is that, while it is difficult to prove that spiral waves exist for systems of reaction-diffusion equations (see, e.g., [10]), it is trivial to produce spirals in the Greenberg-Hastings model. See Figure 1 for a picture of what happens when $\kappa = 3$ and we start with a line of 1s next to a line of 2s in a sea of 3s. Here, and in all of our pictures of this model, excited states are black, rested sites are white, and sites in recovery states are appropriate shades of gray. Things become more interesting and the spirals become more life-like when we follow David Griffeath and generalize the dynamics to use larger neighborhoods $\mathcal{N}_r = \{x : \|x\|_\infty \le r\}$, where $\|x\|_\infty = \max|x_i|$ (one can use other norms but the results are similar), and generalize (ii) to the following.

(ii) If $\xi_n(x) = \kappa$ and at least θ neighbors are excited, then $\xi_{n+1}(x) = 1$; otherwise $\xi_{n+1}(x) = \kappa$.

Figure 2 shows what happens when the range $r = 4$, the number of colors $\kappa = 7$, the threshold $\theta = 9$, and we start from a "randomly chosen" initial state, i.e., each site is independently painted a randomly chosen color. The spirals formed are not unlike those found in the Belousov-Zhabotinsky reaction. (See [12], Chapter 7 for pictures of the chemical reaction.)

It can be argued (see, e.g., [8]) that the theory of cellular automata has a lot to say about the spatio-temporal dynamics of chemically reacting systems, but for the moment we will continue to pursue the subject from a mathematician's perspective. We have an interesting three-parameter family of models, and we will investigate their properties ignoring the relationship between our results and the motivating applications. The main question we will focus on is, "What is the asymptotic behavior of the system starting from a randomly chosen initial state?" Figure 2 shows one possibility — the formation of "spirals". There are also two boring possibilities. The system may "die out", that is, each site may enter the rested state κ after a finite number of steps and then stays there for all time. The reader should note that $\xi(x) \equiv \kappa$ is an absorbing state, or fixed point for the evolution.

Let $\nu = |\mathcal{N}_r|$ be the number of neighbors, and let $\epsilon > 0$. It is easy to prove that, if $\theta = (0.5 + \epsilon)\nu$ and ν is large, then the system dies out. In this case, a large ball on which $\xi(x) = \kappa$ could not become excited even if all the sites outside the ball were excited. The last result is almost certainly true with 0.5 replaced by 1/6, but this seems very difficult to prove. A second uninteresting type of behavior occurs in these systems when θ is too small. Durrett and Griffeath [5] show that, if $\theta = (1 - \epsilon)\nu/2\kappa$ and ν is large, then (i) each site is eventually periodic with period κ and (ii) the limit of $\xi_{m\kappa}$ as $m \to \infty$ looks much like the initial random state. Simulations of the system in this case look like a television set tuned in to a nonexistent channel; we call this the "debris regime."

The aim of the mathematical analysis of these systems is to compute the "phase diagram", i.e., to identify the parameter values that lead to "dying out", "debris", "spirals", and other behaviors. A fairly complete picture is emerging slowly, but the answer is a rather long story, so we will content

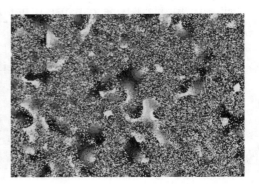

FIGURE 2.

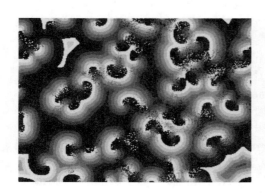

FIGURE 3.

FIGURE 4.

FIGURE 5.

ourselves with illustrating two open problems. Figure 3 shows the system with $r = 5$, $\kappa = 8$, $\theta = 10$ run until it has become periodic (with period 8). Much of the original product measure remains, but many holes have formed and been filled in with "waves." Figure 4 shows what happens when we increase the threshold to 11 and again run the system until it has become periodic. In this situation there is some residual debris from the initial product measure, but most of the screen is covered with spirals. The two pictures are a small part of the experimental evidence that when $\kappa \geq 5$, the transition from debris to spirals occurs at about $0.70\nu/\kappa$, where ν is the number of neighbors. Other experiments suggest that when $\kappa = 3$ or 4 there is no spiral phase; when the debris breaks down, the system dies out.

Figure 5 shows the system with $r = 2$, $\kappa = 5$, $\theta = 4$ run for several hundred units of time. From the picture, it seems that the excited fronts are surviving but are unable to bend enough to make spirals. At first glance, this behavior, which is informally called "macaroni", might seem to be a new phase. However, after some inspired doodling on graph paper, Bob Fisch came up with a "stable periodic object", or s.p.o. for this rule (See Figure 6). An s.p.o. is an assignment of states to a finite set $A \subset \mathbf{Z}^2$ so that, regardless of what occurs outside A, the sites in A are periodic with period κ. Presumably, the existence of an s.p.o. implies that the system becomes periodic with period κ, but this is an open problem. One thing is clear: it would take an enormous computer to have one of these objects appear spontaneously. (The moral of this story is that, to study the asymptotic behavior of these systems, it is not sufficient to just turn on your computer and see what happens. It is necessary to think as well.)

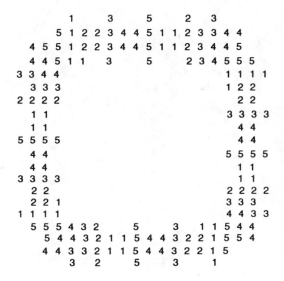

FIGURE 6.

At this point we will move from cellular automata to the more general subject of interacting particle systems. To illustrate the differences we now reformulate the Greenberg-Hastings model in continuous time. The state at time $t \in [0, \infty)$ is a function from $\mathbb{Z}^2$ to $\{1, \ldots, \kappa\}$, and the system evolves as follows.

(a) Sites in state $j < \kappa$ change to $j + 1$ at rate 1;
(b) sites in state κ change to 1 at rate 1 if at least θ neighbors are in state 1.

"At rate 1" here means that the probability of a change between time t and $t + s$ when divided by s converges to 1 as $s \to 0$. It takes a little technique to show that rules (a) and (b) specify a unique process, but it is easy to simulate the system on a computer. We pick a site at random, apply the cellular automaton rules (i) and (ii) at that site, and then repeat the procedure. In the language of cellular automata, we are updating the sites asynchronously (one at a time) rather than synchronously (all at once). This "minor" change makes a large difference in the analysis of these systems. Specifically, s.p.o.s exist for the cellular automata but not for the particle system. The last difference makes the particle system much more difficult to analyze, but we believe that the qualitative features of the phase diagram are the same. One piece of evidence for this is that, if $\theta = (1 - \epsilon)\nu/2\kappa$ and ν is large, then there is an equilibrium state that looks very much like the original product measure (see [4]). It is an open problem to prove the existence of a spiral phase (or even to define it formally), but pictures like Figure 7 indicate that it exists.

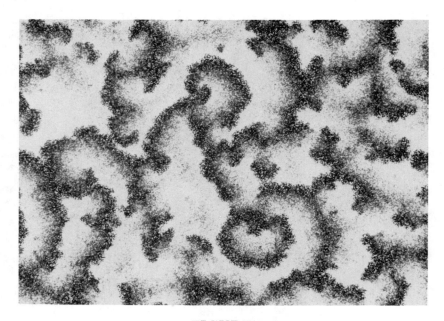

FIGURE 7.

To move toward an application of particle systems, we now introduce a cousin of the Greenberg-Hastings model that has been used to model the spread of epidemics (and forest fires). In this system the state at time $t \in [0, \infty)$ is a function from $\mathbf{Z}^2$ to $\{0, 1, 2\}$, where $0 =$ healthy (a live tree), $1 =$ infected (on fire), and $2 =$ removed (burnt). If you are thinking of a fatal disease, then "removed" here means dead; we prefer to think of an infections disease like the measles where removed means immune. The system evolves as follows.

(*i*)   Sites in state 0 become 1 at a rate equal to the fraction of neighbors in state 1;

(*ii*) sites in state 1 become 2 at rate δ;

(*iii*) sites in state 2 become 0 at rate α.

As in the the Greenberg-Hastings model, one can have a general set of neighbors, but, for simplicity, we will stick to $\mathcal{N} =$ the four nearest neighbors of (0,0). To explain how the system evolves, we note that, if $\lambda = \max\{\delta, \alpha, 1\}$ and U is a random number uniformly distributed on (0,1), then the system may be simulated as follows.

Pick a site x at random;
if x is in state 0, then {pick a neighbor y at random; if (y is in state 1 and $U < 1/\lambda$), change the state of x to 1};
if (x is in state 1 and $U < \delta/\lambda$), change the state of x to 2;
if (x is in state 2 and $U < \alpha/\lambda$), change the state of x to 0.

For an $L \times L$ system, repeat these steps $tL^2\lambda$ times to get to time t. The program is based on the idea that potential changes at site x happen at rate λ; so, for example, to make transitions from 1 to 2 happen at rate δ, we do nothing with probability $1 - \delta/\lambda$.

If we are thinking of forest fires, it is natural to set $\alpha = 0$ and investigate what happens if we start with one burning tree in an otherwise virgin forest. It is not hard to show that if $\delta > 1$ then the fire will always go out because the rate at which we lose a burning tree is always larger than the rate at which a new tree is ignited. (In the best possible situation each burning tree is surrounded by four live trees.) It is more difficult to prove that the opposite can occur: there is a $\delta_c > 0$, so that the forest fire has positive probability of not going out if (and only if) $\delta < \delta_c$. We invite the reader to try to show that $\delta_c > 0$, because attacking this problem is a good way of experiencing first-hand some of the difficulties in the subject. Simulations suggest that the critical value $\delta_c \approx .20$, but you should be happy if you can show that $\delta_c > 10^{-5}$. Figure 8 shows a simulation of the process with $\delta = 0.15$. The states of sites are represented using our Greenberg-Hastings color scheme: black $= 1$, gray $= 2$; white $= 0$. The picture shows one of the basic facts about the forest fire:

when the fire does not go out, the burnt region expands linearly and has an asymptotic shape. For a precise statement and proof of the shape result, as well as a proof of the fact that $\delta_c > 0$, see [1].

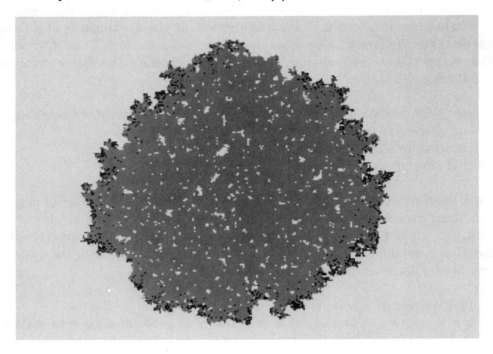

FIGURE 8.

The last result is not a very exciting conclusion about forest fires because most people are interested in putting them out and not just watching them burn. However, analogous results are interesting for the spread of biological populations. Suppose we consider the forest fire with $\alpha = \infty$, or equivalently reformulate the rules as follows.

  (*i*) Sites in state 0 become 1 at a rate equal to the fraction of neighbors in state 1;
  (*ii*) sites in state 1 become 0 at rate δ.

In this case we think of 0 sites as vacant and 1 sites as occupied by a plant, and the resulting system is called the contact process. Again, there is a critical value δ'_c (now ≈ 0.30) so that the species has positive probability of survival starting from a single plant at the origin if and only if $\delta < \delta'_c$; if the species does not die out, the occupied region expands linearly and has an asymptotic shape. For an exposition of these facts, see [3].

Turning now to values $0 < \alpha < \infty$, we will revert to the measles interpretation. At first, the spontaneous mutation of "removed" individuals to susceptible ones may not appear realistic, but it is easy to explain. We are thinking about a population that has a constant size, and hence we combine the deaths and births into one mutation step. To assign values to our parameters, we think of the time unit as one day. Thus, $\delta = 1/14$ (measles lasts two weeks) and $\alpha \approx 4 \times 10^{-5}$ (a 70-year life span is about 25,000 days). These are obviously rough guesses, but the only important things about the numbers are that $\delta < \delta_c$ (the epidemic does not need regrowth to survive) and that α is very, very small.

Since $\delta < \delta_c$ and $\alpha > 0$, it follows from a result of Durrett and Neuhauser [6] that the system has a nontrivial equilibrium in which all three types are present. The last result, like all the other theorems in this article, is for the system on the infinite lattice $\mathbb{Z}^2$, but in this case there are important differences when the system is finite. Figure 9 gives a plot of the density of susceptible and infected individuals versus time for a 100×100 system with $\delta = 0.1$ and $\alpha = 0.04$. (The lower curve is the density of infected individuals.) Although there are fluctuations arising from the fact that we are dealing only with 10,000 sites, the curves drawn indicate that the densities converge exponentially fast to a nontrivial equilibrium. Changing α to 0.01 now gives the radically different graph in Figure 10. The densities oscillate wildly in time and show no signs of settling down to equilibrium. Decreasing α further to 0.005 leads to a rather boring result: the fire dies out when simulated on a 100×100 grid.

Figures 11 and 12 help to explain the oscillations when $\alpha = 0.01$. The first picture shows the system at time 250 when there are several fire fronts. The second picture shows time 350, shortly after the fronts have crashed into each other. Notice that we have a small number of burning trees, but a few of them will burn long enough to produce new fire fronts. When $\alpha = 0.005$ these last few trees go out before the forest has regenerated enough to allow for a new fire to spread across the system. The observations above show promise of explaining an old mystery in epidemiology. It has long been known that there is a critical population size (approximately 250,000) necessary for measles to persist, and that the epidemics show certain periodicities. The parameter values for measles given above indicate that we are in the right ballpark. This approach is currently under investigation.

At this point we have reached the end of our tour. Like a 7-day tour of Europe, we have emphasized sights that can be conveniently packaged together, and we have only begun to scratch the surface. For more systematic introductions to interacting particle systems, see Liggett [9] or Durrett [2]. Fisch, Gravner, and Griffeath [7] give a nice account of the Greenberg-Hastings model. However, the best way to learn about that system is to simulate it and see what happens. If you do not want to program the model yourself, griffeath@math.wisc.edu will happily send you a copy of his program EXCITE! which runs on IBM-compatible PCs equipped with EGA graphics. In connec-

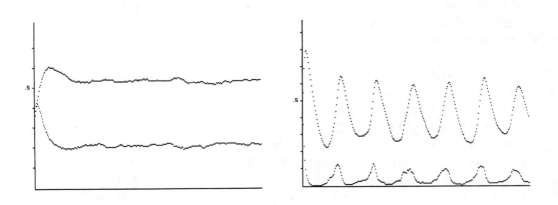

FIGURE 9.   FIGURE 10.

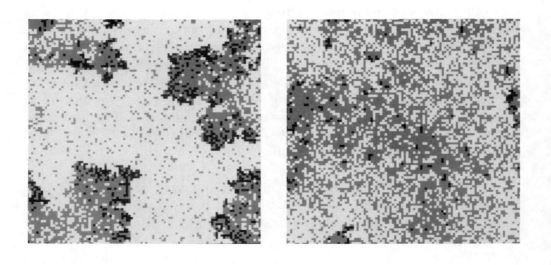

FIGURE 11.   FIGURE 12.

tion with the Greenberg-Hastings model we should mention *Cellular Automata Machines* which is the title of a book by Toffoli and Margolus [11] and the name of a special piece of hardware that plugs into an IBM compatible and allows many cellular automata rules to be simulated on a 256 × 256 lattice with 60 updates a second. David Griffeath's CAM simulations led to many of the conjectures mentioned in this article.

EXCITABLE CELLULAR AUTOMATA: SOME LARGER EXPERIMENTS

David Griffeath

The new computer games that Rick Durrett describes offer windows onto the unbounded universe of a theoretically infinite two-dimensional lattice. However any real-world video array is inevitably finite, necessitating a choice of boundary conditions (wrap-around?, free?, random?,...). How important is this choice, and how accurate a representation of the infinite system do we see on our screen? These are fundamental questions in the study of complex spatially-distributed dynamics. Especially for models such as those Durrett discusses, exhibiting large-scale self-organization, the size of the computer experiment can have a profound impact on system behavior. In particular, if the array is too small then the dynamics may bear little resemblance to the ultimate behavior of the corresponding infinite system.

The best way to understand the dynamics of an infinite system is to use traditional deductive reasoning — theorems and proofs if at all possible — to preclude the existence of distant monsters capable of propagating and taking over the universe. But it is also instructive to simulate our models on as large an array as technically feasible. Since particle systems and cellular automata on large arrays involve an enormous amount of computation, either a super-computer or specialized parallel processing hardware is required for this task. Aided by a CAM-6 cellular automaton machine (see the Toffoli-Margolus reference at the end of Durrett's article), I have produced Color Plates I-IV to illustrate some interesting dynamics involving hundreds of thousands, or even millions of sites.

Plate I illustrates dynamic large-scale spiral formation in a Greenberg-Hastings CA. This specific rule has range 8, threshold 31, and 7 colors. The graphic shows the state of a 640 × 480 system with wrap-around, started from a completely random configuration of the 7 colors, after 50 iterations of the deterministic rule.

Plates II-IV depict systems of more than 3 million sites (2048 × 1536). Plate II shows how, given more room to self-organize and cluster, a "macaroni" rule similar to the one shown in Figure 5 evolves into a state that might better be termed "spaghetti".

Plates III and IV illustrate cyclic systems, close relatives of Greenberg-Hastings models in which every color behaves like the rested state, i.e., updates by contact with a threshold number of its successor. Plate III shows complex nucleation of droplets and spirals in the basic range 1 four nearest-neighbor cyclic cellular automaton that was described in the "Computer Recreations" column of the August 1989 issue of Scientific American.

Plate IV is an example of the turbulent phase for cyclic CA rules: a complex statistical equilibrium of large, finite length-scale characterized by a curious mix of geometrical wave patterns and fine-grained debris. (See [7] for further discussion of this phase.)

References

[1] **Cox, J. T. and Durrett, R.**, Limit theorems for the spread of epidemics and forest fires. *Stoch. Proc. Appl.*, **30**, 171-191, 1988.

[2] **Durrett, R.**, *Lecture Notes on Particle Systems and Percolation*, Wadsworth Pub. Co., Pacific Grove, CA, 1988.

[3] **Durrett, R.**, The contact process: 1974-1989. *Proceedings of the AMS Summer Seminar on Random Media*, American Mathematical Society, Providence, RI, 1991.

[4] **Durrett, R.**, Multicolor particle systems with large threshold and range. *J. Theoretical Probab.*, **5**, 127-152, 1992.

[5] **Durrett, R. and Griffeath, D.**, [in press] Asymptotic behavior of the Greenberg Hastings model.

[6] **Durrett, R. and Neuhauser, C.**, Epidemics with recovery in $d = 2$. *Ann. Applied Probab.*, **1**, 189-206, 1991.

[7] **Fisch, R., Gravner, J. and Griffeath, D.**, Threshold-range scaling for excitable cellular automata, *Statistics and Computing*, **1**, 23-39, 1991.

[8] **Kapral, R.**, Discrete models for chemically reacting systems. *J. Math. Chem.,*, **6**, 113-163, 1991.

[9] **Liggett, T. M.**, *Interacting Particle Svstems*, Springer-Verlag, New York, 1985.

[10] **Murray, J. D.**, *Mathematical Biology*, Springer-Verlag, New York, 1989.

[11] **Toffoli T. and Margolus, N.,** *Cellular Automata Machines*, MIT Press, Cambridge, MA, 1987.

[12] **Winfree, A. T.,** *When Time Breaks Down*, Princeton University Press, Princeton, NJ, 1987.

Chapter 6

CELLULAR AUTOMATA WITH ERRORS:
PROBLEMS FOR STUDENTS OF PROBABILITY

Andrei Toom
Incarnate Word College
San Antonio, Texas

ABSTRACT

This is a survey of some problems and methods in the theory of probabilistic cellular automata. It is addressed to those students who love to learn a theory by solving problems. The only prerequisite is a standard course in probability. General methods are illustrated by examples, some of which have played important roles in the development of these methods. More than a hundred exercises and problems and more than a dozen of unsolved problems are discussed. Special attention is paid to the computational aspect; several pseudo-codes are given to show how to program processes in question.

1. INTRODUCTION

We consider probabilistic systems with local interactions, which may be thought of as special kinds of Markov processes. Informally speaking, they describe the functioning of a finite or infinite system of automata, all of which update their states at every moment of discrete time, according to some deterministic or stochastic rule depending on their neighbors' states.

From statistical physics came the question of uniqueness of the limit as $t \to \infty$ behavior of the infinite system. Existence of at least one limit follows from the famous fixed-point theorem and is proved here as Theorem 1. Systems that have a unique limit behavior and converge to it from any initial condition may be said to forget everything when time tends to infinity, while the others remember something forever. The former systems are called *ergodic*, the latter *non-ergodic*. In statistical physics the non-ergodic systems help us understand conservation of non-symmetry in macro-objects.

Computer scientists prefer to consider finite systems. Although all non-degenerate finite systems converge, some of them converge enormously slower

117

than others. Some converge *fast*, i.e., lose practically all the information about the initial state in a time that does not depend on their size (or grows as the logarithm of their size, according to another definition), whereas some others converge *slowly*, i.e., can retain information during time that grows exponentially in their size. The latter may help to solve the problem of designing reliable systems consisting of unreliable elements.

There is much in common between the discrete-time and continuous-time approaches to interacting processes. American readers are better acquainted with the continuous-time approach, summarized, for example, in [5]. Methods to prove ergodicity are similar in the continuous and discrete time cases. Our Theorem 2 is a simpler version of the well-known result about ergodicity [42]. The same result in continuous time is given as Theorem 4.1 in Chapter 1 of [5].

Methods to prove non-ergodicity, however, have hitherto been better developed for the discrete time case. I think this is because the discrete time case is technically simpler. For example, the theoretical problems of definition provide no essential difficulties in this case. In a fairly general case every question about the discrete-time systems can be reformulated as a question about 'hidden' independent random variables, and we use this possibility throughout this paper.

The theory of interacting systems is a relatively new branch of mathematics where efficient general methods are still in the process of development. In these circumstances it may be useful to pay special attention to individual cases that defy all known methods. Thus we speak of several more or less specific *Examples* which we call special names for reference purposes. Pluses in the following table show which cases of interaction (boxes) between examples (rows) and general themes (columns) are paid attention to here.

themes / *examples*	computer results	attractors and eroders	proofs of ergodicity	proofs of non-ergodicity	critical values	rate of convergence	chaos approximation
percolations	+	+	+	+	+	+	+
flattening	-	+	+	+	-	+	-
NEC-voting	+	+	+	+	-	+	+
1-dim. votings	+	+	+	-	-	-	+
windroses	+	+	+	+	-	-	+
soldiers	+	+	+	-	-	+	-
2-line voting	-	+	+	-	-	-	+

We would like to share with the reader the flavor of connecting theoretical work with computer simulation. Computer results are placed in the first column of our table, as a natural prerequisite for theoretical work. In fact all of them were obtained using the Monte Carlo method, that is, by generation of realizations of the process in question using random numbers. In the vein of this method we shall illustrate the functioning of our systems with pseudo-codes similar to computer programs.

In the sixties and seventies, Ilya Piatetski-Shapiro, who consulted with Roland Dobrushin and Yakov Sinai, challenged members of a seminar of Moscow mathematicians, including me, with results of computer simulations of several carefully chosen examples of interacting random processes with discrete time. He proposed that we prove the observed properties, particularly switches from ergodic to non-ergodic behavior as a result of continuous change of parameters. [2] is the most complete survey of these developments. Some of our examples were first proposed at that seminar. Examples first proposed in [30] and [44], and described as 2-percolation and NEC-voting systems here, moved me to develop a method to prove non-ergodicity [31,32,36,37]. One version of this result is formulated here as Theorem 3.

We prefer to present here results that are explicit and understandable rather than general. More general versions can be found in original papers to which we refer. The only prerequisite for this paper is a standard course in probability. Our potential readers are students who love to learn a theory by solving problems. The idea to write a paper with problems and examples about time-discrete interacting random processes comes from Laurie Snell.

Problems of different levels of difficulty can be found here. **Exercises** are technical; they are for students who strive to become professional mathematicians. **Problems** need more non-trivial reasoning, and some of them were or might have been considered publishable results in the past. I try to refer to papers that contain their solutions, but some have never been published. Some proofs and commentaries are given here; □ marks the end of longer ones. **Unsolved problems** are the most feasible of those whose solutions are unknown to me now. The boundaries between the three certainly are not absolute. After all, our common purpose is to turn every unsolved problem into a series of exercises.

2. SYSTEMS AND DETERMINISTIC CASE

2.1. Systems of neighborhoods

A *system of neighborhoods* or just a *system* is defined as follows. We assume that there is a finite or countable set that we call *space*. We call a system finite if its *space* is finite, and infinite otherwise. The natural number m is called *memory*. (All the examples discussed here belong to the simplest case $m = 1$, but we keep in mind a possibility of interesting examples with $m > 1$.) The set *time* $= \{-m, \ldots, -1, 0, 1, 2, \ldots\}$ is the set of values of the discrete time variable t. The product $V = space \times time$ is called *volume*, whose elements (s, t) are called *points*. Points with $t < 0$ are called *initial* and their set I is called *initial volume*. $V_t = \{(s, t), s \in space\}$ is called t-volume. Any subset $W \subset V$ may be called a *subvolume*.

For every non-initial point w there is a finite sequence $N_1(w), \ldots, N_n(w)$ of points which are called its *neighbors*. The set $N(\cdot)$ of neighbors of a point is called its *neighborhood*. The number n of neighbors is the same for all non-initial points. It is essential that all the neighbors of a point (s, t) have values of their time variable less than t. Sometimes we denote $N_i(s, t) = (n_i(s, t), t - t_i(s, t))$ where $n_i(s, t)$ and $t_i(s, t)$ are suitable functions of s and t and $t_i(s, t) > 0$.

Thus, to specify a *system $\mathcal{S}$* one has to choose *space*, natural numbers m and n and n neighbors of every non-initial point: $\mathcal{S} = \{space,\ m,\ n,\ N(\cdot)\}$.

2.1.1. Exercise. Call neighbors of a point its 1-st degree neighbors. Define a point's k-th degree neighbors as neighbors of its $(k-1)$-th degree neighbors. What is the greatest and the smallest number of a point's k-th degree neighbors in systems with a given n?

Answer: The smallest number is 1. (Or n if we assume that a point's neighbors may not coincide.) The greatest number is n^k.

Comment: For every n there is a *Tree system* (to which we shall refer) in which every point has the greatest possible number n^k of k-th degree neighbors for all k. It can be defined as follows: *space* is the set of vertices of an infinite directed tree, every vertex of which serves as the end point of n edges and as the starting point of one edge. Neighbors of a point (s, t) have their time components equal to $t - 1$ and those space components whence edges go to s.

2.2. Standard systems and uniformity

Let $Z = (\ldots - 1, 0, 1, \ldots)$ be the ring of integers and let Z_S be the ring of residues modulo S. Thus, Z^d is the d-dimensional discrete space and Z_S^d is the d-dimensional torus. Most of our examples belong to the following *standard* case: *space* is Z^d or Z_S^d and there are n *neighbor vectors* $v_1, \ldots v_n \in Z^{d+1}$ such that for all non-initial (s, t) and for all $i = 1, \ldots, n$

$$N_i(s, t) = (s, t) + v_i.$$

(Time components of all the neighbor vectors are negative.) In the standard case it is sufficient to state neighbors of the origin $\mathcal{O} = (\bar{0}, 0)$ of *space $\times$ time*, to define neighbors of all the points, and this is what we shall typically do.

Standard systems are a special case of uniform systems in which 'all automata are equal'. Call a one-to-one map $a : space \mapsto space$ a system's *automorphism* if

$$\forall\, s \in space,\ t \geq 0,\ i = 1, \ldots, n\ :\ N_i(a(s), t) = a(N_i(s, t)).$$

Automorphisms of a system form a group $\mathcal{A}$. Call $\mathcal{A}$ transitive if

$$\forall\, s_1, s_2 \in space\ \exists a \in \mathcal{A}\ :\ a(s_1) = s_2.$$

Call a system *uniform* if it has a transitive group of automorphisms and $t_i(s,t)$ do not depend on s, t. Call a uniform system *commutative* if it has a commutative transitive group of automorphisms. All the systems treated of in our paper are uniform and all except the 2-line voting are commutative. In theory this seems not to be very restrictive. I can mention only one theoretical result [46] where non-uniformity is essential. However, computer simulations sometimes have to be performed on a finite subvolume of Z^d, and in these non-uniform cases one has to specify what is going on at the boundaries.

In uniform systems $n_i(s,t)$ and $t_i(s,t)$ do not actually depend on t, and we shall call them $n_i(s)$ and $t_i(s)$:

$$N_i(s,t) = (n_i(s), \quad t - t_i(s)). \tag{1}$$

Call a uniform system *d-dimensional* if

$$space = Z^d \times S = \{(i,j) \;:\; i \in Z^d, \; j \in S\} \tag{2}$$

where S is finite, and the automorphism group $\mathcal{A}$ contains the subgroup $\mathcal{Z}^d$ whose elements are shifts at any constant vector $V \in Z^d$:

$$\mathcal{Z}^d = \{z \;:\; (i,j) \mapsto (i+V, j), \; V \in Z^d\}. \tag{3}$$

2.2.1. Exercise. Prove that any standard system is *d*-dimensional and commutative.

2.2.2. Exercise. Prove that the tree system, which we defined when answering the exercise 2.1.1, is commutative but non-standard.

Comment: The group of all automorphisms of a tree system is not commutative. But it has transitive commutative subgroups.

2.2.3. Exercise. In a commutative system every map $n_i \;:\; space \mapsto space$ is an automorphism, where $i = 1, \ldots, n$ and $n_i(\cdot)$ is that of formula (1).

Proof: Choose some $s_0 \in space$. We can choose $a_i \in \mathcal{A}$ such that $a_i(s_0) = n_i(s_0)$. Now let us prove that $a_i(s) = n_i(s)$ for any $s \in space$. Take $a \in \mathcal{A}$ such that $s = a(s_0)$. Then

$$a_i(s) = a_i(a(s_0)) = a(a_i(s_0)) = a(n_i(s_0)) = n_i(a(s_0)) = n_i(s) \qquad \square$$

2.2.4. Exercise. For a commutative system prove that $\mathcal{A}$ contains such a subgroup $\mathcal{A}_0$ that for any s, $s' \in space$, there is exactly one $a \in \mathcal{A}_0$ such that $a(s) = s'$.

Comment: For any $s \in space$ denote the subgroup $\mathcal{A}_s = \{a \in \mathcal{A} : a(s) = s\}$. Generally, $\mathcal{A}_s$ may depend on s, but in the commutative case it does not, and $\mathcal{A}_0 = \mathcal{A}/\mathcal{A}_s$.

Using this exercise, we can choose a *reference element* $s_0 \in space$ that defines the one-to-one correspondence between $\mathcal{A}_0$ and *space* by the rule $a \mapsto a(s_0)$.

2.3. States and configurations

Let us call elements of *space automata*. Every automaton at any time has the same finite set X_0 of possible states. We may write $X_0 = \{0, \ldots, k\}$. The simplest case $X_0 = \{0, 1\}$ alone provides so many unsolved problems that we shall enjoy it as much as possible. To every point $(s, t) \in V$ there corresponds a variable $x_{s,t} \in X_0$ which represent the state of automaton s at time t. Set $X = X_V = X_0^V$ is the configuration space, whose elements are called configurations. Any configuration consists of *components*, that is, elements of A corresponding to all points. 'All a-s' will mean any configuration, all components of which equal the same element $a \in X_0$. $X_I = X_0^I$ stands for the set of *initial conditions*, that is configurations on I. Elements of the configuration space $X_t = X_0^{V_t}$ at time t are called t-configurations. Generally, to any subvolume W there corresponds its configuration space $X_W = S^W$, whose elements are called configurations on W. Also $X^m = X_0^{space \times [1,m]}$ will be used when denoting operators.

We call a point w a *point of difference* between two configurations s and s' iff $s_w \neq s'_w$. *The set of difference* between two configurations is the set of their points of difference. Call a configuration a *finite perturbation* of another configuration if their set of difference is finite.

2.4. Deterministic systems

To turn a system into a *deterministic system* we need to choose a set X_0 of states of every point and a *transition function* $\mathrm{tr} : X_0^n \mapsto X_0$. This function gives the state of a non-initial point using states of its neighbors as arguments.

For any deterministic system we define a map $\mathrm{TR} : X_I \mapsto X$ as follows: given an initial configuration x_I, first, the initial components of $\mathrm{TR}(x_I)$ are the same as those of x_I, and, second, all the non-initial components of $y = \mathrm{TR}(x_I)$ are determined inductively according to the formula $y_w = \mathrm{tr}(y_{N(w)})$. We call $\mathrm{TR}(x_I)$ the *trajectory* resulting from the initial condition x_I.

To any deterministic system $\mathcal{D}$ there corresponds a *(deterministic) operator* $D : X^m \mapsto X^m$ defined in the following way: the result of application D to a configuration $x = (x_{s,t})$, $s \in space$, $t \in [1, m]$ has components

$$D(x)_{s,t} = \begin{cases} x_{s,t+1} & \text{if } t < m, \\ \mathrm{tr}(x_{N(s,t)}) & \text{if } t = m. \end{cases}$$

(If $m = 1$, the upper line is not actually used.) Call a configuration $x \in X^m$ *fixed* for an operator D (and thereby for the corresponding system $\mathcal{D}$) if $D(x) = x$.

2.4.1. Exercise. Prove: If x is fixed then components of x and $\mathrm{TR}(x)$ do not depend on t.

2.4.2. Exercise. Prove that any deterministic system $\mathcal{D}$ and operator D are uniform in the following sense: For any automorphism $\mathcal{A}$ of the system, the corresponding one-to-one maps of X and X^m commute with $\mathcal{D}$ and D respectively.

2.4.3. Problem. For any commutative d-dimensional system with *space* given by formula (2), prove that if components of an initial state do not depend on j, then components of the resulting trajectory also do not depend on j.

Sketch of proof: From exercise 2.2.3, every n_i is an automorphism. Call $\mathcal{H} : \mathcal{A} \mapsto \mathcal{Z}^d$ the homomorphism of the commutative transitive group $\mathcal{A}$ of the system's automorphisms to the subgroup $\mathcal{Z}^d$ of shifts (3). Thus $\mathcal{H}(n_i)$ is a shift at some vector V_i. Now let $z(w)$ stand for the first coordinate of any point $w \in V$ in the (2) representation of *space*. The statement in question is assured by the fact that the difference $z(n_i(s)) - z(s)$ does not depend on s. This is because this difference equals V_i. $\square$

This exercise shows that commutative d-dimensional systems boil down to standard ones if we apply them only to configurations whose components do not depend on j.

2.5. Pseudo-codes

Since this article emphasizes a computer approach, we illustrate the functioning of our systems with pseudo-codes of imaginary programs which model them. These pseudo-codes do not obey strict syntactical rules of any actually existing programming language; they explain the algorithm. *space* and the range of t are assumed to be finite whenever we write a pseudo-code. Lines of every pseudo-code are numbered for reference purposes. For example, the functioning of a deterministic system can be expressed by the following pseudo-code:

$$
\begin{aligned}
&1 \quad \text{for } t = -m \text{ to } -1 \text{ do} \\
&2 \qquad \text{for all } s \in Space \text{ do} \\
&3 \qquad\qquad x_{s,t} \leftarrow x_{s,t}^{initial} \\
&4 \quad \text{for } t = 0 \text{ to } t_{max} - 1 \text{ do} \\
&5 \qquad \text{for all } s \in Space \text{ do} \\
&6 \qquad\qquad x_{s,t} \leftarrow \mathrm{tr}(x_{N(s,t)})
\end{aligned}
\tag{4}
$$

Here lines 1–3 assign the initial configuration and lines 4–6 calculate t_{max} subsequent t-configurations.

In our imagination, values of $x_{s,t}$ for all $s \in space$ are assigned simultaneously, at one moment t. However, in the actual programming this is not essential, and practically the line 'for all $s \in space$ do' can be implemented using d nested cycles if $space$ is d-dimensional.

The well-known cellular automata (see, for example, [1,7]) are standard deterministic systems with $m = 1$ and $space = Z^d$.

2.5.1. Exercise. Write a pseudo-code for the well-known Game of Life, that is, a cellular automaton with $space = Z^2$ and $X_0 = \{0,1\}$. Here neighbors of a cell are its eight nearest neighbors. A cell in the state 0 is called 'dead' and a cell in the state 1 is called 'alive'. The transition rule is: a dead cell becomes alive at the next moment only if it has exactly three live neighbors and an alive cell stays alive only if it has two or three live neighbors.

3. MEASURES

3.1. Measures and topology

For any configuration space X_W we shall consider the set $\mathcal{M}_W$ of probabilistic, i.e., normed measures on the σ-algebra generated by cylinder sets in the product-space X_W. Most often we deal with $\mathcal{M}_V$, which we call simply $\mathcal{M}$. Call a measure *degenerate* if it has a zero value on at least one cylinder set. Let δ_s stand for the δ-measure concentrated on one configuration s. A product-measure on any configuration space $X_W = X_0^W$ is a product of measures for coordinates. A *mixture* of several normed measures is their linear combination with non-negative coefficients whose sum equals 1. A measure is called uniform if it is fixed for any map on $\mathcal{M}$ corresponding to an automorphism.

Let the *weak topology* correspond to convergence on all the cylinder sets and let the *strong topology* correspond to convergence on all the elements of the σ-algebra generated by them. The following exercises show why the weak topology is typically used when treating infinite interactive systems:

3.1.1. Exercise. Prove that, if W is finite, the weak and strong topologies on $\mathcal{M}_W$ coincide, but in the infinite case they are different.

3.1.2. Exercise. Let $B(p)$ stand for the uniform product-measure on the product-space $X = \{0,1\}^Z$ in which every component equals 1 with probability p, independently of others. Prove that when $p \to 0$, the measure $B(p)$ tends to the measure concentrated in the configuration 'all zeroes' in the weak topology, but not in the strong topology.

1)   For any configuration space X_0^W, where X_0 is compact (e.g., finite) and W is finite or countable, $\mathcal{M}_W$ is compact in the weak topology.

2)   For any configuration space X_0^W, where $|X_0| > 1$ and W is infinite, $\mathcal{M}_W$ is not compact in the strong topology.

For any two normed measures μ_1 and μ_2 on a measurable space X, call a measure on $X \times X$ a *coupling* of μ_1 and μ_2 if its marginals are μ_1 and μ_2. Coupling of measures is the simplest version of couplings. Below we shall speak of couplings of random systems [5,4]; [16,24] are some of the newest papers on coupling.

3.2. A distance between measures

As long as we are interested only in convergence vs. non-convergence, the weak topology is sufficient for us. But as soon as we want to speak about how fast systems converge, we need a distance between measures. For any coupling c of μ_1 and μ_2 on a configuration space $X_W = X_0^W$, let

$$rift(c) = \sup\{c(x_w^1 \neq x_w^2),\ w \in W\},$$

where $x^1 = (x_w^1)$ and $x^2 = (x_w^2)$ stand for the first and second components in $X^1 \times X^2 = X_W \times X_W$. Define $dist(\mu_1, \mu_2)$, i.e. the *distance* between μ_1 and μ_2, as the infimum of $rift(c)$ over all the couplings c of μ_1 and μ_2.

3.2.1. Exercise. Prove: Any value of $rift(\cdot)$ or $dist(\cdot)$ belongs to $[0, 1]$.

3.2.2. Exercise. Let $\mu_1, \mu_2 \in \mathcal{M}_W$ where $X_W = \prod_{w \in W} X_w$. Let $\mu_1|_w$ and $\mu_2|_w$ stand for projections of μ_1 and μ_2 to X_w. Prove:

$$dist(\mu_1, \mu_2) = \sup\{dist(\mu_1|_w, \mu_2|_w)\ :\ w \in W\}.$$

3.2.3. Exercise. Let S be a cylinder subset of $X_W = \{(x_w),\ w \in W\}$ indicated by a condition involving some r components $x_1, \ldots, x_r$. Prove:

$$\forall\, \mu_1, \mu_2 \in \mathcal{M}_W\ :\ |\mu_1(S) - \mu_2(S)| \leq r \cdot dist(\mu_1, \mu_2).$$

3.2.4. Problem. Prove that $dist$ is a metric, that is

1) $dist(\mu_1, \mu_2) \geq 0$.

2) $dist(\mu_1, \mu_2) = dist(\mu_2, \mu_1)$.

3) $dist(\mu_1, \mu_2) = 0 \iff \mu_1 = \mu_2$.

4) $dist(\mu_1, \mu_3) \leq dist(\mu_1, \mu_2) + dist(\mu_2, \mu_3)$.

Proof: We leave proof of 1) and 2) to the reader and 3) follows from exercise 3.2.3. The following proof of 4) for the finite case was proposed by Eugene Speere. Take any μ_1, μ_2 and μ_3 on X_W. Let c_{12} be any coupling of μ_1 and μ_2 and c_{23} be any coupling of μ_2 and μ_3. All we have to do is to present a coupling c_{13} of μ_1 and μ_3 for which

$$rift(c_{13}) \leq rift(c_{12}) + rift(c_{23}). \tag{5}$$

Take the following measure c_{123} on $X^1 \times X^2 \times X^3 = X_W^3$:

$$c_{123}(x^1, x^2, x^3) = \begin{cases} 0 & \text{if } \mu_2(x^2) = 0; \\ c_{12}(x^1, x^2) \times c_{23}(x^2, x^3)/\mu_2(x^2) & \text{otherwise.} \end{cases}$$

Since μ_2 is a marginal of c_{23}, the sum of $c_{23}(x^2, x^3)$ over x^3 gives $\mu_2(x^2)$. Therefore the sum of $c_{123}(x^1, x^2, x^3)$ over x^3 gives $c_{12}(x^1, x^2)$. Analogously, the sum of $c_{123}(x^1, x^2, x^3)$ over x^1 gives $c_{23}(x^2, x^3)$. Thus c_{123} is a normed measure and the following is a coupling of μ_1 and μ_3:

$$c_{13}(x^1, x^3) = \sum_{x^2} c_{123}(x^1, x^2, x^3).$$

Thus defined c_{13} fits our claim (5) because

$$\forall w \in W : c_{13}(x_w^1 \neq x_w^3) \leq c_{12}(x_w^1 \neq x_w^2) + c_{23}(x_w^2 \neq x_w^3). \qquad \square$$

3.2.5. Exercise. Extend this proof to the infinite case.

Call a coupling c of μ_1 and μ_2 their *fine coupling* if $rift(c) = dist(\mu_1, \mu_2)$.

3.2.6. Exercise. Let W be finite. Prove that our metric *dist* is continuous both in the weak and strong topology and that $\mathcal{M}_W$ is compact w.r.t. our metric. Prove also that in this case for every two measures there exists a fine coupling.

3.2.7. Exercise. Let W be infinite and $|X_0| > 1$. Prove that $\mathcal{M}$ is not compact with *dist* as a metric. Prove also that *dist* is not continuous in the weak topology, because there is a μ and a sequence $\mu_1, \mu_2, \ldots$ which tends to μ, such that $dist(\mu_n, \mu)$ does not tend to 0.

3.2.8. Unsolved problem. Is it true that for any two normed measures on X_W there exists a fine coupling?

Comment: The difficulty lies in non-compactness of $\mathcal{M}$. However, I can not present measures for which there is no fine coupling.

3.2.9. Exercise. Prove that fine coupling may be non-unique.

Proof: For example, let some x_w certainly equal 0 in μ_1 and certainly equal 1 in μ_2. Then *rift* of any coupling of these measures equals 1, and all their couplings are fine.   □

The following are some cases when fine coupling exists and sometimes is unique, even with an infinite W:

3.2.10. Exercise. If $\mu_1 = \mu_2$ then their fine coupling c_{fine} exists and is unique. Prove this and write an explicit formula for $c_{fine}(S)$ for any cylinder set $S \subset X_W \times X_W$.

3.2.11. Exercise. Let μ_1 be an arbitrary measure on X_W and let μ_2 be a δ-measure δ_y concentrated on one configuration $y \in X_W$. Then μ_1 and μ_2 have only one coupling at all. Therefore their fine coupling exists and is unique and $dist(\mu_1, \mu_2) = \sup\{\mu_1(x_w \neq y_w) : w \in W\}$.

3.2.12. Problem. Let $|W| = 1$, i.e., $X_W = X_0$, and μ_1 and μ_2 be any measures on X_W. Denote $m(a) = \min(\mu_1(a), \mu_2(a))$ for all $a \in X_0$. Then:

1)
$$
\begin{aligned}
dist(\mu_1, \mu_2) &= \max\{\mu_1(S) - \mu_2(S) : S \subset X_0\} \\
&= \max\{\mu_2(S) - \mu_1(S) : S \subset X_0\} \\
&= \sum_{a \in X_0} (\mu_1(a) - m(a)) = \sum_{a \in X_0} (\mu_2(a) - m(a)).
\end{aligned}
$$

2)   The set of fine couplings of μ_1 and μ_2 is non-empty and coincides with the set of those couplings of μ_1 and μ_2 which have the form $c_{fine} = c_0 + c_1$ where

$$
c_0(x_1, x_2) = \begin{cases} m(x_1) & \text{if } x_1 = x_2, \\ 0 & \text{otherwise.} \end{cases}
$$

and c_1 is any measure on $X_0 \times X_0$, for which

$$
\begin{cases} \forall\, x_1 : \sum_{x_2 \in X_0} c_1(x_1, x_2) = \mu_1(x_1) - m(x_1), \\ \forall\, x_2 : \sum_{x_1 \in X_0} c_1(x_1, x_2) = \mu_2(x_2) - m(x_2). \end{cases} \tag{6}
$$

Comment: One example of c_1 which satisfies (6) is

$$
c_1^*(x_1, x_2) = (\mu_1(x_1) - m(x_1)) \times (\mu_2(x_2) - m(x_2)),
$$

and the resulting $c_{fine}^* = c_0 + c_1^*$ is Vaserstein's *-operation which was defined in [42] for any measurable X_0.

3.2.13. Problem. Let μ_1 and μ_2 be product-measures on the product-space X_W:

$$\mu_1 = \prod \mu_1^s, \quad \mu_2 = \prod \mu_2^s, \quad s \in space.$$

Then any product of fine couplings of μ_1^s and μ_2^s is a fine coupling of μ_1 and μ_2.

3.2.14. Problem. Let $X_0 = \{0, \ldots, k\}$, let $X = X_0^W$, and let $\mathcal{B}(p)$ stand for the product-measure on the product-space X, every factor of which is the same measure $p = (p_0, \ldots, p_k)$ on X_0. Prove that for any $\mathcal{B}(p)$ and $\mathcal{B}(q)$ there exists a fine coupling and

$$dist(\mathcal{B}(p), \mathcal{B}(q)) = \max\{p(S) - q(S) \ : \ S \subset X_0\}.$$

Clearly, there are many ways to define a distance between measures. The distance we describe here seems to be useful at least in the present context. The facts that it is not continuous in the weak topology and that $\mathcal{M}$ is not compact seem to be an inevitable trade-off for relevance to problems about how quickly or slowly large interactive systems converge. The idea of this definition of distance was present in [42] (in the form of that definition on p. 50 which speaks of estimate α_k). However, the $*$-operation proposed in [42], generally speaking, does not provide a fine coupling. For example, the $*$-operation applied directly to any different uniform product-measures on X^W with $|X| \geq 2$ and infinite W has *rift* equal to 1.

4. RANDOM SYSTEMS

We shall define *random systems* using (every time the same) mutually independent *hidden* variables h_w for all points $w \in V$, everyone of which is uniformly distributed on $[0, 1]$; thus we have the *hidden* product-measure η on the hidden configuration space $H = [0, 1]^V$ of the hidden variables. In pseudo-codes the same idea is expressed in generating and using random numbers *rnd*. We assume that every call of *rnd* in a pseudo-code generates a random number uniformly distributed on $[0, 1]$ that is independent from all the past events, including past calls of *rnd*.

4.1. Special case: $X_0 = \{0, 1\}$

All we need to define a random system in this case is a function $p_0 : X_0^n \mapsto [0, 1]$, the conditional probability of a point to be in the state 0 given states of its neighbors. A point's probability to be in the state 1 is $p_1(\cdot) = 1 - p_0(\cdot)$. Values of $p_0(\cdot)$ and $p_1(\cdot)$ are called *transition probabilities*. If at least one of them equals 0, we call the random system *degenerate*. The following pseudo-code describes functioning of an arbitrary random system with $X_0 = \{0, 1\}$ and a finite *space*:

```
1    for t = −m to −1 do
2        for all s ∈ space do
3            x_{s,t} ← x_{s,t}^{initial}
4    for t = 0 to t_{max} − 1 do
5        for all s ∈ space do
6            if rnd < p_0(x_{N(s,t)}) then x_{s,t} ← 0 else x_{s,t} ← 1
```
$$(7)$$

Here lines 1–3 assign initial values and lines 4–6 calculate a realization of the process during t_{max} time steps. In mathematical terms this means that we define our process as the measure on X induced by any initial measure and the hidden measure η with the following map from $X_I \times H$ to X:

$$\begin{cases} \forall\, s \in space,\ t \in [-m, -1] : x_{s,t} = x_{s,t}^{initial}, \\ \forall\, s \in space,\ t \in [\ 0,\ \infty) : x_{s,t} = \begin{cases} 0 & \text{if } h_{s,t} < p_0(x_{N(s,t)}), \\ 1 & \text{otherwise.} \end{cases} \end{cases} \quad (8)$$

4.2. General case: $X_0 = \{0, \ldots, k\}$

Take a function p from X_0^n to the set of normed measures on X_0 and call its values $p(a \mid b_1, \ldots, b_n)$ *transition probabilities*; they serve as conditional probabilities of $x_w = a$ if $x_{N(w)} = (b_1, \ldots, b_n)$. [Denotations $p_0(\cdot)$ and $p_1(\cdot)$, which we used in the case $X_0 = \{0, 1\}$, now are substituted by $p(0|\cdot)$ and $p(1|\cdot)$.] If at least one of the transition probabilities equals 0, we call the random system *degenerate*. As soon as parameters $p(a \mid b_1, \ldots, b_n)$ are given, we can define our process as a measure $proc(\mu_I) \in \mathcal{M}$ induced by any initial measure μ_I and the hidden measure η with the following map:

$$\begin{cases} \forall\, s \in space,\ t \in [-m, -1] : x_{s,t} = x_{s,t}^{initial}, \\ \forall\, s \in space,\ t \in [\ 0,\ \infty) : x_{s,t} = \\ \quad \begin{cases} 0 & \text{if } h_{s,t} \leq p(0|x_{N(s,t)}), \\ i & \text{if } \sum_{j=0}^{i-1} p(j|x_{N(s,t)}) < h_{s,t} \leq \sum_{j=0}^{i} p(j|x_{N(s,t)}) \text{ for all } i > 0. \end{cases} \end{cases} \quad (9)$$

4.2.1. Exercise. Write a pseudo-code for a random system with $X_0 = \{0, 1, \ldots, k\}$.

4.2.2. Exercise. Let a random system be given. Prove that a mixture of two processes is a process too.

4.2.3. Exercise. Prove that, if a sequence of processes of a random system has a limit (in the weak topology), this limit is also a process for the random system.

4.2.4. Exercise. For any deterministic system with a transition function $\text{tr}(\cdot)$ let us define the corresponding random system with the same *space*, *m* and neighborhoods as follows:

$$p(a_w | b_{N(w)}) = \begin{cases} 1 & \text{if } a_w = \text{tr}(b_{N(w)}), \\ 0 & \text{otherwise.} \end{cases}$$

Prove that for any deterministic initial condition the resulting process is concentrated in the resulting trajectory.

4.3. Deterministic systems with noise

Here we consider random systems which result from deterministic ones by adding random noise.

First assume $X_0 = \{0,1\}$. In this case we need only two non-negative parameters ε and δ where $\varepsilon + \delta \le 1$. Take the process induced by the hidden measure η with the following map from H to X:

$$\begin{cases} \forall s \in space, \ t \in [-m, -1] : x_{s,t} = x_{s,t}^{initial}, \\ \forall s \in space, \ t \in [\ 0, \ \infty) : x_{s,t} = \begin{cases} 0 & \text{if } h_{s,t} < \varepsilon, \\ 1 & \text{if } h_{s,t} > 1 - \delta, \\ \text{tr}(x_{N(s,t)}) & \text{otherwise.} \end{cases} \end{cases} \quad (10)$$

We shall say that this system results from the deterministic one by adding an ε-δ noise. To obtain the corresponding pseudo-code it is sufficient to add the following two lines to the pseudo-code (4):

```
7              h ← rnd
8              if h < ε then x_{s,t} ← 0 else if h > 1 − δ then x_{s,t} ← 1    (11)
```

The resulting pseudo-code (4) and (11) imitates a process in which every automaton first follows the deterministic rule, then makes a two-way random error: turns 0 with probability ε and turns 1 with probability δ. We call the ε-δ noise *symmetric* if $\varepsilon = \delta$, *biased* if $\varepsilon \ne \delta$, and *one-way* or *degenerate* if $\varepsilon = 0$ or $\delta = 0$.

4.3.1. Exercise. Show that for any $\text{tr}(\cdot)$, ε, and δ there exists $p(\cdot)$ such that the process (8) is the same as the process (10).

Now the general case $X_0 = \{0, \ldots, k\}$. We need non-negative parameters ε_i for all $i \in X_0$ whose sum does not exceed 1. The process is induced by any initial measure and the hidden measure η with the following map:

$$
\begin{cases}
\forall\, s \in space,\ t \in [-m, -1] : x_{s,t} = x_{s,t}^{initial}, \\
\forall\, s \in space,\ t \in [\ 0,\ \infty) : x_{s,t} = \\
\quad \begin{cases}
0 & \text{if } h_{s,t} < \varepsilon, \\
i & \text{if } \sum_{j=0}^{i-1} \varepsilon_j \leq h_{s,t} < \sum_{j=0}^{i} \varepsilon_j, \text{ for } i > 0, \\
tr(x_{N(s,t)}) & \text{otherwise.}
\end{cases}
\end{cases}
\tag{12}
$$

To obtain the corresponding pseudo-code we may add the following lines to the pseudo-code (4):

7	$h \leftarrow rnd$	
8	$i \leftarrow 0;\ \theta \leftarrow \varepsilon_0$	
9	while $i < k$ and $\theta < h$ do	(13)
10	$\quad i \leftarrow i + 1;\ \theta \leftarrow \theta + \varepsilon_i$	
11	if $\theta \geq h$ then $x_{s,t} \leftarrow i$	

The resulting pseudo-code (4) and (13) imitates a process in which every automaton first follows the deterministic rule, after which the noise turns it into the state i with probability ε_i for all $i \in A$. We call the noise *degenerate* if some $\varepsilon_i = 0$.

4.3.2. Exercise. Show that for any $tr(\cdot)$ and ε_i, $i \in A$ there exists $p(\cdot)$ such that the process (9) is the same as the process (12).

4.4. Definitions of ergodicity, slow and fast convergence

By definition, *time-shift* of the volume V maps any (s,t) to $(s, t+1)$. *Time-shifts* $T : X \mapsto X$ and $T : M \mapsto M$ are defined in the same vein:

$$\forall\, x \in X,\ s \in space,\ t \in time\ :\ T(x)_{s,t} = x_{s,t+1},$$

$$\text{and } \forall\, S \subseteq X\ :\ T(\mu)(S) = \mu(T(S)).$$

4.4.1. Exercise. Prove that if μ is a process of a random system then $T(\mu)$ also is a process of the same system.

Call a process μ *invariant* for a system if $T(\mu) = \mu$.

Theorem 1. Any random system has at least one invariant process.

Proof: Take any process μ_1 and for all $n = 2, 3, \ldots$ define

$$\mu_n = \frac{1}{n} \sum_{i=0}^{n-1} \mathcal{T}^n(\mu_1). \tag{14}$$

Since all $\mathcal{T}^n(\mu_1)$ are processes (exercise 4.4.1), all their mixtures are processes too (exercise 4.2.2). Since $\mathcal{M}$ is compact (exercise 3.1.3), any sequence in $\mathcal{M}$ has at least one limit point (may be more). From exercise 4.2.3 every limit point of μ_n is an invariant process. □

Definition. Say that a random system is *ergodic* if it has only one invariant process μ_{inv} and for all its processes μ the limit $\lim_{n\to\infty} \mathcal{T}^n(\mu)$ exists and equals μ_{inv}.

4.4.2. Exercise. Does a non-ergodic random system exist which has only one invariant process?

Answer: Yes. Take a deterministic system with *space* consisting of one element 0, with $N(0, t)$ consisting of one point $(0, t - 1)$, in which $X_0 = \{0, 1\}$ and $\mathrm{tr}(a) = 1 - a$.

4.4.3. Unsolved problem. Does there exist a non-degenerate non-ergodic random system which has only one invariant process?

4.4.4. Exercise. Prove that any non-degenerate finite random system is ergodic.

4.5. Patterns

Let us say that all those standard systems with the same d, m, and $N(\mathcal{O})$ belong to one *pattern*. In the same vein, a deterministic pattern is a quadruple $(d, m, N(\mathcal{O}), \mathrm{tr}(\cdot))$ and a random pattern is a quadruple $(d, m, N(\mathcal{O}), p(\cdot))$. This allows us to compare behavior of the infinite system and of finite systems with various S belonging to one pattern $(d, m, N(\mathcal{O}))$. Let μ_{inv} stand for the invariant process of any system that has only one invariant process. For any random pattern let us define $\rho(S, t)$ as the supremum of $dist(\mathcal{T}^t(\mu), \mu_{inv})$ over all processes μ of the finite system with *space* $= Z_S^d$ that belongs to the given pattern.

Let us say that a pattern *converges fast* if there is a positive constant $q < 1$ such that

$$\forall\, S, t\ :\ \rho(S, t) \leq q^t.$$

Let us say that a pattern *converges slowly* if there are $\varepsilon > 0$ and $Q > 1$ such that

$$\forall\, S, t\ :\ t \leq Q^S \Longrightarrow \rho(S, t) \geq \varepsilon.$$

4.5.1. Unsolved problem. It is conjectured that any random pattern converges fast iff the infinite system is ergodic and it converges slowly iff the infinite system is non-ergodic. Is it always true?

Comment: Note that most computer simulations refer directly to finite systems. (Typically they imitate functioning of some finite system using the Monte Carlo method.) Thus we certainly need to know (or assume) some relation between finite and infinite systems whenever we interpret results of computer simulations as telling us something about ergodicity or non-ergodicity of infinite systems.

4.6. Operators

It is possible to define ergodicity in the following equivalent way. Let $\mathcal{M}^m$ stand for the set of normed measures on X^m. To any random system there corresponds a linear *operator* $P : \mathcal{M}^m \mapsto \mathcal{M}^m$ whose definition is based on the following assumption: application of P to a δ-measure concentrated in a state $y = (y_1, \ldots, y_m) \in X^m$ is a product of the following measures pertaining to components:

$$\text{the distribution of } x_{s,t} \text{ is } \begin{cases} \delta_{y_{s,t+1}} & \text{if } t < m, \\ p(y_{N(s,t)}) & \text{otherwise} \end{cases}.$$

(If $m = 1$, the upper line is not used.) Call a measure μ *invariant* for an operator P if $P(\mu) = \mu$.

4.6.1. Exercise. Prove that any operator has at least one invariant measure.

Proof: analogous to proof of Theorem 1.

Let us say that operator P of a random system is *ergodic* if it has only one invariant measure μ_{inv} and for all μ the limit $\lim_{n\to\infty} P^n(\mu)$ exists and equals μ_{inv}.

4.6.2. Exercise. Prove that a random system is ergodic iff its operator is ergodic.

4.7. Monotony

We shall use the signs $\preceq$ and $\succeq$ to speak about (perhaps, partial) ordering of sets. For example, any set S of real numbers is said to be *naturally ordered* if $\forall i, j \in S : i \preceq j \iff i \leq j$. Call a function f from one ordered set to another *monotonic* if $s \preceq s' \implies f(s) \preceq f(s')$.

Let X_0 be ordered (for example, naturally). Then any configuration space X_0^W is partially ordered by the rule $s \preceq s' \iff \forall i \in W : s_i \preceq s_i'$. Call a deterministic system *monotonic* if

$$\forall s, s' \in X_I : s \preceq s' \implies \text{TR}(s) \preceq \text{TR}(s').$$

4.7.1. Exercise. Prove that a deterministic system is monotonic iff its transition function is monotonic.

For any two measures μ_1, $\mu_2 \in \mathcal{M}_W$ on any configuration space X_W we say that $\mu_1 \preceq \mu_2$ iff $\mu_1 f \le \mu_2 f$ for any monotonic function $f : X_W \mapsto R$.

Call a random system *monotonic* if

$$\forall \mu_1, \mu_2 \in \mathcal{M}_I \; : \; \mu_1 \preceq \mu_2 \Longrightarrow proc(\mu_1) \preceq proc(\mu_2).$$

Call an operator P *monotonic* if

$$\forall \mu_1, \mu_2 \in \mathcal{M}^m \; : \; \mu_1 \preceq \mu_2 \Longrightarrow P(\mu_1) \preceq P(\mu_2).$$

4.7.2. Exercise. Prove that a random system is monotonic iff its operator is monotonic.

4.7.3. Exercise. Prove that, if a map (9) from $\mathcal{M}_I \times H$ to X is monotonic ($[0, 1]$ naturally ordered), then the resulting random system is monotonic.

4.7.4. Exercise. Any order can be reversed by turning $\preceq$ into $\succeq$ and vice versa. Prove that this reversal keeps all our kinds of monotony.

4.7.5. Problem. Let A be ordered and have a smallest element $a_{\min}$ and a largest element $a_{\max}$. Thus X has the smallest state $s_{\min} =$'all $a_{\min}$' and the largest state $s_{\max} =$'all $a_{\max}$'. Then for any monotonic operator P the limits $\lim_{t \to \infty} P^t(s_{\min})$ and $\lim_{t \to \infty} P^t(s_{\max})$ exist. These limits are equal iff P is ergodic.

5. PROOFS OF ERGODICITY AND FAST CONVERGENCE

5.1. Case $X_0 = \{0, 1\}$

To prove ergodicity and fast convergence of a system, it is convenient to use its *coupling* (see [4] and [5]) which we shall first illustrate by the following pseudo-code:

```
1     for t = -m to -1 do
2           for all s ∈ space do
3                 x_{s,t} ← x_{s,t}^{initial}
4                 y_{s,t} ← y_{s,t}^{initial}
5                 m_{s,t} ← 0
6     for t = 0 to t_{max} - 1 do
7           for all s ∈ space do
8                 x_{s,t} ← f(x_{N(s,t)}, rnd)
9                 y_{s,t} ← f(y_{N(s,t)}, rnd)
10                m_{s,t} ← min(m_{N_1(s,t)}, ..., m_{N_n(s,t)})        (15)
11                h ← rnd
12                if h < ε then
13                      x_{s,t} ← 0
14                      y_{s,t} ← 0
15                      m_{s,t} ← 1
16                else if h > 1 - δ then
17                      x_{s,t} ← 1
18                      y_{s,t} ← 1
19                      m_{s,t} ← 1
```

Let us first ignore all the lines that deal with the values $m_{s,t}$ and concentrate our attention on those which deal with $x_{s,t}$ and $y_{s,t}$. Then we see that we are modelling simultaneously two processes of the same system, using for them a common source of random noise. In both processes every automaton every time does the following: first, due to line 8 or 9, it assumes some value that depends in a random way on its neighbors and, second, it makes a random error, becoming 0 with probability ε due to line 13 or 14 and becoming 1 with probability δ due to line 17 or 18. (We assume that $\varepsilon + \delta \leq 1$.)

A coupling c of k random systems with a common S is a random system with the same S system of neighborhoods, $X_0(c) = X_0(1) \times X_0(2) \times \cdots \times X_0(k)$, and the transition probability which is a coupling (in the sense of coupling of measures defined in section 3.1) of their transition probabilities. Note that according to this definition the coupled systems need not to be identical; it is sufficient for them to correspond to the same S system of neighborhoods, that is, to have the same $space$, m, and $N(\cdot)$. Couplings typically used in the literature are couplings of a system with itself in our terms. The pseudocode (15) describes coupling of a given system with itself and with another system that deals with the values $m_{s,t}$ (about which we comment below). We assume that the function $f : X_0^n \times [0, 1] \mapsto X_0$ is chosen in such a way that each marginal process coincides with a certain given one. This assumption is substantiated below in the form of exercise 5.1.4.

5.1.1. Exercise. Prove that marginals of any process of a coupling of several random systems are processes of the coupled random systems.

Generally, any system has many different couplings. But only those, where points of difference die out fast enough for any pair of initial conditions, help to prove ergodicity. To check this, our pseudo-code assumes that at every point w we have a special *mark* m_w which may equal 0 or 1. This is to mark loss of the memory (when we are sure about it). Let us say that a point w is *marked* if $m_w = 1$. Initially all the points are unmarked (line 5) and become marked (lines 15 and 19) whenever they are assigned values which certainly do not depend on the prehistory and therefore cannot be points of difference.

Let us estimate the percentage U_t of unmarked points at time t in this process. Let k_i stand for the number of the origin's neighbors whose time coordinate equals $-i$. (Note that $k_1 + \cdots + k_m = n$.) The percentage of unmarked points that proliferate into V_t as a result of line 10 does not exceed $k_1 U_{t-1} + \cdots + k_m U_{t-m}$. Then $\varepsilon + \delta$ of them die as a result of lines 15 and 19. Thus

$$U_{-m} = \cdots = U_{-1} = 1, \qquad U_t \le (1 - \varepsilon - \delta)(k_1 U_{t-1} + \ldots + k_m U_{t-m}).$$

Therefore $U_t \le \tilde{U}_t$ where

$$\tilde{U}_{-m} = \cdots = \tilde{U}_{-1} = 1, \qquad \tilde{U}_t = (1 - \varepsilon - \delta)(k_1 \tilde{U}_{t-1} + \ldots + k_m \tilde{U}_{t-m}).$$

5.1.2. Exercise. Prove that $\tilde{U}_t \to 0$ iff $(1 - \varepsilon - \delta)(k_1 + \cdots + k_m) < 1$, that is $1 - \varepsilon - \delta < 1/n$. Prove also that if $\tilde{U}_t$ tends to 0, then it does so exponentially.

5.1.3. Problem. For any (finite or infinite) system modelled by the pseudo-code (15), prove that dying out of unmarked points guarantees ergodicity and fast convergence.

Now to apply our results to a general process with $X_0 = \{0, 1\}$. This is done by the following:

5.1.4. Exercise. Prove that for any two processes of one random system described by the pseudo-code (7) and formula (8), there is a coupling which can be represented by the pseudo-code (15) with a suitable f and

$$\varepsilon = \min\{p_0(x_{N(s,t)}), \ x_{N(s,t)} \in X_{N(s,t)}\},$$
$$\delta = \min\{p_1(x_{N(s,t)}), \ x_{N(s,t)} \in X_{N(s,t)}\}.$$

5.1.5. Problem. Prove that if $X_0 = \{0, 1\}$ and

$$\max\{p_0(x_{N(s,t)}), \ x_{N(s,t)} \in X_{N(s,t)}\} - \min\{p_0(x_{N(s,t)}), \ x_{N(s,t)} \in X_{N(s,t)}\} < 1/n$$

then the system is ergodic and converges fast.

Comment: This follows from 5.1.2–5.1.4.

5.2. General case $X_0 = \{0, \ldots, k\}$

Let $\mathbf{P}$ stand for the set of functions from A^n to the set of normed measures on A. Let a δ-function $\delta_a \in \mathbf{P}$ stand for a constant function that maps any argument to the same δ-measure on A concentrated in some $a \in X_0$. To prove ergodicity of a random system, it makes sense to represent its transition probability p as a mixture of elements of $\mathbf{P}$, in which δ-functions have the maximal possible sum of coefficients. Let $\Sigma(p)$ stand for this maximal possible sum.

5.2.1. Exercise. Prove that $\Sigma(p) = 1 - \max\{|p(B|s_1) - p(B|s_2)|,\ s_1, s_2 \in X_0^n,\ B \subset X_0\}$.

5.2.2. Exercise. Prove that if $1 - \Sigma(p) < 1/n$ then the system is ergodic and converges fast.

Based on these two exercises, the following holds:

Theorem 2. If $\max\{|p(B|s_1) - p(B|s_2)|,\ s_1, s_2 \in X_0^n,\ B \subset X_0\} < 1/n$ then the random system is ergodic and converges fast.

Comment: See a similar result in [42].

6. PERCOLATION SYSTEMS

6.1. General percolation

Since marking proved useful, it makes sense to examine it as such. This leads us to our first examples. The following pseudo-code is obtained from (15) by deleting everything which pertained to $x_{s,t}$ and $y_{s,t}$ and denoting $\theta = \varepsilon + \delta$:

```
1    for t = -m to -1 do
2        for all s ∈ space do
3            m_{s,t} ← 0
4    for t = 0 to t_{max} - 1 do                              (16)
5        for all s ∈ space do
6            m_{s,t} ← min(m_{N_1(s,t)}, ..., m_{N_n(s,t)})
7            if rnd < θ then m_{s,t} ← 1
```

The process described by this code can be called a *percolation process* because of the following interpretation. Assume that some fluid is supplied to all the initial points. Every non-initial point is connected with its neighbors by pipes which pass the fluid in one direction: to the point from its neighbors. But in every non-initial point there is a tap which is closed with probability θ

independently from all the other taps. Then a point is unmarked iff the fluid percolates to this point.

Lines 1–3 in this pseudo-code set the initial condition 'all zeroes'. Instead of this, we can take an arbitrary initial condition, and thereby obtain an arbitrary *random percolation system*. Lines 4–6 describe the *deterministic percolation system* and line 7 adds the one-way random noise.

6.1.1. Exercise. Prove that any percolation system is monotonic.

Note that the measure concentrated in 'all ones' is an invariant process for any percolation system. Therefore, ergodicity of a percolation system amounts to a tendency to 'all ones' from any initial state.

6.1.2. Exercise. Let P stand for the *percolation operator*, i.e., operator of the percolation system. Prove that any percolation system is ergodic iff $P^n(\text{'all zeroes'})$ tends to 'all ones' when $n \to \infty$.

Let us discuss ergodicity of a percolation system $\mathcal{P}(\theta)$ as depending on θ with a given *space*, m, and neighborhoods. We are interested in *critical* values θ^* that separate ergodicity from non-ergodicity of $\mathcal{P}(\theta)$. If a percolation system is ergodic for all positive θ, we say that $\theta^* = 0$ is its only critical value.

6.1.3. Exercise. Prove that any finite percolation system is ergodic for all $\theta > 0$.

6.1.4. Exercise. Prove that any standard percolation system in which all the neighbor vectors are collinear is ergodic for all $\theta > 0$.

6.1.5. Exercise. Prove that $\theta_1 < \theta_2 \implies \mathcal{P}(\theta_1) \preceq \mathcal{P}(\theta_2)$.

6.1.6. Exercise. Prove that any percolation system $\mathcal{P}(\theta)$ has only one critical value.

6.1.7. Exercise. Let two percolation processes $proc_1(\theta)$ and $proc_2(\theta)$ be defined with the same *space*, m, and initial condition, and let $N_1(\mathcal{O}) \subset N_2(\mathcal{O})$. Prove that $\forall\, \theta\ :\ proc_1(\theta) \succeq proc_2(\theta)$ and that $\theta_1^* \leq \theta_2^*$.

6.1.8. Problem. Prove that a standard infinite percolation system has $\theta^* > 0$ iff it has at least two non-collinear neighbor vectors.

Comment: Due to the preceding exercise, it is sufficient to prove $\theta^* > 0$ for 2-percolation (see below) which is the simplest percolation process.

6.1.9. Unsolved problem. It is conjectured that percolation systems are ergodic at the critical value of θ whenever it is positive. Is this true?

Comment: Some infinite percolation systems, including 2-percolation, are proved to be ergodic at the critical point, see [8,9,12,13,23].

6.2. 2-Percolation

The 2-percolation system has $d = 1$, $m = 1$, $n = 2$ and $N(0,0) = \{(-1,-1),(0,-1)\}$. The following pseudo-code models 2-percolation on Z_S:

```
1    for all s ∈ Z_S do
2        x_{s,-1} ← 0
3    for t = 0 to t_{max} − 1 do
4        for all s ∈ Z_S do
5            x_{s,t} ← min(x_{s-1,t-1}, x_{s,t-1})
6            if rnd < θ then x_{s,t} ← 1
```
$$(17)$$

    This system, in its finite and infinite versions, was first introduced in [30] and received much attention. A proof of its non-ergodicity for small θ by the well-known contour method can be found in [31] or [2].

6.3. Different rates of convergence of finite percolation systems

For percolation systems the following special way is used to speak about how fast or slowly they converge: Let $T(\theta, S)$ stand for the mean time when a given standard finite percolation system with $Space = Z_S$ and a given θ first reaches the state 'all ones', if it started from the state 'all zeroes'.

6.3.1. Exercise. Prove that, if θ is large enough (say $\theta > 1 - 1/n$), then the percolation pattern converges fast and $T(\theta, S)$ grows logarithmically as a function of S.

6.3.2. Problem. Prove that, if θ is small enough, then the percolation pattern converges slowly and $T(\theta, S)$ grows exponentially as a function of S.

Comment: See [31] where it is proved for 2-percolation. For other percolation patterns it follows from monotony.

6.3.3. Unsolved problem. For any percolation pattern the following critical values may be defined:

$\theta_\infty^* = \theta^*$ — the boundary between ergodicity and non-ergodicity of the infinite system,

θ_{fast}^* — infimum of those values of θ with which finite systems converge fast,

θ_{slow}^* — supremum of those values of θ with which finite systems converge slowly,

θ_{log}^* — infimum of those values of θ for which $M(\theta, S)$ grows logarithmically,

θ_{exp}^* — supremum of those values of θ for which $M(\theta, S)$ grows exponentially.

It is conjectured that for any percolation pattern all of these critical values are equal. Is it true?

Comment: This conjecture is not proved (or disproved) even for 2-percolation. However, some relations between these critical values can be stated.

6.3.4. Problem. Prove for all percolation patterns:

$$\theta^*_{\text{slow}} \leq \left\{ \begin{array}{c} \theta^*_{\text{exp}} \\ \theta^*_{\infty} \end{array} \right. \leq \theta^*_{\text{log}} \leq \theta^*_{\text{fast}}.$$

Comment: See [2], p. 72–84 where several critical values and relations between them are discussed.

7. NON-ERGODICITY AND SLOW CONVERGENCE

Theorem 2 has shown that any 'random enough' system is ergodic. Thus, to be non-ergodic, a system has to be 'deterministic enough'. We shall prove non-ergodicity of some systems obtained by 'spoiling' a deterministic system with a small random noise. Naturally, properties of the deterministic system are essential with such approach, and this is what we start with.

7.1. Attractors and eroders

Let X_0 have an element called 0. Let an *island* stand for a finite perturbation of 'all zeroes'. Call a deterministic system an *eroder* if for any initial island x_I the corresponding trajectory $\text{TR}(x_I)$ is also an island.

7.1.1. Exercise. Prove that in any eroder $\text{TR}(\text{'all zeroes'}) = \text{'all zeroes'}$.

We concentrated our attention on 'all zeroes' for simplicity; generalization is easy. Say that an initial configuration s_I is an *attractor* for a deterministic system, or *attracts* it, if for any finite perturbation s'_I of s_I the trajectory $\text{TR}(s'_I)$ is a finite perturbation of $\text{TR}(s_I)$. For example, a system is an eroder iff it is attracted by the initial configuration 'all zeroes'. Note that according to our definition, any finite perturbation of an attractor is an attractor too. So it is better to speak of *bunches* of attractors, a bunch being a class of equivalence, with two initial conditions equivalent iff their set of difference is finite. Let us call a bunch an *attractor* iff all of its elements are attractors.

7.1.2. Exercise. Prove that, if a bunch contains an attractor, it is an attractor.

7.1.3. Exercise. Prove that in any deterministic percolation system the only attractors are those in the bunch of 'all zeroes'.

7.1.4. Exercise. Can a finite deterministic system have more that one bunch of attractors?

7.1.5. Unsolved problem. Can an infinite standard monotonic deterministic system with $X_0 = \{0, 1\}$ have a non-periodic fixed attractor?

Comment: The answer is unknown even in the one-dimensional case with $m = 1$.

Our interest in attractors is motivated by the following consideration. It seems plausible (and sometimes it is true) that, if an initial configuration is an attractor, the system remains in the vicinity of the resulting trajectory even in the presence of a small random noise. Hence it is a good idea to construct systems that are attracted by more than one bunch, and to see how they behave with a small random noise added. In the next section we give exact definitions.

7.2. Stable trajectories

As before, $\mathcal{M} = \mathcal{M}_V$ stands for the set of normed measures on $X = X_0^V$. To every initial configuration s there corresponds $\mathcal{M}^s \subset \mathcal{M}$, which consists of those measures whose projection to I is concentrated in s. To every value of the parameter $\varepsilon \in [0, 1]$ there corresponds $\mathcal{M}_\varepsilon \subset \mathcal{M}$ defined as follows: a measure μ belongs to $\mathcal{M}_\varepsilon$ if

$$\text{for all finite } W \subset V \ : \ \mu(\forall \, w \in W \ : \ x_w \neq \text{tr}(x_{N(w)})) \leq \varepsilon^{|W|},$$

where $|W|$ is the cardinality of W. Finally, $\mathcal{M}_\varepsilon^s = \mathcal{M}_\varepsilon \cap \mathcal{M}^s$. An initial condition s and the resulting trajectory $\text{TR}(s)$ are termed *stable* if

$$\lim_{\varepsilon \to \infty} \sup\{\mu(x_w \neq \text{TR}(s)_w) : \mu \in M_\varepsilon^s, \ w \in V\} = 0.$$

This definition always makes sense: the supremum makes sense because the set $\mathcal{M}_\varepsilon^s$ is non-empty, as it contains the measure concentrated in $\text{TR}(s)$, and the limit makes sense because the set $\mathcal{M}_\varepsilon^s$ decreases when ε decreases.

7.2.1. Problem. Prove that a finite system can not have more than one stable initial configuration.

To show that our definition is non-trivial, i.e., the set $\mathcal{M}_\varepsilon^s$ is rich, we propose the following exercise that constructs explicitly a multi-parametric subset of $\mathcal{M}_\varepsilon^s$.

7.2.2. Exercise. Let us have a parameter $\varepsilon_w \leq \varepsilon$ for every non-initial point w and let the hidden measure η induce a measure $\mu \in \mathcal{M}$ with the following map. For all initial w we set $x_w = s_w$ and for all non-initial w the value of x_w is defined inductively as follows: $x_w = \text{tr}(x_{N(w)})$ if $\eta_w \leq 1 - \varepsilon_w$ and is assigned an arbitrary value otherwise. Prove that any measure μ defined in this way belongs to $\mathcal{M}_\varepsilon^s$. Show also that measures of this sort do not exhaust $\mathcal{M}_\varepsilon^s$.

7.2.3. Unsolved problem. Consider all infinite standard deterministic monotonic systems with $X_0 = \{0, 1, \ldots, k\}$ naturally ordered. Formulate checkable equivalent criteria for their initial condition '*all zeroes*': 1) to be an attractor, 2) to be stable.

Comment: The answer is unknown now even for the case with $m = 1$, $d = 2$ and $|X_0| = 3$. Criteria for arbitrary d, $X_0 = \{0, 1\}$, and $m = 1$ were given in [33]. The same criteria generalized for any m were given in Theorem 6 of [36] and are repeated here as Theorem 3. Criteria for $d = 1$ and arbitrary k and m were given in [21]. The monotony condition is essential; it is known that, without demanding monotony, the problem of discerning eroders is algorithmically unsolvable even with $m = k = 1$ [28]. Some similar proofs of unsolvability are given in [34].

7.2.4. Exercise. Prove that, if the initial condition '*all zeroes*' is stable, then it is an attractor under the conditions of problem 7.2.3. Prove also that the opposite is not true, and therefore the items 1) and 2) of the problem 7.2.3 are not equivalent to each other.

The next problem highlights the difficulty of problem 7.2.3. Let us call: *diameter* of an island — the greatest Euclidean distance between its non-zero components; *lifetime* of an initial condition y — the greatest time coordinate of non-zero components of $\mathrm{TR}(y)$. Of course, a system is an eroder iff lifetime of any initial island is finite. Finally, let us define a function $T(D)$ — the greatest lifetime of initial islands whose diameters do not exceed D.

7.2.5. Problem. Assume conditions of problem 7.2.3. Then:

1. For any system with $k = 1$ there is such a constant C that

$$T(D) \leq C \cdot (D + 1).$$

2. There is a system with $k = 2$ for which

$$T(D) \geq C^D - const$$

   where $C = const > 1$.

7.3. A theorem about attraction and stability

The following theorem treats standard deterministic systems with $space = Z^d$. In this case the volume $V = space \times time$ is a half of the $d + 1$-dimensional integer space. Let us consider V as a subset of the continuous space R^{d+1}. Let $conv(S)$ stand for the convex hull of any set S in R^{d+1}. For any set $S \subset R^{d+1}$ and any number k we define:

$$kS = \{ks : s \in S\}, \quad ray(S) = \bigcup \{kS : k \geq 0\}.$$

We call $ray(S)$ *ray* of S. Let a *zero-set* stand for any subset $z \subseteq N(\mathcal{O})$ for which

$$(\forall w \in z : x_w = 0) \implies \text{tr}(x_{N(\mathcal{O})}) = 0.$$

Let $\sigma(\mathcal{D})$ stand for the intersection of rays of convex hulls (in the continuous space) of all the zero-sets of $\mathcal{D}$.

Theorem 3. Let $\mathcal{D}$ be any monotonic standard deterministic system with $X_0 = \{0, 1\}$ and *space* $= Z^d$. The following four conditions are mutually equivalent:

1) $\mathcal{D}$ is an eroder.

2) $\sigma(\mathcal{D}) = \{\mathcal{O}\}$.

3) There exist homogeneous linear functions $L_1, \ldots, L_r : R^{d+1} \mapsto R$ (where $r \leq \nu + 2$) such that $L_1 + \cdots + L_r \equiv 0$, and for every $i = 1, \ldots, r$, the set $\{w \in N(\mathcal{O}) : L_i(w) \geq 1\}$ is a zero-set.

4) The initial condition '*all zeroes*' is stable.

From now on $r(\mathcal{D})$ stands for the minimal value of r for which condition 3) holds for an eroder $\mathcal{D}$.

7.3.1. Exercise. Prove that 2-percolation is an eroder and find $L_1, \ldots, L_r$ with which condition 3) is fulfilled.

Answer: With $r = 2$, $L_1(s, t) = 2s - t$, $L_2(s, t) = -2s + t$.

We use Theorem 3 to prove non-ergodicity (in fact non-uniqueness of the invariant measure), in the following way.

7.3.2. Exercise. For any k construct a deterministic system $\mathcal{D}$ that has at least k different fixed attractors, for which there is $\varepsilon > 0$ such that any random system resulting from $\mathcal{D}$ by adding a random noise, all of whose parameters ε_i are less than or equal to ε, has at least k different invariant processes.

Comment: See [36].

7.4. About the proof of Theorem 3

The most difficult part to prove is that 1), 2), or 3) implies 4). This is nontrivial even for particular examples. For 2-percolation this proof boils down to examination of a directed planar percolation that has been described in [31] or survey [2]. The case $r = 2$ (which is the smallest possible value of r) is more difficult than 2-percolation, but still is essentially simpler than the general case. In this case 'contours' are used also, but there is no percolation interpretation.

For any system with $r > 2$ there is a combinatorial general proof, but the topological objects it involves are more complicated than contours, because they ramify. The most general versions of this proof are in [36] and [37]. It seems that the best way to understand their main idea is to read [26] where the proof is reworded for the NEC-voting (for which $r = 3$). There is another method [14] to obtain similar results, but we do not discuss it here.

Here we prove only one part of Theorem 3, namely that 2) implies 3), to show how this proof uses the theory of convex sets. Assume that $\sigma(D) = \{\mathcal{O}\}$. For any finite set $S \subset R^{d+1}$ the set $ray(conv(S))$ can be represented as an intersection of several halfspaces. (A halfspace is a set in R^{d+1}, where some homogeneous linear function is non-negative.) Apply this to zero-sets, and you have a finite list of zero-halfspaces whose intersection is $\{\mathcal{O}\}$. (A zero-halfspace is a halfspace whose intersection with $N(\mathcal{O})$ is a zero-set.) For everyone of these zero-halfspaces we introduce a linear function f_i which is non-positive on it and only on it. We know that the origin is the only point in R^{d+1} where all f_i are non-positive. This allows us to apply Theorem 21.3 of [6] (a variant of Helly's theorem) to the hyperplane $\Pi = \{(s, t) : t = -1\}$, restrictions $f_i|_\Pi$ of our functions to Π and any non-empty closed convex set $C \subseteq \Pi$. We take $C = \Pi$. Then there exist such non-negative real numbers λ_i, at most $d + 1$ of which are positive, that for some $\varepsilon > 0$.

$$\forall\, w \in C : \sum_i \lambda_i f_i(w) \geq \varepsilon.$$

Since the left part is linear and bounded from below by a positive constant on C, it has to be a positive constant on C:

$$\forall\, w \in C : \sum_i \lambda_i f_i(w) = \delta = \text{const} \geq \varepsilon.$$

Hence

$$\forall\, w \in R^{d+1} : \sum_i \lambda_i f_i(w) = -\delta t.$$

Then the functions

$$L_i = -\left(t + \frac{n\lambda_i f_i}{\delta}\right)$$

fit our claim.   □

7.4.1. Exercise. Prove that $\sigma(D)$ equals the intersection of convex hulls of rays of *minimal* zero-sets. We call a zero-set *minimal* iff any its proper subset is not a zero-set. This fact simplifies checking of condition 2) for particular systems.

7.4.2. Exercise. Reformulate Theorem 3 as a criterion for attraction and stability of 'all ones'.

7.4.3. Exercise. Simplify conditions 2) and 3) of Theorem 3 for the case $m = 1$.

7.4.4. Problem. Prove that item 4) implies 1), 2), and 3).

7.4.5. Problem. Prove that items 1), 2), and 3) are equivalent.

7.4.6. Problem. Formulate and prove an analogue of Theorem 3 for all standard monotonic deterministic systems with $X_0 = \{-1, 0, 1\}$. In other words, prove that in all of these systems the initial state 'all zeroes' is stable iff it is an attractor, and give a criterion for that.

7.4.7. Problem. Generalize Theorem 3 to any d-dimensional commutative system with $X_0 = \{0, 1\}$. See [38].

7.5. Example: flattening

This example illustrates an application of Theorem 3. It is a deterministic system with $d = 2$, $m = 1$ with ε-δ noise added. First let us define a function of four arguments:

$$F(a, b, c, d) = \max(\min(a, b), \min(c, d)).$$

The following pseudo-code shows neighborhoods and transition probabilities of this system:

```
1    for all (i, j) ∈ Z²_S do
2        x_{i,j,-1} ← x^{initial}_{i,j}
3    for t = 0 to t_max − 1 do
4        for all (i, j) ∈ Z²_S do                                    (18)
5            x_{i,j,t} ← F(x_{i,j,t−1}, x_{i,j−1,t−1}, x_{i−1,j,t−1}, x_{i−1,j−1,t−1})
6            if rnd < ε then x_{i,j,t} ← 0 else
7            if rnd > δ then x_{i,j,t} ← 1
```

The infinite flattening has the same pattern and transition probabilities and is an eroder. The condition 3) is fulfilled with $r = 2$, $L_1(i, j, t) = 2i - t$, $L_2(i, j, t) = -2i + t$.

Theorem 3 guarantees that for ε and δ small enough the infinite version of this system is non-ergodic. More specifically, Theorem 3 proves the following:

$$\forall\, (i, j) \in Z^2 : x^{initial}_{i,j} = 0 \implies \sup_{t,\varepsilon} proc(x_{i,j,t} = 1) \to 0 \text{ when } \delta \to 0.$$

$$\forall\, (i, j) \in Z^2 : x^{initial}_{i,j} = 1 \implies \sup_{t,\delta} proc(x_{i,j,t} = 0) \to 0 \text{ when } \varepsilon \to 0.$$

(Both of these probabilities do not actually depend on i or j.)

7.5.1. Exercise. For which values of ε and δ does Theorem 2 prove fast convergence of the finite version of flattening?

7.5.2. Exercise. Prove that in the flattening systems both bunches of 'all zeroes' and 'all ones' are attractors.

7.5.3. Problem. Prove that although the deterministic flattening has infinitely many fixed initial conditions, it has only two bunches of attractors: those of 'all zeroes' and 'all ones'.

7.5.4. Problem. Prove that the finite version of flattening converges slowly with small enough, but positive, ε and δ.

Comment: This can be proved by a method analogous to that of [11]. The general proof is in [39,40].

7.5.5. Unsolved problem. Does flattening have any attractors besides bunches of 'all zeroes' and 'all ones'?

7.6. Quasi-Stability

In the section 4.4 we have defined slow convergence as a finite analogue of non-ergodicity. In the same vein we can define quasi-stability as a finite analogue of stability. Take some $a \in A$. Take a pattern in which 'all a-s' is a trajectory and call this trajectory *quasi-stable* for the given pattern if there exist $\varepsilon > 0$ and $Q > 1$ such that for the processes resulting from the initial condition 'all a-s',

$$\forall S, t \; : \; t \leq Q^S \implies \mathrm{Prob}(x_{s,t} \neq a) \leq \varepsilon.$$

7.6.1. Exercise. Prove that if an ergodic system has two non-equivalent quasi-stable trajectories, it converges slowly.

7.6.2. Problem. Prove that 'all zeroes' is quasi-stable for all patterns for which the conditions of Theorem 3 hold.

Comment: [11] proved this for the NEC-voting. The general proof is in [39,40].

8. STANDARD VOTINGS

For any odd n, the Boolean function *voting* with n arguments equals 1 iff the majority of its arguments are ones. Equivalently, *voting* equals 0 iff the majority of its arguments are zeros. Let us consider the following systems that use *votings*: a deterministic voting is a deterministic system that has *voting* as its transition function, and a random voting results from deterministic voting by adding the ε-δ noise. The first computer simulations of random votings were done in [44] for the symmetric case $\varepsilon = \delta$.

8.1. Symmetric votings

We call a standard voting *symmetric* if $N(\mathcal{O})$ is symmetric with respect to $ray(\bar{0}, -1)$.

8.1.1. Exercise. Prove that all deterministic symmetric votings are not eroders and have $\sigma = ray(\bar{0}, -1)$.

8.1.2. Exercise. Prove that deterministic symmetric votings have no fixed attractors.

8.1.3. Problem. Prove that all random symmetric votings are ergodic if $\varepsilon = 0$ and $\delta > 0$ (or $\varepsilon > 0$ and $\delta = 0$).

Comment: A similar statement is proved as Proposition 1 in [35].

8.1.4. Exercise. For which values of ε and δ does Theorem 2 prove ergodicity of random Symmetrical votings?

8.1.5. Unsolved problem. Prove that all one-dimensional votings with a non-degenerate noise are ergodic.

Comment: Computer simulations [44] of one-dimensional symmetric votings with
$$N(\mathcal{O}) = \{(s, -1) \ : \ s = -1, \ 1, \ 0\} \tag{19}$$
showed ergodicity in any non-degenerate case. It is very plausible that in fact all non-degenerate one-dimensional votings are ergodic. However, a rigorous proof [22] exists now only for the continuous-time analogue of (19).

Let us call a symmetrical voting a *windrose* if its neighbor vectors are non-coplanar.

8.1.6. Unsolved problem. Prove that all windroses are non-ergodic with small enough $\varepsilon = \delta$.

8.1.7. Unsolved problem. Prove ergodicity of some windrose with some positive values of ε and δ.

Comment: Both of the last problems are suggested by computer simulations [44,10]. However, both are unproved even for particular cases, including the following simplest two-dimensional ones:

5-windrose has $n = 5$ and

$$N(\mathcal{O}) = \{(s, -1) \ : \ s = (0,0), (-1,0), (1,0), (0,-1), (0,1)\}.$$

9-windrose has $n = 9$ and

$$N(\mathcal{O}) = \{(i, j, -1) \ : \ i, j = -1, 0, 1\}.$$

8.2. NEC and other non-symmetric votings

One of non-symmetric votings, the NEC-voting, has attracted special attention. Here NEC means north, east, center. It has $N(\mathcal{O}) = \{(s, -1) \ : \ s = (0,0), (1,0), (0,1)\}$. NEC-voting was first introduced in [44] with a symmetric noise, and the results of computer simulation showed that it is non-ergodic if $\varepsilon = \delta$ is small enough. Now this non-ergodicity is proved for any small enough ε and δ; it follows from our Theorem 3 because NEC-voting is an eroder. Condition 3) is fulfilled with

$$r = 3, \quad L_1(i,j,t) = -3i - t, \quad L_2(i,j,t) = -3j - t, \quad L_3(i,j,t) = 3i + 3j + 2t.$$

8.2.1. Exercise. Prove that in the deterministic NEC-voting both bunches of 'all zeroes' and 'all ones' are attractors.

8.2.2. Problem. Prove that, although deterministic NEC-voting has infinitely many fixed initial conditions, it has only two bunches of attractors, those of 'all zeroes' and 'all ones'.

8.2.3. Unsolved problem. Does NEC-voting have any attractors besides bunches of 'all zeroes' and 'all ones'?

8.2.4. Problem. Consider any voting with $m = 1$ and $n = 3$ whose neighbor vectors are non-coplanar. Prove that it is ergodic or non-ergodic whenever the NEC-voting with the same values of ε and δ is.

8.2.5. Problem. Use NEC-voting to construct deterministic systems with at least m stable fixed trajectories for any natural m. In more detail: For any finite set of periodic configurations in $space = Z^d$, propose a system for which all of the corresponding fixed configurations are stable trajectories. (You can even propose a monotonic system with this property.) See [36].

8.2.6. Problem. Take any standard voting in which all the neighbor vectors are pairwise non-collinear. Prove that it is not an eroder iff there exists such a neighbor vector V that all the other neighbor vectors form pairs, the two vectors of each pair having the opposite V-direction projections to *space*.

Comment: Importance of non-symmetry in a similar context was discussed in [29].

9. ONE-DIMENSIONAL CONSERVATORS

Let a *forget-me-not* stand for a non-ergodic non-degenerate process. For $d > 1$ forget-me-nots certainly exist, flattening and NEC-voting, for example. The question whether one-dimensional forget-me-nots exist was very intriguing for years. The *positive rates conjecture*, proposed by several authors, claimed that all non-degenerate one-dimensional random systems are ergodic (see for example [5], Chapter 4, section 3). Now this seems to be refuted: after some preliminary work [46,25], Péter Gács proposed an elaborate construction [19], which presents a one-dimensional forget-me-not, but his construction is not yet examined sufficiently, and we shall not discuss it here.

Our purpose is more modest. Note that, from the practical point of view, it is not always necessary to remember forever; it may be sufficient to keep information for a finite but long time. So it seems worthwhile to look for deterministic systems that converge slowly in presence of a small noise. With this idea in mind, the term 'conservator' and first examples of conservators were proposed in [20]. A *conservator* is a deterministic system with at least two different fixed initial attractors. The following is one of the simplest conservators.

9.1. Soldiers

This is a standard system in which

$$N(\mathcal{O}) = \{(s, -1) \; : \; s = -3, -1, \; 0, \; 1, \; 3\}.$$

Let us call every point of Z a *a soldier*. Every soldier has only two possible states, -1 and 1. The transition function equals

$$\text{tr}(a_{-3}, a_{-1}, a_0, a_1, a_3) = \begin{cases} -1 & \text{if } a_0 = 1, \; a_1 = a_3 = -1, \\ 1 & \text{if } a_0 = -1, \; a_{-1} = a_{-3} = 1, \\ a_0 & \text{otherwise.} \end{cases}$$

9.1.1. Exercise. Prove that the soldiers system is non-monotonic, whether we assume $-1 \prec 1$ or $-1 \succ 1$.

9.1.2. Exercise. Prove that the soldiers system is symmetric in the following sense. If we define a one-to-one map $a : X \mapsto X$ by the rule $(a(x))_{s,t} = -x_{-s,t}$, then $\forall\, x_I :\ a(\mathrm{TR}(x_I)) = \mathrm{TR}(a(x_I))$.

9.1.3. Exercise. Prove that if soldiers has only one invariant process μ_{inv} then

$$\mu_{inv}(x_i = -1) = \mu_{inv}(x_i = 1) = 1/2.$$

9.1.4. Problem. Prove that the soldiers system is a conservator. In more detail: prove that both 'all ones' and 'all minus ones' are attractors. (From symmetry these two facts are equivalent.)

Comment: This was first proved in [20] and reinforced in [17].

9.1.5. Unsolved problem. Is it true that 'all ones' and 'all minus ones' are the only non-equivalent attractors for soldiers system?

Comment: [17] proved that 'all ones' and 'all minus ones' are the only fixed attractors of soldiers. See also [27].

9.1.6. Unsolved problem. Prove ergodicity of soldiers with the one-way ε-δ noise in which $\varepsilon = 0$ or $\delta = 0$.

9.2. 2-Line Voting

This is a non-standard voting with 3 neighbors. Here $space = Z \times \{1, -1\}$, i.e., automata form two parallel rows and are indexed (i, j), where $i \in Z$ and $j \in \{1, -1\}$, and

$$N(i, j, t) = \{(s, t - 1)\ :\ s = (i - 2j, j), (i - j, j), (i, -j)\}.$$

9.2.1. Exercise. Show that 2-line voting is uniform, but non-commutative.

9.2.2. Exercise. Prove that both 'all zeroes' and 'all ones' are attractors for 2-line voting.

9.2.3. Problem. Prove that 2-line voting is ergodic in the presence of a one-way noise, which turns zeros into ones with probability ε and never turns ones into zeros (or vice versa).

Comment: This can be proved by the method of [35]. Due to monotonicity, it it sufficient to take only 'all zeroes' and 'all ones' as initial states, because all the others are between them.

9.2.4. Problem. Has 2-line voting any other fixed attractors besides the bunches of 'all zeroes' and 'all ones'?

Answer: No. To prove it, note that there are only six fixed states: 1) 'all zeroes'; 2) 'all ones'; 3) zeroes if $j = 1$, ones if $j = -1$; 4) zeroes if $j = -1$, ones if $j = 1$; 5) zeroes if i is even, ones if i is odd; 6) zeroes if i is odd, ones if i is even.

9.2.5. Unsolved problem. Has 2-line voting any other attractors besides the bunches of 'all zeroes' and 'all ones'?

9.3. Relaxation time

The first results of computer simulation of soldiers (and similar systems) with a random noise were reported in [20]. If the simulation of soldiers began in 'all ones' (or in 'all minus ones'), the system remained in the vicinity of this state for a long time (in fact, all the computer time that was available to the authors). Moreover, the system approached one of these states if started from a chaotic initial condition. A recent computer simulation [17] of soldiers shows ergodicity for all non-degenerate cases (which indeed is very plausible), but it suggests some non-trivial dependence of the rate of convergence on ε.

In more detail: [17] defines the 'relaxation time' as the mean time when the percentage of ones first gets into the range $(S/2 - \sqrt{S},\ S/2 + \sqrt{S})$ if we started from the state 'all minus ones'. This definition is convenient for all systems with $X_0 = \{-1, 1\}$ and the kind of symmetry which was shown for soldiers in exercise 9.1.2. The computer results of [17] seem to show that the relaxation time for soldiers does not depend on S (which corresponds to fast convergence in our sense), but grows unusually fast when $\varepsilon \to 0$. Authors of [17] claim that it is asymptotic to $\sim e^{const/\varepsilon}$ as $\varepsilon \to 0$. This is enormously greater than that for $\mathcal{D}_{\equiv}$ (see the next exercise). This suggests that the soldiers system, although ergodic, effectively checks dissent when ε is small enough.

9.3.1. Exercise. Let $\mathcal{D}_{\equiv}$ stand for the identity standard deterministic system with $space = Z$, in which automata never change their states. Show that the relaxation time for $\mathcal{D}_{\equiv}$ with an ε-δ noise is $O(1/(\varepsilon + \delta))$ when ε and δ tend to 0.

9.3.2. Unsolved problem. How do relaxation times of various Votings, especially of 2-line voting, behave as functions of ε when $\varepsilon \to 0$?

Conjecture: One-dimensional votings with $N(\mathcal{O}) = \{(i, -1),\ i \in [-R,\ R]\}$ have relaxation times that behave as $O((\varepsilon + \delta)^{-(R+1)})$ and the relaxation time of 2-line voting behaves like that of soldiers.

9.4. Further problems

9.4.1. Problem. Prove that there are no commutative one-dimensional monotonic conservators with $|X_0| = 2$.

Sketch of Proof: Assume the contrary and come to a contradiction. Formula (2) here becomes

$$space = Z \times S = \{(i,j) \ : \ i \in Z, \ j \in S\}.$$

Define two initial configurations:

$$L(i,j) = \begin{cases} 0, & \text{if } i < k \\ 1, & \text{otherwise} \end{cases} \quad \text{and} \quad R(i,j) = \begin{cases} 1, & \text{if } i < k \\ 0, & \text{otherwise.} \end{cases}$$

Since components of L and R do not depend on j, components of $TR(L)$ and $TR(R)$ also do not depend on j (Exercise 2.4.3). Then prove that

$$\min\{i \ : \ TR(L)_{i,j,t} = 1, \ t = T\} = V_{left} \cdot T + O(1),$$

$$\min\{i \ : \ TR(R)_{i,j,t} = 0, \ t = T\} = V_{right} \cdot T + O(1).$$

Here V_{left} and V_{right} are constants that may be called velocities of movement of the boundaries between zeroes and ones (the technique of velocities was developed in [21]). The following three statements complete the proof:

1)   If $V_{left} < V_{right}$ then the bunch of 'all ones' is the only attractor.

2)   If $V_{left} > V_{right}$ then the bunch of 'all zeroes' is the only attractor.

3)   If $V_{left} = V_{right}$ then there are no attractors.   □

Note that we did more than the problem asked; we described completely all attractors rather than only fixed ones.

Thus, to have a conservator we need either a greater-than-one dimension, or non-monotonicity, or non-commutativity, or $|X_0| > 2$. In fact, in each case there is a conservator. For greater dimensions, it is flattening or NEC-voting. For non-monotonity it is the soldiers system. For non-commutativity it is the 2-line voting. An example for $|X_0| > 2$ is the subject of the following problem.

9.4.2. Problem. Propose a one-dimensional standard monotonic deterministic system with $|X_0| = 3$ which is a conservator.

Comment: You can take $X_0 = \{-1, 0, 1\}$ and arrange an interaction such that both 'all ones' and 'all minus ones' will be attractors. You can also make this system symmetric in a sense similar to soldiers.

Note. [18] and [41] present different one-dimensional systems, which display some forms of stability.

10. CHAOS APPROXIMATION

Let $m = 1$ and let $\mathcal{M}^1$ be the set of normed measured on X_0^{Space}. Define a map $chaos : \mathcal{M}^1 \mapsto \mathcal{M}^1$ as follows. Any $\mu \in \mathcal{M}^1$ given, $chaos(\mu)$ is that product-measure, whose projections to all X_s, $s \in space$ coincide with those of μ. Thus, $chaos(\mathcal{P})$, that is, $chaos$ applied after $\mathcal{P}$, is the new operator that hopefully approximates $\mathcal{P}$. The chaos approximation is an exact solution for the tree system that we defined when answering the exercise 2.1.1. We shall illustrate the chaos approximation by two examples.

10.1. Percolations

Take any percolation operator $\mathcal{P}(\theta)$ with $m = 1$ and take the measure concentrated in 'all zeroes' as the initial condition. Let $P_t(\theta)$ and $\tilde{P}_t(\theta)$ stand for the percentage of ones at the t-th step in the measures $\mathcal{P}^t('all\ zeroes')$ and $(chaos(\mathcal{P}))^t('all\ zeroes')$. The sequence of $\tilde{P}_t(\theta)$ satisfies the simple iteration formula

$$\tilde{P}_{t+1}(\theta) = \theta + (1 - \theta)\tilde{P}_t^n(\theta).$$

This allows us to examine the behavior of $\tilde{P}_\infty(\theta) = \lim_{t \to \infty} \tilde{P}_t(\theta)$ and to compare it with the behavior of $P_\infty(\theta) = \lim_{t \to \infty} P_t(\theta)$.

10.1.1. Exercise. Prove that when θ grows from 0 to 1, $\tilde{P}_\infty(\theta)$ starts at 0 as $\theta + o(\theta)$, grows monotonically and becomes equal to 1 at the critical value $\theta_{chaos}^* = 1 - \frac{1}{n}$.

This exercise shows that, crude as it is, the chaos approximation's behavior may be similar to that of the original process.

10.1.2. Unsolved problem. Let $\mathcal{P}$ be a percolation operator. Let us apply $chaos$ after every T steps of $\mathcal{P}$. Does the critical point of $chaos(\mathcal{P}^T)$ tend to that of $\mathcal{P}$ when $T \to \infty$?

10.2. Votings

It is instructive to examine the chaos approximation for random votings. As before, $\tilde{P}_t$ stands for the percentage of ones at t-step. The iteration formula for $n = 3$ is

$$\tilde{P}_{t+1} = \varepsilon + (1 - \varepsilon - \delta)(\tilde{P}_t^3 + 3\tilde{P}_t^2(1 - \tilde{P}_t)).$$

Let us compare two sequences generated by this formula with two different initial conditions $\tilde{P}_0 = 0$ and $\tilde{P}_0 = 1$. We must discriminate the case when the limits of these sequences are equal, from the case when they are different.

10.2.1. Exercise. Show that these limits are equal if $\varepsilon + \delta$ is close enough to 1 and are different if ε and δ are small enough.

We see that behavior of the chaos approximation of a random voting with 3 neighbors is similar to that of NEC-voting, but different from one-dimensional votings, which seem to be ergodic in all non-degenerate cases.

10.2.2. Exercise. Write and examine the iteration formula for the chaos approximation of a voting with 5 neighbors and compare it with the behavior of the 5-windrose.

References

Books and surveys

[1] Cellular Automata, Theory and Experiment / edited by Howard Gutowitz. 1st MIT Press ed. Imprint: Cambridge, Mass. : MIT Press, 1991. Series: Special issues of physica D.

[2] **Toom, A. L., Vasilyev, N. B., Stavskaya, O. N., Mityushin, L. G., Kurdyumov, G. L. and Pirogov, S. A.,** Discrete Local Markov Systems, *Stochastic Cellular Systems: Ergodicity, Memory, Morphogenesis,* Nonlinear Science: theory and applications, Ed. by R. Dobrushin, V. Kryukov and A. Toom, Manchester University Press, 1990.

[3] **Durrett, Richard,** Lecture Notes on Particle Systems and Percolation. Wadsworth & Brooks/Cole Statistics/Probability Series, 1988.

[4] **Griffeath, David,** Coupling Methods for Markov Processes, *Studies in Probability and Ergodic Theory; Advances in Mathematics, Supplementary Studies,* **2,** 1–43, Academic Press, New York, 1978.

[5] **Liggett, Thomas M.,** *Interacting Particle Systems,* Springer-Verlag, N. Y., 1985.

[6] **Rockafellar, R. T.,** *Convex Analysis,* Princeton Univ. Press, 1970.

[7] Theory and Application of Cellular Automata (including selected papers 1983–1986), **Wolfram, Stephen,** a.o. *Advanced series on complex systems,* **1,** World Scientist, 1986.

Papers

[8] **Barsky, David J., Grimmett, Geoffrey and Newman, Charles M.,** Dynamic Renormalization and Continuity of the Percolation Transition in Ortants, *Spatial Stochastic Processes,* K. Alexander and J. Watkins, Eds., Birkhauser, Boston, 37–55, 1991.

[9] **Barsky, David J., Grimmett Geoffrey and Newman, Charles M.,** Percolation in Half-Spaces: Equality of Critical Densities and Continuity of the Percolation Probability, *Prob. Theory and Related Fields,* **90(1),** 111–148, 1991.

[10] **Bennett, Charles H. and Grinstein, G.,** Role of Irreversibility in Stabilizing Complex and Nonergodic Behavior in Locally Interacting Discrete Systems, *Phys. Rev. Letters,* **55(7),** 657–660, 1985.

[11] **Berman, Piotr and Simon, Janos,** Investigations of Fault-Tolerant Networks of Computers, *ACM Symp. on Theory of Computing,* **20,** 66–77, 1988.

[12] **Bezuidenhout, Carol and Grimmett, Geoffrey,** The Critical Contact Process Dies Out, *Annals of Probability,* **18(4),** 1462–1482, 1990.

[13] **Bezuidenhout, Carol and Gray, Lawrence,** Critical Attractive Spin Systems, To appear in *Annals of Probability.*

[14] **Bramson, Maury and Gray, Lawrence,** A Useful Renormalization Argument, *Random walks, Brownian motion and interacting particle systems,* Festschrift for F. Spitzer. Ser. Progress in Probability, **28,** Rick Durrett & Harry Kesten, Eds. Birkhauser, 113–152, 1991.

[15] **Bramson, Maury and Neuhauser, Claudia,** Survival of One Dimensional Cellular Automata under Random Perturbations, *Annals of Probability,* **22(1),** 244–263, 1994.

[16] **Chen, Mu Fa,** On Coupling of Jump Processes, *Chinese Ann. Math.,* Ser. B **12(4),** 385–399, 1991.

[17] **Gonzaga de Sa, Paula and Maes, Christian,** The Gacs-Kurdyumov-Levin Automaton Revisited, *Journal of Stat. Physics,* **67,** 507–522, 1992.

[18] **Evans, M. R., Foster, D. P., Godrèche, C. and Mukamel D.,** Asymmetric exclusion model with two species: spontaneous symmetry breaking. Submitted to *Journal of Stat. Physics.*

[19] **Gács, Péter,** Reliable Computation with Cellular Automata, *Journal of Computer and System Sciences,* **32(1),** 15–78, 1986.

[20] **Gács, Péter, Kurdjumov, George and Levin, Leonid,** One-Dimensional Homogeneous Media, which Erode Finite Islands, *Problems of Information Trasmission,* **14(3),** 223–226, 1978.

[21] **Galperin, Gregory,** One-Dimensional Local Monotone Operators With Memory, *Soviet Math. Dokl.,* **17(3),** 688–692, 1976.

[22] **Gray Lawrence F.,** The Positive Rates Problem for Attractive Nearest Neighbor Spin Systems on Z, *Z. Wahrscheinlichkeitstheorie verw. Gebiete*, **61**, 389–404, 1982.

[23] **Grimmett G. and Marstrend, J. M.,** The Supercritical Phase of Percolation is Well-Behaved, *Proc. Royal Soc. London*, Ser A, **430**, 439–457, 1990.

[24] **Kopylov, I. G.,** Some Properties of a Markovian Coupling, *Moscow Univ. Comput. Math. Cybernet.*, **4**, 58–63, 1991.

[25] **Kurdyumov, George,** An Example of a Nonergodic One-Dimensional Homogeneous Random Medium with Positive Transition Probabilities, *Soviet Math. Dokl.*, **19(1)**, 211–214, 1978.

[26] **Lebowitz, Joel L., Maes, Christian and Speer, Eugene R.,** Statistical Mechanics of Probabilistic Cellular Automata, *Journal of Stat. Physics*, **59(1–2)**, 117–168, 1990.

[27] **Li, Wentian,** Non-Local Cellular Automata, 1991 Lectures in Complex Systems, SFI Studies in the Sciences of Complexity, Lect. Vol. IV, Eds. L. Nadel & D. Stein, Addison-Wesley, 1992.

[28] **Petri, N. V.,** The unsolubility of the problem of discerning of annuling iterative nets, *Researches in the Theory of Algorithms and Mathematical Logic*, Moscow, Nauka, 1979 (in Russian).

[29] **Pippenger, Nicholas,** Symmetry and Self-Repair. To appear in *Journal of Computer and System Science*.

[30] **Stavskaya, O. N. and Piatetski-Shapiro, I. I.,** [Uniform networks of spontaneously active elements.] *Problemy Kibernetiki*, **20**, 91–106, 1968 (in Russian).

[31] **Toom, Andrei,** A Family of Uniform Nets of Formal Neurons, *Soviet Math. Dokl.*, **9(6)**, 1338–1341, 1968.

[32] **Toom, Andrei,** Non-ergodic Multidimensional Systems of Automata, *Problems of Information Transmission*, **10**, 239–246, 1974.

[33] **Toom, Andrei,** Monotonic Binary Cellular Automata, *Problems of Information Transmission*, **12(1)**, 33–37, 1976.

[34] **Toom, Andrei and Leonid, Mityushin,** Two Results Regarding Non-computability for Univariate Cellular automata, *Problems of Information Transmission*, **12(2)**, 135–140, 1976.

[35] **Toom, Andrei,** Unstable Multicomponent Systems, *Problems of Information Transmission*, **12(3)**, 220–225, 1976.

[36] **Toom, Andrei,** Stable and Attractive Trajectories in Multicomponent Systems, *Multicomponent Random Systems,* (R. L. Dobrushin, Ya. G. Sinai, Eds.) Advances in Probability and Related Topics, **6,** Dekker, 549–575, 1980.

[37] **Toom, Andrei,** Estimations for measures that describe the behavior of random systems with local interaction, *Interactive Markov Processes and their Application to the Mathematical Modelling of Biological Systems.* Pushchino, 21–33, 1982 (in Russian).

[38] **Toom, Andrei,** On Reliable Simulation in Real Time, *Preprints of the 1-st World Congress of the Bernoulli Society for Mathematical Statistics and Probability Theory,* **2,** 676, 1986.

[39] **Toom, Andrei,** On Critical Phenomena in Interacting Growth Systems, Part I: General. *Journal of Stat. Physics,* v. 74, nn. 1–2, 91–109, 1994.

[40] **Toom, Andrei,** On Critical Phenomena in Interacting Growth Systems, Part II: Bounded Growth. *Journal of Stat. Physics,* v. 74, nn. 1–2, 111–130, 1994.

[41] **Toom, Andrei,** Simple 1-Dimensional Systems with Super-Exponential Relaxation Times. Submitted to *Journal of Stat. Physics.*

[42] **Vaserstein, Leonid,** Markov Processes over Denumerable Products of Spaces, Describing Large Systems of Automata, *Problems of Information Transmission,* **5(3),** 47–52, 1969.

[43] **Vaserstein, Leonid and Leontovitch, Andrei,** Invariant Measures of Certain Markov Operators Describing a Homogeneous Random Medium, *Problems of Information Transmission,* **6(1),** 61–69, 1970.

[44] **Vasilyev, N. B., Petrovskaya M. B. and Piatetski-Shapiro, I. l.,** Simulation of Voting with Random Errors, *Automatica i Telemekhanika,* **10,** 103–107, 1969 (in Russian).

[45] **Vasilyev, N. B. and Pyatetskii-Shapiro, I. I.,** The Classification of One-Dimensional Homogeneous Networks, *Problems of Information Transmission,* **7(4),** 340–346, 1971.

[46] **Zirelson, Boris S.,** Non-Uniform Local Ineraction Can Produce 'Far-Range Order' in a One-Dimensional System, *Theor. Probability Appl.,* **21 (3),** 681–683, 1976.

Chapter 7

METROPOLIS-TYPE MONTE CARLO SIMULATION ALGORITHMS AND SIMULATED ANNEALING

Basilis Gidas*
Division of Applied Mathematics
Brown University

ABSTRACT

In this article we describe the mathematical framework(s) and properties of Metropolis-type Monte Carlo simulation algorithms, and simulated annealing. The simulation algorithms are based on Markov processes — typically, homogeneous Markov chains or diffusions. Simulated annealing is a class of *stochastic global optimization algorithms* obtained by modifying the simulation algorithms via one or more time-dependent control parameters. The spirit of the article is expository. We describe the most useful variants of the algorithms, and give the main results concerning the convergence and other properties of the algorithms. In addition, we address some key practical issues associated with the implementation of the algorithms, and indicate current efforts for designing *non-local or multiresolution* variations of the algorithms. Finally, we posed one of the most important open problems: how does the performance of the algorithms depend on the problem-size?

1. INTRODUCTION

Metropolis-type simulation algorithms are instances of (*dynamic*) Monte Carlo procedures based on Markov processes — typically homogeneous Markov chains or homogeneous diffusion processes. They have been important in computational studies of statistical mechanics systems (where the basic algorithms were introduced first), quantum field theories, image processing and computer vision, and various applications of statistics. Simulated annealing is a *stochastic* global optimization algorithm (more precisely, a class of such algorithms), obtained by modifying the Metropolis-type simulation algorithms via one or more *control parameters* that change as the algorithm evolves. This

*Research partially supported by ARO Contracts DAAL03-90-G-0033, DAAL03-86-K-0171, and ONR Contract N00014-88-K-0289

modification leads to nonhomogeneous Markov processes (chains or diffusions). Simulated annealing has found a host of applications in such diverse areas as low-temperature behavior of statistical physics systems (e.g., finding the *"ground states,"* i.e., states with globally minimum energy), image analysis and computer vision tasks, combinatorial optimization (e.g., VLSI routing and placement, graph partitioning, traveling salesman problem), code design for communication systems, multiprocessor load-balancing, neural networks, and statistics (e.g., cluster analysis).

In this article we will describe the mathematical framework(s) of the simulation algorithms and simulated annealing, and present the main results concerning convergence and other properties of the algorithms. We will also expose some key practical issues associated with the implementation of the algorithms, state some open problems, and indicate the two most important directions in the future study of the algorithms: (i) The *complexity* of the algorithms, i.e., their performance as the problem-size becomes larger and larger and (ii) the design, within the same framework, of *non-local* or *multiresolution* algorithms for improving the efficiency of the basic algorithms (see Sections 2.2.3 and 2.4). In the remainder of this introduction, we will indicate the roots and intuitive motivation of the Metropolis-type simulation algorithms and simulated annealing.

The basic Metropolis et al. [71] algorithm was introduced for studying numerically properties of statistical mechanics systems at *equilibrium*. Consider, for example, a large but finite system of N interacting molecules (e.g., gas molecules) confined in a region (typically a box) $\Omega_0 \subset I\!\!R^3$. Let $x = \{x_i \in \Omega_0 : i = 1, \ldots, N\}$ be the molecular positions — a *state* or *configuration* of the system. The set of possible states is the *state* (or *configuration*) *space* $\Omega = \Omega_0^N \subset I\!\!R^{3N}$. If the system is in thermal equilibrium with its surroundings, at temperature T, then the behavior of the system is determined by the *Gibbs distribution*

$$\pi(x) = \frac{1}{Z} e^{-\frac{1}{kT} U(x)}, \qquad x \in \Omega, \tag{1.1a}$$

$$Z = \int_\Omega e^{-\frac{1}{kT} U(x)} d^{3N} x, \tag{1.1b}$$

where k is the Boltzmann constant, and $U(x)$ is the (potential) *energy* of the system at state $x \in \Omega$. The normalizing constant Z is called the partition function; because of the large dimensionality of Ω, and because $U(x)$ is typically a complicated function, the partition function Z is computationally intractable in applications.

A problem of central interest in the study of systems described by (1.1), for which the Metropolis algorithm was invented, is the evaluation of *"ensemble averages"* or *"equilibrium expectations"*

$$< f > \equiv \int_\Omega f(x) \pi(dx) \tag{1.2}$$

for various functions ("*observables*") $f(x)$, e.g., $f(x) = U(x)$.

Another important question for physical systems with energy function $U(x)$ is to find the globally minimum energy states or *ground states* of the system, i.e., to find the set

$$\underline{\Omega} = \{x \in \Omega : U(x) = \inf[U(y) : y \in \Omega]\}. \tag{1.3}$$

Simulated annealing, introduced by Kirpatrick et al.[62] and independently by Cerny [14], is designed to solve global optimization problems like this.

The applications mentioned in the first paragraph above lead to problems with similar mathematical structure: There is an underlying measurable space Ω — the *state* or *configuration space* — and either a probability measure π on Ω, or real-valued function $U(x)$ on Ω. A typical question of interest in the former case is the (numerical) computation of averages as in (1.2). In the latter case, one is interested in finding the global minima of $U(x)$. Simulation algorithms (to be described in Section 2) are aimed at the former and related problems, while simulated annealing (to be described in Section 3) is aimed at the latter problem and variants of it. The state space Ω may be finite, countable, or continuous (compact or non-compact), but in most applications Ω is very large, which reflects the fact that the problems have very many *degrees of freedom*. Also, most natural problems are intrinsically *non-linear*, i.e., $\pi(x)$ or $U(x)$ are highly complex.

Mathematically, the problem of estimating averages, such as in (1.2), is a problem of numerical integration. But because of the high dimensionality of Ω and the complexity of $\pi(x)$ (or $U(x)$), it is a very difficult one. Traditional deterministic methods of integration, such as Simpson's rule, are highly inefficient in high dimensions. For example, Simpson's rule in d-dimensions with n nodal points has an error of the order of $n^{-\frac{4}{d}}$ (for smooth integrands), which is unsatisfactorily large for large d. Monte Carlo integration [46] is typically more efficient than deterministic methods in high dimensions. In fact, we will see that Monte Carlo procedures have an error proportional to $n^{-\frac{1}{2}}$, independent of dimensions. Although better than the error in deterministic methods, the error is still large. But the problems we are concerned with are very difficult (typically NP-complete [73]) due to the fact that they have very many degrees of freedom and are highly non-linear and non-perturbative.

The precise principles of Monte Carlo simulation and integration will be given in Section II. In brief, one generates random samples $X(1), X(2), \ldots$ from $\pi(x)$, and then estimates expectations like those in (1.2), by the *time* or "*ergodic*" *averages*

$$\overline{f} = \frac{1}{n} \sum_{t=1}^{n} f(X(t)). \tag{1.4}$$

There are at least two basic procedures for generating the samples. The first class of procedures (called *static* Monte Carlo methods [86]) generates *statistically independent* samples $X(1), X(2), \ldots$ from π. These methods are quite useful for distributions on not-very-high dimensional spaces, but are unfeasible for Gibbs or other high dimensional distributions. The second class of methods (called *dynamic* Monte Carlo [86]) designs a discrete or continuous time stochastic process $\{X(t)\}_{t\geq 0}$ — typically a homogeneous Markov chain or diffusion — with state space Ω and π as its unique stationary distribution. Then one simulates the stochastic process starting from an initial configuration. The Metropolis algorithm is a specific recipe for constructing a Markov chain. For countable Ω and π given by (1.1) (integration replaced by summation) the recipe is briefly the following: Choose an arbitrary irreducible and symmetric stochastic matrix $Q = \{q(x, y) : x, y \in \Omega\}$, and an initial configuration $X(0) = x^{(0)}$. Then construct a discrete-time Markov chain $\{X(t) : t \geq 0\}$ by the rule: If at time $t \geq 0$, $X(t) = x$, then propose a move $x \mapsto y$ according to Q, and set

- $X(t+1) = y$ with probability 1, if
  $\Delta E \equiv \frac{1}{kT}[U(y) - U(x)] \leq 0$

  (1.5)

- $X(t+1) = y$ with probability
  $\exp\{-\Delta E\}$, if $\Delta E > 0$.

In Section 2.2, we will see how Q is chosen in practice. A more general version of the Metropolis algorithm was given by Hastings [47], and a very different approach was introduced by Geman and Geman [35]. These procedures (to be described in Section 2.2) apply also to the cases when Ω is a continuous state space (compact or non-compact). In these cases, one may also design stochastic processes $\{X(t)\}$ having π as an invariant measure. These processes are diffusions determined through stochastic differential equations, the prototype of which is the *Langevin equation* (see Section 2.3). For statistical mechanics systems, the process $\{X(t)\}$ may be thought as a *stochastic time evolution* for a given system. But this evolution need not correspond to any real physical dynamics, such as those of *non-equilibrium statistical mechanics* (although the Langevin equation does have a physical interpetation; see Section 2.3). Rather, it is simply a stochastic numerical algorithm, and it is designed to be as computationally efficient as possible.

Next we come to the global optimization problem (1.3). In the applications we are concerned with, the function $U(x)$, assumed to be bounded below, has very many local minima. Traditional *deterministic* global optimization methods [25], typically based on heuristic techniques that try to enumerate all local and global minima, require a prior knowledge of various properties of the cost functions; have a formidable computational cost; and often do not guarantee success. Classical "randomization" global optimization algorithms [79] are by and large variants of the so-called multistart technique; they do in general converge with probability one, but in practice are extremely inefficient for high dimensional Ω's and cost functions U with large number of local minima. The family of algorithms that is close in spirit to, but technically very different from, simulated annealing is local search algorithms [73, 60] based on stepwise improvements in the value of the cost function by exploring neighborhood systems.

The intuitive basis of simulated annealing [62, 14, 35] lies in the following observations: As $T \downarrow 0$, the Gibbs distribution (1.1) concentrates on the global minima $\underline{\Omega}$ of $U(x)$. Thus the ground states and other states near them in energy control (roughly speaking) the low-temperature behavior of a system described by (1.1), a central problem in low temperature physics. Experimentally, the ground states are reached by a procedure known as *chemical annealing*, whereby one first "melts" a substance and then cools it *slowly*, being careful to pass especially slowly through the "freezing" temperature (if any!); certain crystals, for example, are obtained this way. If the temperature is lowered too *abruptly*, then the system may end up not in a ground state, but in a nearby "metastable" state, i.e., in a local but not global minimum (this often is called the "*adiabatic*" effect). On the other hand, if the temperature is lowered *too slowly*, then the system will indeed approach the ground state(s), but may do so extremely slowly (this is called the "*supercooling*" effect). The optimal speed of cooling — the *annealing schedule* — is determined by the competition between the two effects.

The basic idea of [62, 63] (see also Pincus [75]) was to mimic mathematically the experimental process of chemical annealing. To this end, one regards the global minima of an arbitrary cost function $U(x)$ on Ω as the ground states of an *imaginary* physical system with energy function $U(x)$, and associates with it a Gibbs distribution of the form (1.1) with *time dependent* temperature $T = T(t)$. Then one uses one of the simulation methods mentioned above to reach "steady state", or *equilibrium*, at each of a sequence of decreasing temperatures $T_1 \geq T_2 \geq T_3 \geq \ldots$, converging to zero as $t \to +\infty$. More precisely, starting with an initial configuration $X(0)$, generate samples $X(1), X(2), \ldots, X(\tau_1)$ at temperature T_1; $X(\tau_1 + 1), X(\tau_1 + 2), \ldots, X(\tau_2)$ at temperature T_2, and so on. The discrete-time process $\{X(t) : t \geq 0\}$ generated in this way is the *annealing process* (typically an inhomogeneous Markov chain or diffusion). The pair (T_t, τ_t) is called the *annealing* or *cooling schedule* and it is chosen so that the process $\{X(t)\}$ converges, as $t \to +\infty$, weakly to a law supported on the global minima $\underline{\Omega}$ of $U(x)$:

$$P\{X(t) \in \underline{\Omega}\} \to 1 \text{ as } t \to +\infty. \tag{1.6}$$

The determination of the *optimal* annealing schedule is controlled by the adiabatic and supercooling effects mentioned above.

Various types of annealing processes and their mathematical properties will be given in Section 3. The first rigorous result on the convergence of simulated annealing was established by Geman and Geman [35]. In [38], annealing was treated as a special case of non-homogeneous Markov chains. The optimal annealing schedule for the convergence criterion (1.6) was first determined by Hajek [44] (finite Ω).

The study of simulated annealing and Metropolis-type algorithms has involved, in addition to the theory of homogeneous and non-homogeneous Markov chains and diffusion processes, large deviation theory, spectral analysis of operators, singular perturbation theory, and (for certain problems) stochastic partial differential equations. References to topics not covered in this article will be given in Sections 2 and 3; see [33], [1], [86], and [45] for a more complete list of references on the applications and theoretical studies of the algorithms.

2. METROPOLIS-TYPE MONTE CARLO SIMULATION ALGORITHMS

In Subsection 2.1 we summarize the key aspects of the general theory of dynamic Monte Carlo algorithms via Markov chains, with special emphasis on the concept of "relaxation" time. Subsection 2.2 contains the Metropolis-Hasting algorithms and the Gibbs sampler. Simulation algorithms via the Langevin equation and its generalizations are treated in Subsection 2.3. And in Subsection 2.4, we briefly review the Swendsen-Wang algorithm.

2.1   Dynamic Monte Carlo simulation via Markov chains[†]

We will assume, for simplicity, that the state space Ω is countable, and we will concentrate on discrete-time Markov chains (MCs). The theory can easily be adapted to continuous-time MCs, and much of it can formally be extended to general state spaces, replacing sums by integrals and matrices by kernels, although the actual theory in this case is technically much harder [72].

Let $\{X_t\}_{t \geq 0}$ be a discrete-time, homogeneous Markov chain (MC) on Ω, with (1-step) transition probability matrix

$$P = \{p(x, y) : x, y \in \Omega\}$$

$$p(x, y) = P\{X_t = y | X_{t-1} = x\}, t \geq 1$$

and intial distribution

$$\mu = \{\mu(x) : x \in \Omega\}, \ \ \mu(x) = P\{X_0 = x\}.$$

The t-step, $t = 0, 1, 2, \ldots$, transition probabilities (the matrix elements of P^t) will be denoted by

$$p(t; x, y) \equiv P\{X_{t+n} = y | X_n = x\}, \ \ \forall n \geq 0$$

$$p(0; x, y) = \delta_{xy}$$

and the components of the probability μP^t at time t by

$$p(t, x) = (\mu P^t)(x) = P\{X_t = x\}, \ \ t \geq 0.$$

[†]The presentation of this section has greatly been influenced by Sokal's lecture notes [86].

A probability measure $\pi = \{\pi(x) : x \in \Omega\}$ on Ω is called a *stationary distribution* (= *invariant distribution*) for the matrix P if $\pi P = \pi$, i.e.,

$$\sum_x \pi(x)p(x,y) = \pi(y), \quad \forall y \in \Omega.$$

Note that stationary distributions satisfy $\pi P^t = \pi$, for all $t \geq 0$.

We will assume that the MC (i.e., P) is *irreducible* [in the sense that for any two states $x, y \in \Omega$ there exists an $n = n(x,y) \geq 0$ such that $p(n; x, y) > 0$], and that it has period $d \geq 1$. It is well-known [82] that the state space Ω of an irreducible MC with period d can be decomposed into disjoint subsets $G_1, G_2, \ldots, G_d$, such that $\Omega = \cup_{\alpha=1}^d G_\alpha$, and with the property that the chain moves cyclically around these subsets, in the sense that $p(n; x, y) = 0$ whenever $x \in G_\alpha$, $y \in G_\beta$ with $\beta - \alpha \neq n \pmod{d}$. If $d = 1$, the MC (or P) is called *aperiodic*.

Dynamic Monte Carlo simulation algorithms are based on the following well-known

Theorem 2.1 *Let $P = \{p(x,y)\}$ be an irreducible transition probability matrix with period $d \geq 1$. If P has a stationary probability distribution π, then π is unique and $\pi(x) > 0$ for all $x \in \Omega$. Furthermore for $m = 0, 1, 2, \ldots, d-1$*

$$\lim_{t \to +\infty} p(td + m; x, y)$$

$$= \begin{cases} d\pi(y), & \text{if } x \in G_\alpha, y \in G_\beta \text{ with } \beta - \alpha = m(\mathrm{mod}\, d) \\ \\ 0, & \text{if } x \in G_\alpha, y \in G_\beta \text{ with } \beta - \alpha \neq m(\mathrm{mod}\, d) \end{cases} \tag{2.1}$$

for all $x, y \in \Omega$. In particular

$$\lim_{n \to +\infty} \frac{1}{n} \sum_{t=1}^n p(t; x, y) = \pi(y), \quad \text{for all } x, y \in \Omega. \tag{2.2}$$

If, in addition, P is aperiodic then

$$\lim_{t \to +\infty} p(t; x, y) = \pi(y), \quad \text{for all } x, y \in \Omega. \tag{2.3}$$

Remarks:

1) From the point of view of simulation algorithms, one is given a probability distribution $\pi = \{\pi(x) : x \in \Omega\}$ and wants to design an irreducible transition probability matrix $P = \{p(x,y) : x, y \in \Omega\}$ having π as its stationary distribution (necessarily unique). This will be the subject of Subsection 2.2.

2) Equation (2.3) [resp. (2.2)] says that the probability $p(t, x) = (\mu P^t)(x)$ (resp. its Cesaro mean) converges to $\pi(x)$ *irrespective of the initial distribution*

μ. Under the conditions of Theorem 2.1, one can establish other properties, such as a strong law of large numbers, a central limit theorem, and an iterated logarithm law [20, Section I.16]. Of special interest to us will be the convergence in probability (or in some stronger sense) of the time (or "ergodic") averages

$$\overline{f}^{(n)} = \frac{1}{n} \sum_{t=1}^{n} f(X_t) \qquad (2.4a)$$

to the *equilibrium expectations*

$$< f > \equiv < f >_\pi \equiv \sum_x \pi(x) f(x). \qquad (2.4b)$$

Next we address the long-time behavior of $\{X_t\}$, a subject of special significance for the feasibility and *efficiency* of simulation algorithms. We will consider two (related) aspects of the long-time behavior: a) the "rate of convergence to equilibrium," i.e., the rate at which μP^t converges to π as $t \to +\infty$, and b) the rate of convergence of $\overline{f}^{(n)}$ to $< f >$. The former is characterized by a constant (depending on P) called the *relaxation time*, while the latter is characterized by another constant that describes the size (properly normalized) of the *fluctuations* of the long-time averages $\overline{f}^{(n)}$, and is called the *fluctuation constant* (or *integrated relaxation time* [86]). To make things precise we need some formalism.

Let $\ell_2(\pi)$ be the Hilbert space of complex-valued functions on Ω with inner product

$$< f, g > \equiv < f, g >_\pi \equiv \sum_x \pi(x) f^*(x) g(x),$$

where f^* denotes complex conjugation. The norm of $f \in \ell_2(\pi)$ will be denoted by $\|f\|$ or $\|f\|_\pi$. We think of P^t, $t \geq 1$, as an operator on $\ell_2(\pi)$ acting as follows

$$(P^t f)(x) = \sum_y p(t; x, y) f(y).$$

Let Π be the operator on $\ell_2(\pi)$ defined by

$$(\Pi f)(x) = \sum_y \pi(y) f(y) \equiv < f > \equiv < f >_\pi .$$

Clearly Π is the matrix with rows the row-vector $\pi = \{\pi(x) : x \in \Omega\}$. It is easily seen that Π is an orthogonal projection on $\ell_2(\pi)$ with range the constant functions.

The following theorem [83] is an extension of the Perron-Frobenius Theorem.

Theorem 2.2 *Let P be the transition probability matrix of an irreducible MC with (necessarily unique) stationary distribution π. Then P as an operator on $\ell_2(\pi)$ has the properties:*

a) *P is a contraction; in particular its spectrum lies in the closed unit disk.*

b) *1 is a **simple** eigenvalue of P with corresponding eigenspace the constant functions; in particular $P\Pi = \Pi P = \Pi$. Furthermore, 1 is also a simple eigenvalue of the adjoint P^* of P, with eigenvector π.*

c) *If P is aperiodic, then 1 is the only eigenvalue of P and P^* on the unit circle.*

Let $\sigma(P - \Pi)$ denote the spectrum of $P - \Pi$, and let

$$r(P - \Pi) = \sup\{|\lambda| : \lambda \in \sigma(P - \Pi)\}$$

be the *spectral radius* of $P-\Pi$ (note that $P-\Pi$ is equal to P on the orthogonal complement of constant functions). By Theorem 2.2a, $r(P - \Pi) \leq 1$. If P is periodic (i.e., $d \geq 2$) then $r(P - \Pi) = 1$, since the dth roots of unity are eigenvalues of P [58]. If P is aperiodic and Ω is finite then $r(P - \Pi) < 1$ (by Theorem 2.2c), but if P aperiodic and Ω is infinite, then we may have $r(P - \Pi) = 1$. The quantity

$$\tau_0 = \frac{1}{-\log r(P - \Pi)} \tag{2.5}$$

is called the *relaxation time*. We will see below that the "rate of convergence to equilibrium" can be bounded in terms of τ_0. Clearly $0 \leq \tau_0 \leq +\infty$, with $\tau_0 = +\infty$ if $r(P - \Pi) = 1$; hence for fast convergence we like to choose Ps so that $r(P - \Pi) < 1$.

The spectral radius $r(P - \Pi)$ is given by the following *spectral radius formula* [76, p.192]:

$$r(P - \Pi) = \lim_{t \to +\infty} \|(P - \Pi)^t\|^{\frac{1}{t}} = \inf_{t \geq 1} \|(P - \Pi)^t\|^{\frac{1}{t}} \tag{2.6}$$

where the operator norm $\|\cdot\|$ is the operator norm on $\ell_2(\pi)$. If P is self-adjoint on $\ell_2(\pi)$, then $r(P - \Pi) = \|P - \Pi\|$. The next proposition [87] gives a very useful characterization of the spectral radius. Let $f \in \ell_2(\pi)$, and consider the autocorrelation (more precisely, autocovariance) function

$$C_f(|t - s|) = E_\pi\{[f(X_s) - <f>] \cdot [f(X_t) - <f>]\} \tag{2.7a}$$

$$= E_\pi\{f(X_s)f(X_t)\} - (<f>)^2$$

where $E_\pi\{\cdot\}$ denotes expectation with initial measure π. A straightforward computation gives (recall that $< \cdot, \cdot >$ denotes the inner product on $\ell_2(\pi)$)

$$C_f(|t - s|) = <f, (P^{|t-s|} - \Pi)f> = <f, (P - \Pi)^{|t-s|}f> \tag{2.7b}$$

$$= <f, P^{|t-s|}(I - \Pi)f> = <f, (I - \Pi)P^{|t-s|}(I - \Pi)f> .$$

The last three equalities are obtained by using the properties $(P - \Pi)^n = P^n - \Pi$ *and* $P^n \Pi = \Pi P^n = \Pi$ *for* $n \geq 1$. Let

$$\tau_{0,f} = \lim_{t \to +\infty} \sup \left(-\frac{1}{t} \log |C_f(t)|\right)^{-1}. \tag{2.8}$$

Proposition 2.1 [87]

$$\frac{1}{-\log r(P - \Pi)} \equiv \tau_0 = \sup_{f \in \ell_2(\pi)} \tau_{0,f}.$$

As a consequence of this proposition, we have, asymptotically as $|t-s| \to +\infty$,

$$C_f(|t - s|) \sim \exp\left\{-\frac{|t - s|}{\tau_0}\right\}.$$

This means that τ_0 measures the time it takes for $f(X_s)$ and $f(X_t)$ to become statistically independent (approximately); in fact we will see next that τ_0 bounds from above the rate at which μP^t converges to π. To characterize this rate, we need to define a *distance* between μP^t and π, i.e., a criterion that measures deviation from equilibrium. Throughout this article we will use either distance in the ℓ_2-*sense*

$$d(\mu P^t, \pi) \equiv \sup_{\|f\|_\pi \leq 1} |\mu P^t f - \pi(f)| = \left\|\frac{\mu P^t}{\pi} - 1\right\|_{\ell_2(\pi)} \tag{2.9a}$$

or distance in *total variation* [80, Chapter 6]

$$\|\mu P^t - \pi\|_{\mathrm{var}} \equiv \sum_x |(\mu P^t)(x) - \pi(x)|. \tag{2.10a}$$

One easily derives

$$d(\mu P^t, \pi) \leq \|P^t - \Pi\| d(\mu, \pi) \tag{2.9b}$$

and

$$\|\mu P^t - \pi\|_{\mathrm{var}} \leq \|P^t - \Pi\| d(\mu, \pi). \tag{2.10b}$$

By the spectral radius formula, we have asymptotically as $t \to +\infty$,

$$\|P^t - \Pi\| \sim (r(P - \Pi))^t = e^{-\frac{t}{\tau_0}} \tag{2.11}$$

with equality for all t if P is self-adjoint on $\ell_2(\pi)$. This together with (2.9b) and (2.10b) makes precise the statement that the relaxation time τ_0 bounds from above the convergence to equilibrium. From the algorithmic point of view, τ_0 bounds the number of Monte Carlo iterations that should be discarded at the beginning of a run, before the system reaches equilibrium.

Remark: Often convergence in the ℓ_2-sense is formulated in terms of $\|P^t f - <f>\|_\pi$. Since

$$\|P^t f - <f>\|_\pi \le \|P^t - \Pi\| \|f - <f>\|_\pi, \qquad (2.12)$$

we see that τ_0 also controls the convergence of this norm. A much stronger mode of convergence is *uniform convergence*, where the ℓ_2 norm on the left-hand side of (2.12) is replaced by the uniform norm, and $\|f - <f>\|_\pi$ on the right-hand side is replaced by some constant depending on f.

Next we turn to the convergence of the long-time averages (2.4a) and the *fluctuation constant*. Let P be irreducible, and π, Π as before. By Theorem 2.2, the matrix

$$Z = (I - (P - \Pi))^{-1} \qquad (2.13a)$$

is a well-defined operator on $\ell_2(\pi)$; if 1 is *not* an isolated eigenvalue of P, then Z is unbounded, but it is densely defined. Z is called the *fundamental matrix* of the MC [61]. From (2.2) one easily deduces that

$$Z = I + \lim_{n \to +\infty} \sum_{k=1}^{n} \frac{n-k}{n} (P^k - \Pi), \qquad (2.13b)$$

and if P is aperiodic

$$Z = I + \sum_{k=1}^{+\infty} (P^k - \Pi). \qquad (2.13c)$$

Let $D(Z) \subseteq \ell_2(\pi)$ denote the *quadratic form domain* [76, p.277] of Z (if Z is bounded then $D(Z) = \ell_2(\pi)$). Then we have the following theorem.

Theorem 2.3 *Let Z, $D(Z)$ be as above, and $E_\pi\{\cdot\}$ denote expectation with initial distribution π. If $f, g \in D(Z)$, then*

$$\lim_{n \to +\infty} n E_\pi \{ [\frac{1}{n} \sum_{t=1}^{n} f^*(X_t) - <f^*>] \cdot [\frac{1}{n} \sum_{s=1}^{n} g(X_s) - <g>] \} = <f, \frac{C + C^*}{2} g> \qquad (2.14a)$$

and

$$\lim_{n \to +\infty} n E_\pi \{ |\frac{1}{n} \sum_{t=1}^{n} f(X_t) - <f>|^2 \} = <f, Cf> \qquad (2.14b)$$

where

$$C = 2Z - I - \Pi. \qquad (2.15)$$

Proof: Using the notation of (2.4a), we have

$$nE_\pi\{[\bar{f}^{*(n)} - <f^*>][\bar{g}^{(n)} - <g>]\}$$

$$= \frac{1}{n}\sum_{t,s=1}^n E_\pi\{[f^*(X_t) - <f^*>][g(X_s) - <g>]\}$$

$$= \frac{1}{n}\sum_{t=1}^{n-1}\sum_{s=t+1}^n <f, (P^{s-t} - \Pi)g>$$

$$+ \frac{1}{n}\sum_{s=1}^{n-1}\sum_{t=s+1}^n <g^*, (P^{t-s} - \Pi)f^*>$$

$$+ <f,g> - <f^*><g>$$

$$= <f,g> + \sum_{k=1}^{n-1}\frac{n-k}{n} <f, (P^k - \Pi)g>$$

$$+ <g^*, f^*> + \sum_{k=1}^{n-1}\frac{n-k}{n} <g^*, (P^k - \Pi)f^*>$$

$$- <f,g> - <f^*><g>$$

which converges as $n \to +\infty$ to $<f, (Z + Z^* - I - \pi)g>$; this gives (2.14a), which in turn implies (2.14b).

Remarks:

1) For finite Ω, Theorem 2.3 can be found in [56, Section 4.4.4]. For self-adjoint P and countable Ω see [10]. See also [20, p.97].

2) Note that the limiting operator C may be written as

$$C = (I + P)(I - P)^{-1}(I - \Pi). \qquad (2.16)$$

3) Theorem 2.3 implies in particular that the long-time average of f converges in probability to $<f>$, with *fluctuations* of size $2\tau_{1,f}n^{-\frac{1}{2}}$, asymptotically as $n \to +\infty$, where $\tau_{1,f}$ is the *fluctuation constant* defined by

$$\tau_{1,f} = \frac{1}{2}\frac{<f, Cf>}{<f, (I - \Pi)f>} \qquad (2.17)$$

$$= \frac{1}{2} + \sum_{t=1}^{+\infty}\frac{C_f(t)}{C_f(0)} \qquad (2.18)$$

where $C_f(t)$ is the autocovariance function (2.7). Because of (2.18), $\tau_{1,f}$ is also called the *integrated relaxation time* [86]. The factor $\frac{1}{2}$ in (2.17) is inserted so that $\tau_{1,f} \approx \tau_{0,f}$ [see (2.8)], if

$$C_f(t) \sim \exp\{-\frac{|t|}{\tau}\}$$

with τ very large.

4) Clearly $2\tau_{1,f}$ measures the *statistical error* or fluctuations in Monte Carlo measurements of the equilibrium expectation $< f >$. If the samples $X_1, X_2, X_3, \ldots$ were statistically independent, then

$$nE_\pi\{(\frac{1}{n}\sum_{t=1}^{n} f(X_t) - < f >)^2\} = < f, (I - \Pi)f > .$$

Hence Theorem 2.3 says that, in dynamic Monte Carlo algorithms, the variance of $\overline{f}^{(n)}$ is $2\tau_{1,f}$ times larger than it is in independent sampling algorithms (static Monte Carlo algorithms). Put differently, the number of "effectively independent samples" in a run of length n is approximately $n/2\tau_{1,f}$. This role of $\tau_{1,f}$ should be contrasted with the role of τ_0 which measures the number of iterations that should be discarded at the beginning of a run. For many systems, $\tau_{1,f}$ and $\tau_{0,f}$ are of the same order of magnitude, but this is *not* true in general; for example, in statistical mechanics systems at (or near) critical points ("second-order" phase transition points), $\tau_{1,f}$ and $\tau_{0,f}$ are of different orders of magnitude. (See [86] for an interesting discussion of this and related issues.)

The theory we have developed so far simplifies considerably if the transition probability matrix P is self-adjoint on $\ell_2(\pi)$. In the next subsection, we will see that the matrices in the Metropolis-type algorithms are typically either self-adjoint or are "built" up out of self-adjoint Ps. It is easily verified that P is self-adjoint on $\ell_2(\pi)$, i.e., $< f, Pg >_\pi = < Pf, g >_\pi$ for all $f, g \in \ell_2(\pi)$, iff P satisfies

$$\pi(x)p(x,y) = \pi(y)p(y,x), \text{ for all } x, y \in \Omega. \tag{2.19}$$

This implies, in particular, that π is a stationary distribution. In the physics literature, property (2.19) is called the *detailed-balance condition*; while in the mathematics literature a MC (i.e., P) that satisfies (2.19) is called *reversible* [61, Section 5.3]. By Theorem 2.2, the spectrum $\sigma(P)$ of a self-adjoint P lies in $[-1, 1]$, in $(-1, 1]$ if P is aperiodic, and in $[0, 1]$ if P is non-negative.

For self-adjoint Ps, the spectral radius $r(P - \Pi)$ and the relaxation times $\tau_0, \tau_{0,f}, \tau_{1,f}$ have a more convenient representation. Let

$$\mu_2 \equiv \mu_2(P - \Pi) \equiv \sup \sigma(P - \Pi),$$

$$\underline{\mu} \equiv \underline{\mu}(P - \Pi) \equiv \inf \sigma(P - \Pi).$$

Then

$$r(P - \Pi) = \max\{|\underline{\mu}|, \mu_2\} \tag{2.20}$$

which induces another representation of τ_0 (see (2.5)). If P has a discrete spectrum and we order its eigenvalues so that $\mu_1 = 1 > \mu_2 \geq \mu_3 \geq \ldots \geq -1$, then the μ_2 in (2.20) is the "second" eigenvalue. Let

$$\rho_f(|t|) = \frac{C_f(|t|)}{C_f(0)} = \frac{< f, (P - \Pi)^{|t|} f >}{< f, (I - \Pi)f >} \tag{2.21}$$

be the *autocorrelation function.* The spectral theorem [76] applied to the self-adjoint operator $P - \Pi$ implies that, for any $f \in \ell_2(\pi)$, there exist $\underline{\mu}_f$ and $\overline{\mu}_f$ and a non-negative spectral density $d\sigma_f(\mu)$ supported on $[\underline{\mu}_f, \overline{\mu}_f] \subseteq [\underline{\mu}, \mu_2] \subseteq [-1, 1]$, so that [by (2.21)]

$$\rho_f(|t|) = \int_{\underline{\mu}_f}^{\overline{\mu}_f} \mu^{|t|} d\sigma_f(\mu), \tag{2.22}$$

and by (2.18) [or (2.16) and (2.17)]

$$\tau_{1,f} = \frac{1}{2} \int_{\underline{\mu}_f}^{\overline{\mu}_f} \frac{1+\mu}{1-\mu} d\sigma_f(\mu), \tag{2.23}$$

and clearly

$$\tau_{0,f} = (-\log \max(|\underline{\mu}_f|, \overline{\mu}_f))^{-1}. \tag{2.24}$$

It follows from (2.23)

$$\frac{1}{2}\frac{1+\rho_f(1)}{1-\rho_f(1)} \leq \tau_{1,f} \quad \leq \frac{1}{2}\frac{1+\overline{\mu}_f}{1-\overline{\mu}_f}$$

$$\leq \frac{1}{2}\frac{1+\exp\{-1/\tau_{0,f}\}}{1-\exp\{-1/\tau_{0,f}\}} \tag{2.25}$$

$$\leq \frac{1}{2}\frac{1+\mu_2}{1-\mu_2} \leq \frac{1}{2}\frac{1+\exp\{-1/\tau_0\}}{1-\exp\{-1/\tau_0\}}.$$

The first inequality is obtained from Jensen's inequality applied to the convex function $(1 + \mu)/(1 - \mu)$, while the other inequalities are obtained by the monotonicity of the same function. Another useful inequality is obtained from (2.22):

$$\rho_f(|t|) \geq (\rho(1))^{|t|}, \quad \text{for even } t's.$$

This holds also for odd ts if $P \geq 0$, i.e., if $d\sigma_f(\mu)$ is supported in $[0, \overline{\mu}_f]$.

An interesting mathematical problem, which is also very important in practical aspects of simulation algorithms, is the problem of deriving *upper and lower bounds* for τ_0, $\tau_{1,f}$ (and $\tau_{0,f}$). For fast convergence and small statistical fluctuations, we like τ_0 and $\tau_{1,f}$ to be small (at least bounded); hence the significance of the upper bounds is clear. But lower bounds are also useful, since if these bounds are very large then convergence is definitely slow. Typically, it is much harder to establish upper bounds than lower bounds. The inequalities in (2.25) yield lower bounds (often quite useful as we will see in Section 2.2) for τ_0 and $\tau_{0,f}$ in terms of $\rho_f(1)$, provided that $\rho_f(1) > 0$ [always $-1 \leq \rho_f(1) \leq 1$]; one always has a lower bound of $\tau_{1,f}$ in terms of $\rho_f(1)$, but this bound is useless if $\rho_f(1)$ is near -1. These lower bounds are extremely important if $\rho_f(1)$, for *some* f, is near $+1$ (in this case τ_0, $\tau_{1,f}$ are large).

Note that deriving upper (resp. lower) bounds for τ_0 is equivalent to deriving upper (resp. lower) bounds for the spectral radius $r(P - \Pi)$. The problem of deriving sharp upper or lower bounds for $r(P - \Pi)$ is very difficult if P is not self-adjoint, and in fact it is quite difficult even if P is self-adjoint.

In the self-adjoint case, if $\mu_2 > 0$ then μ_2 yields a lower bound for $r(P - \Pi)$ (and hence τ_0); furthermore, if P has no spectrum near -1 (more precisely, if $\mu_2 > |\underline{\mu}|$), then deriving upper (resp. lower) bounds for τ_0 is equivalent to deriving upper (resp. lower) bound for μ_2. Also, from (2.23) we see that for the statistical fluctuations (measured by $2\tau_{1,f}$), the spectrum near -1 is not important; it is the spectrum near $+1$ (i.e., μ_2) that is essential. For these reasons most of the studies have addressed the derivation of upper and lower bounds for μ_2. Next, we survey the main mathematical tools that have been used for this problem.

Let

$$\lambda_2 = 1 - \mu_2. \tag{2.26}$$

This represents the distance between the simple eigenvalue $\lambda_1 = 0$ of $I - P$ and the rest of its spectrum (the spectrum of $I - P$ lies in $[0, 2]$). In the physics literature λ_2 is called the *mass gap* of $I - P$; if Ω is finite and P aperiodic, then $\lambda_2 > 0$, but if Ω is infinite then we may have $\lambda_2 = 0$ even if P is aperiodic. The mass gap λ_2 has the minimax representation

$$\lambda_2 = \inf_f \frac{< f, (I - P)f >}{< f, (I - \Pi)f >} = \inf_f [1 - \rho_f(1)], \tag{2.27}$$

where the inf is over non-constant functions $f \in \ell_2(\pi)$. The ratio in (2.27) is called the *Rayleigh quotient*. The numerator has the following useful representation

$$< f, (I - P)f > = \frac{1}{2} \sum_{x,y} \pi(x) p(x, y) |f(x) - f(y)|^2 \tag{2.28}$$

for any real or complex $f \in \ell_2(\pi)$; the derivation of (2.28) is straightforward (the formula does not hold if P is not self-adjoint, unless f is real).

Note that upper (resp. lower) bounds of μ_2 correspond to lower (resp. upper) bounds of λ_2. It is clear from (2.27) that it is much harder to derive lower bounds for λ_2 than to derive upper bounds. Indeed, to derive an *upper* bound for λ_2, it suffices to choose a suitable function ("observable") f and estimate its Rayleigh quotient; this is called the *variational* or *Rayleigh-Ritz method*; the difficulty here is, of course, how to find a "good" function f so that $1 - \rho_f(1)$ is near its infimum. On the other hand, to derive lower bounds for λ_2, one needs to understand the behavior of the Rayleigh quotient for all $f \in \ell_2(\pi)$. In sub-Section 2.2.3, we will use the Rayleigh-Ritz methods to derive upper bounds, and (2.27) to compare the λ_2s for various Ps.

The mathematical techniques for deriving lower and upper bounds for λ_2 are quite different. For upper bounds, in addition to the Rayleigh-Ritz variational method mentioned above, there is a *minimum hitting-time argument* introduced in [88]. The problem of establishing lower bounds for λ_2 is similar to the problem of deriving lower bounds for the second eigenvalue of the Laplacian on Riemannian manifolds, or of Schrödingertype operators [84, 85, 48]. A powerful method for the Laplacian problem was proposed by Cheeger [15] in 1970; a general version of Cheeger's inequality for Markov chains and Markov processes with killing was proven by Lawler and Sokal [64]. Diaconis and Stroock [24] (finite Ω) established sharper bounds than those obtained from Cheeger's inequality; their method uses Poincaré type inequalities and arguments of Jurrum and Sinclair. (See [24] for more details on the methodology and history for lower bounds of λ_2.)

2.2   Metropolis-type dynamics for simulation

In this Section we will assume that we are given a probability distribution π on some state space Ω, and wish to choose an irreducible transition probability matrix P having π as its stationary measure. In Subsection 2.2.1 we describe the strategy suggested in [71] and extended in [47]. Subsection 2.2.2 describes the Gibbs sampler [35]. In subsection 2.2.3 we briefly compare various dynamics, and derive lower bounds for the relaxation time and fluctuation constant for some systems. As in Section 2.1, we will assume that Ω is countable and $\pi(x) > 0$ for all $x \in \Omega$ (the case when π is zero on some subset of Ω will be treated in Section 3 together with simulated annealing). The matrices we will construct are either self-adjoint [on $\ell_2(\pi)$] or are built up out of self-adjoint matrices $P_1, \ldots, P_n$ in one of two ways:

a) *Random Sampling.* Let $P = \sum_{i=1}^{n} \lambda_i P_i$ with $\lambda_i \geq 0$, $i = 1, 2, \ldots, n$, and $\sum_{i=1}^{n} \lambda_i = 1$. If π is a stationary distribution for $P_1, \ldots, P_n$, then it is so for P; and if $P_1, \ldots, P_n$ are self-adjoint, then so is P. [Algorithmically, sampling with P amounts to choosing *randomly* from $P_1, \ldots, P_n$ according to the weights $\{\lambda_i\}$; hence the name random sampling].

b) *Sequential* or *Systematic Sampling.* Let $P = P_1 \cdots P_n$. If π is a stationary distribution for $P_1, \cdots, P_n$, then it is so for P; but P need not be self-adjoint (reversible) even if all $P_1, \ldots, P_n$ are. [Algorithmically, sampling with P amounts to performing *sequentially* = "*systematically*" the operations $P_1, \ldots, P_n$; hence the name sequential or systematic sampling].

2.2.1   Metropolis-Hasting dynamics

The basic idea for designing "Metropolis" dynamics is the following: Choose an arbitrary irreducible transition probability matrix $Q = \{q(x, y) : x, y \in \Omega\}$, and a function $\alpha(x, y)$, $0 \le \alpha(x, y) \le 1$, for each pair $x \ne y$, $x, y \in \Omega$. Then define $P = \{p(x, y) : x, y \in \Omega\}$ by

$$p(x, y) = q(x, y)\alpha(x, y) , \quad \text{for } x \ne y, \tag{2.29a}$$

and

$$p(x, x) = 1 - \sum_{y \ne x} p(x, y) = q(x, x) + \sum_{y \ne x} q(x, y)[1 - \alpha(x, y)]. \tag{2.29b}$$

This has the interpretation: If the chain $\{X_t\}_{t \ge 0}$ at some t is at state x, i.e., $X(t) = x$, then generate a *proposed* move $x \to y$ according to Q; then accept y as the state of the chain at time $t + 1$ with probability $\alpha(x, y)$, or reject it with probability $1 - \alpha(x, y)$, i.e., set

$$X(t+1) = \begin{cases} y & \text{with probability} \quad \alpha(x, y), \\ X(t) = x & \text{with probability} \quad 1 - \alpha(x, y). \end{cases}$$

Because of its role, Q is called the *proposal matrix*; later we will see how it is chosen in practice. P is reversible iff

$$\frac{\alpha(x, y)}{\alpha(y, x)} = \frac{\pi(y)q(y, x)}{\pi(x)q(x, y)}, \quad \text{for all } x \ne y.$$

This is satisfied if we choose

$$\alpha(x, y) = F\left(\frac{\pi(y)q(y, x)}{\pi(x)q(x, y)}\right), \quad x \ne y \tag{2.29c}$$

with a function $F : [0, \infty] \to [0, 1]$ satisfying

$$F(\xi) = \xi F\left(\frac{1}{\xi}\right), \quad \text{for all } \xi \ge 0. \tag{2.30}$$

Observe that such an F need to be specified only for $0 \le \xi \le 1$. In fact, the set of such functions is in 1-1 correspondence with the set of function $G(\zeta)$ defined on [0,1] and satisfying

$$0 \le \frac{G(\zeta)}{1 + \zeta} \le 1, \quad \text{for } 0 \le \zeta \le 1.$$

The correspondence between F and G is

$$F(\xi) = \frac{\xi}{1 + \xi} G(\min\{\xi, \frac{1}{\xi}\}) , \quad \text{for all } \xi \ge 0.$$

Also, there is a 1-1 correspondence between the functions F and the set of functions $K : [0, +\infty] \to [0, 1]$ satisfying $K(t) \le e^{-t}$ for all $t \ge 0$; explicity

$$F(\xi) = \begin{cases} K(-\log \xi), & \text{if } 0 \le \xi \le 1 \\ \xi K(\log \xi), & \text{if } \xi > 1. \end{cases}$$

The matrices P given by (2.29a)–(2.29c) will be referred to as the *Metropolis-Hasting dynamics* or *sampling*, or *algorithm*). The original [71] *Metropolis dynamics* corresponds to the choice

$$F_M(\xi) = \min(\xi, 1).$$

Note that $F(\xi) \leq F_M(\xi)$ for all $\xi \geq 0$ and all F satisfying (2.30); thus F_M is the *maximal* function satisfying (2.30). The dynamics corresponding to

$$F_B(\xi) = \frac{\xi}{1 + \xi}$$

is called the *Barker dynamics*. The choice

$$F_\gamma(\xi) = \frac{\xi}{1+\xi}[1 + 2(\frac{1}{2}\min\{\xi, \frac{1}{\xi}\})^\gamma], \quad \gamma \geq 1$$

"interpolates" between the Metropolis ($\gamma = 1$) and Barker ($\gamma = +\infty$) choices. For symmetric Qs [i.e., $q(x, y) = q(y, x)$], the Metropolis matrix reads, for $x \neq y$

$$p(x, y) = \begin{cases} q(x, y), & \text{if} \quad \pi(y) \geq \pi(x) \\ \\ q(x, y)\frac{\pi(y)}{\pi(x)}, & \text{if} \quad \pi(y) < \pi(x) \end{cases} \qquad (2.31)$$

and the Barker matrix

$$p(x, y) = q(x, y)\frac{\pi(y)}{\pi(y) + \pi(x)}, \quad y \neq x. \qquad (2.32)$$

In both cases (and always), $p(x, x)$ is given by (2.29b).

If Q is reversible with respect to π, then $\alpha(x, y) \equiv 1$ and $P = Q$. Thus the Metropolis-Hasting procedure may be thought as a way of modifying an arbitrary matrix Q into a matrix P that satisfies detailed-balance relative to π. The irreducibility or aperiodicity of P (or lack thereof) needs to be verified separately for each dynamics. The irreducibility of Q does not guarantee the irreducibility of P. It is easily seen, however, that the Ps in (2.31) and (2.32) are irreducible iff Q is irreducible (provided that $\pi(x) > 0$ for all $x \in \Omega$); this is true also of the general Metropolis and Barker dynamics (i.e., for non-symmetric Q). Furthermore, if Ω is finite and π non-constant, then (2.31) and (2.32) are also aperiodic. To see this choose $x \in \Omega$ so that $\pi(x) \geq \pi(y)$ for all $y \in \Omega$; then for the Metropolis dynamics (2.31)

$$p(x, x) = q(x, x) + \sum_{y \neq x} q(x, y)[1 - \frac{\pi(y)}{\pi(x)}] > q(x, x) \geq 0$$

and similarly for the Barker dynamics (2.32). The condition that Q be symmetric for aperiodicity can be relaxed in various ways; for example, it suffices to satisfy

$$\alpha(x)q(x, y) = \alpha(y)q(y, x) \qquad (2.33)$$

for some probability distribution $\alpha(x) > 0$ for all $x \in \Omega$.

As we mentioned in the Introduction, Gibbs distributions play a major role in many applications; these distributions are of the form [compare with (1.1)]

$$\pi(x) = \frac{1}{Z}e^{-H(x)}, \quad Z = \sum_x e^{-H(x)} \qquad (2.34)$$

with some *energy* (or *Hamiltonian*) function $H(\cdot)$ on Ω so that $Z < +\infty$. Often (e.g., in problems of image processing and computer vision, statistical mechanics systems, quantum field theories, and other areas), $H(x)$ is built up from "local" energy terms — the *interactions* (see, for example, the Ising model in subsection 2.2.3); this is very convenient in the implementation of the algorithms, but it plays no special role in the general theory. Note that the ratios $\pi(y)/\pi(x)$ that enter into the Metropolis-Hasting dynamics do not involve the *partition function* Z, a computationally intractable object. For Gibbs distributions, the Metropolis dynamics with a symmetric Q [i.e., (2.31)] reads

$$p(x,y) = q(x,y)e^{-[H(y)-H(x)]^+}, \quad \text{for } x \neq y \qquad (2.35)$$

where $u^+ = \max(u,0)$. This leads to the sampling rule stated in (1.5).

How do we choose the proposal matrix Q? In applications, most of the choices are based on the notion of *neighborhood systems*. Suppose for a moment that we are given an irreducible Q. Let

$$\mathcal{N}(x) = \{y \in \Omega : y \neq x, q(x,y) > 0\}$$

$[\mathcal{N}(x)$ may be empty for some states $x \in \Omega]$. The elements of $\mathcal{N}(x)$ are called the *neighbors of x*, $\mathcal{N}(x)$ a neighborhood of x, and $\mathcal{N} = \{\mathcal{N}(x) : x \in \Omega\}$ a *neighborhood system* on Ω. If Q is symmetric, then $\mathcal{N}$ is a *symmetric* neighborhood system in the sense that $x \notin \mathcal{N}(x)$, and $y \in \mathcal{N}(x)$ iff $x \in \mathcal{N}(y)$. Now we reverse the procedure: We choose a (not necessarily symmetric) neighborhood system $\mathcal{N} = \{\mathcal{N}(x) : x \in \Omega\}$ [by definition $x \notin \mathcal{N}(x)$]; then Q is chosen so that $q(x,y) > 0$ iff $y \in \mathcal{N}(x)$, and $q(x,x) = 1 - \sum_{y \neq x} q(y,x)$. Such a Q, no matter what its exact form, is irreducible if the neighborhood system $\mathcal{N}$ has the property:

Strong irreducibility: For each pair $x \neq y$ there exists a sequence of states $x_1 = x, x_2, x_3, \ldots, x_{n-1}, x_n = y$, so that $x_{l+1} \in \mathcal{N}(x_l)$ for $1 \leq l \leq n-1$.

Here is an example of an explicit Q: Suppose that Ω is finite and $\mathcal{N}$ has the strong irreducibility property; then define

$$q(x,y) = \frac{1}{|\mathcal{N}(x)|}\mathbb{1}_{\{y \in \mathcal{N}(x)\}} = \begin{cases} \frac{1}{|\mathcal{N}(x)|}, & \text{if } y \in \mathcal{N}(x) \\ \\ 0, & \text{else.} \end{cases} \qquad (2.36)$$

This Q is irreducible; furthermore, if $\mathcal{N}$ is symmetric as well, then Q is reversible [i.e., (2.33) holds] relative to

$$\alpha(x) = \frac{|\mathcal{N}(x)|}{\sum_x |\mathcal{N}(x)|}.$$

In particular, Q is symmetric if $|\mathcal{N}(x)|$ is independent of x.

How do we choose the neighborhood system $\mathcal{N}$? In many applications the state space Ω has the form

$$\Omega = \Omega_0^S \tag{2.37}$$

where S is a finite set whose elements will be called "*sites*" or "*vertices*," and will be denoted by $i, j, k, \ldots$; Ω_0 is a countable state space called the *single-site state space* (more generally, Ω has the form $\Omega = \Pi_{i \in S}\Omega_i$, with possibly different single-site state spaces $\Omega_i, i \in S$). In the Ising model (see subsection 2.2.3) $\Omega_0 = \{-1, 1\}$. In some applications, S is a subset of $\mathbb{Z}^d, d \geq 1$. More generally, S is the set of vertices of a graph $\mathcal{G}$; the vertices S of $\mathcal{G}$ index the underlying samples (states) $x = \{x_i : i \in S\}$, whereas the "*bonds*" of $\mathcal{G}$ capture [33, 36, 35] the *interactions* among the individual components of x. In the remaininder of this subsection we will assume that Ω has the form (2.37). We will use the notation: If A is a subset of Ω, then

$$x_A = \{x_i : i \in A\}, \quad {}_i x = x_{S \setminus \{i\}}$$
$$\Omega_A = \Omega_0^A, \quad A^c = S \setminus A. \tag{2.38}$$

If Ω_0 is finite, $2 \leq |\Omega_0| < +\infty$ (we always assume $|S| < +\infty$), then the following neighborhood systems and dynamics are commonly used: Let $A_1, \ldots, A_N$ be distinct subsets of S (not necessarily disjoint) so that no subset is contained in another subset, and

$$S = A_1 \cup A_2 \cup \cdots \cup A_N. \tag{2.39}$$

Clearly $N \leq |S|$. For example, each A_i may contain a single site $i \in S$, in which case $N = |S|$, and $|A_i| = 1, i \in S$. Now define the neighborhoods

$$\mathcal{N}_i(x) \equiv \mathcal{N}_{A_i}(x) = \{y \in \Omega : y_{A_i} \neq x_{A_i}, y_{A_i^c} = x_{A_i^c}\}, i = 1, \ldots, N.$$

Clearly $|\mathcal{N}_i(x)| = |\Omega_{A_i}| - 1$. Set $\ell_i = |\Omega_{A_i}| - 1$, and define for $y \neq x$

$$q_i(x, y) \equiv q_{A_i}(x, y) = \frac{1}{\ell_i}\mathbb{1}_{\{y \in \mathcal{N}_i(x)\}} = \begin{cases} \frac{1}{\ell_i}, & \text{if } y \in \mathcal{N}_i(x) \\ 0, & \text{else,} \end{cases} \tag{2.40}$$

and set $q_i(x, x)$ so that $\sum_y q_i(x, y) = 1$. Note that $Q_i = \{q_i(x, y)\}$ is symmetric (not irreducible); hence the acceptance probability (2.29c) depends on $\pi(y)/\pi(x)$ only. The transition probability matrix

$$p_i(x, y) \equiv p_{A_i}(x, y) \equiv p_{A_i}^{(F)}(x, y) = q_i(x, y)F(\frac{\pi(y)}{\pi(x)}), \quad x \neq y \tag{2.41}$$

for any F satisfying (2.30) is reversible relative to π (but not irreducible). Note that $p_i(x, y)$ changes only the x_{A_i} components of $x = \{x_i : i \in S\}$. A *full dynamics* (or full *sweep*) is obtained either by *random* or *sequential* sampling, i.e., we define:

Random sampling (RF):

$$P \equiv P^{(F)} = \frac{1}{N} \sum_{i=1}^{N} P_i. \tag{2.42}$$

Sequential or systematic sampling (SF):

$$P \equiv P^{(F)} = P_{i_1} P_{i_2} \cdots P_{i_N} \tag{2.43}$$

where $A_{i_1}, \ldots, A_{i_N}$ is a deterministic ordering (*"visiting scheme"*) of $A_1, \ldots, A_N$. The matrix (2.42) is reversible relative to π, but (2.43) is not (although it has π as a stationary distribution). It is easily verified that the random dynamics (2.42) is always irreducible. But the sequential dynamics (2.43) is not always irreducible. For example, consider the Metropolis dynamics (i.e., $F = F_M$) for the one-dimensional Ising model (see subsection 2.2.3) with zero-external field, N sites, and periodic boundary conditions (i.e., $i = 1, 2, \ldots, N$ sites on the circle). For N odd, say $N = 2n + 1$, choose a single-site visitation scheme corresponding to the lexicographic ordering of sites $i = 1, 2, \ldots, 2n+1$; the full-sweep Metropolis dynamics is $P = P_1 P_2 \ldots P_{2n+1}$. This dynamics is *not* irreducible; indeed, the states $+ - + \cdots - +$ (alternating $+$ and $-$) and $+ + \cdots + +$ (all sites $+$) do not communicate. For $N = 2n$ with n also *even*, choose a single-site visitation-scheme visiting the odd sites $1, 2, \ldots, 2n - 1$ first, and then the even sites $2, 4, 6, \ldots, 2n$. The corresponding full-sweep Metropolis dynamics $P = P_1 P_3 \cdots P_{2n-1} P_2 P_4 \cdots P_{2n}$ is *not* irreducible; indeed the states $+ + - - + + - - \cdots + + - -$ (alternating pairs of $++$ and $-$) and $+ + \cdots + +$ (all sites $+$) do not communicate. For $N = 2n$ with n *odd*, choose the single-site visitation scheme with corresponding full-sweep dynamics $P = P_1 P_{n+1} P_2 P_{n+2} P_3 P_{n+3} \cdots P_n P_{2n}$. This dynamics is *not* irreducible; indeed, the state $+ + \cdots + - - \cdots -$ (the first n sites $+$, and the last n sites $-$) and $+ + + \cdots +$ (all sites $+$) do not communicate. The above dynamics have obvious analogues in two or higher dimensions, and the non-irreducibility still holds.

The proposal matrix Q for the reversible dynamics (2.42) is given by

$$q(x, y) = \frac{1}{N} \sum_{i=1}^{N} \frac{1}{\ell_i} q_i(x, y), \quad \text{for } x \neq y,$$

and it is clearly symmetric; hence (2.41) reads

$$p(x, y) = p^{(F)}(x, y) = q(x, y) F\left(\frac{\pi(y)}{\pi(x)}\right), \quad \text{for } x \neq y.$$

Remarks:

1) In practice, the *"visitation scheme"* $A_1, \ldots, A_N$ (distinct A_ks) is chosen so that each set contains a small number of sites. If each set contains only one site (in which case $N = |S|$), then the dynamics (2.42) and (2.43) are called *single-site ("updating") dynamics*; otherwise they are called *multiple-site ("updating") dynamics*. If $|\Omega_0| = 2$, then single-site dynamics are called *single-flip dynamics*.

2) If the A_ks contain a single site, then the following *exchange dynamics* are used in some applications: For each pair $i, j \in S, i \neq j$, define the neighborhoods

$$\mathcal{N}_{ij}(x) = \{y \in \Omega : y_i = x_j, y_j = x_i, {}_i y =_i x\}$$

where ${}_i x$, ${}_i y$ are defined in (2.38). Then one proceeds to define $q_{ij}(x, y)$ and $p_{ij}(x, y)$ as in the single-site dynamics. Again, the full dynamics are constructed either through random or sequential updating. Note that the states generated by the exchange dynamics preserve the sum $\sum_{i \in S} x_i$. Hence these dynamics are not irreducible, but they become irreducible (at least for $F = F_M$ and $F = F_B$) when restricted to the subspace of Ω of fixed $\sum_{i \in S} x_i$.

3) If instead of the "visitation scheme" (2.39) we consider subsets $\{A_k : k = 1, 2, 3, \ldots\}$ so that $A_k \subset S$ for all k and

$$S = \cup_{k=m}^{+\infty} A_k, \quad \text{for all } m \geq 1, \tag{2.44}$$

then we may construct a *non-homogeneous* Markov chain with 1-step transition matrices $P_1, P_2, \ldots$, where P_k is defined as in (2.41), and the n-step transition matrix is $P_1 P_2 \ldots P_n$. Although non-homogeneous, this chain also converges to π (for several choices of F). This type of dynamics is also referred to as *sequential sampling*.

2.2.2 Gibbs sampler

The dynamics described in this subsection were introduced in [35] where the name Gibbs sampler appears. The Gibbs sampler provides a different approach than that of Metropolis-Hastings, although it coincides with the Barker dynamics in some cases (see below).

We will assume that Ω has the form (2.37) with S and Ω_0 finite (extensions to countable or continuous Ω_0 are straightforward, at least formally). If x and y are states in Ω, and A is a subset of S, we will denote by $x_A y_{A^c}$ the state that equals x_A on A and y_{A^c} on $A^c = S \backslash A$. Furthermore, if $A \subset S$ then

$$\pi(x_A | x_{A^c}) = \frac{\pi(x)}{\sum_{z_A} \pi(z_A x_{A^c})} \tag{2.45}$$

is the conditional probability ("*local characteristic*") of $x_A \in \Omega_A$ given $x_{A^c} \in \Omega_{A^c}$.

We fix a "visitation scheme" $A_1, \ldots, A_N$ as in (2.39), and define for each A_i, $i = 1, \ldots, N$,

$$p_i(x, y) \equiv p_{A_i}(x, y) = \mathbb{1}_{\{y_{A_i^c} = x_{A_i^c}\}} \pi(y_{A_i} | x_{A_i^c}) \tag{2.46}$$

for all $x, y \in \Omega$. It is easily verified that each matrix $P_i = P_{A_i} = \{p_i(x, y) : x, y \in \Omega\}$ is reversible relative to π (but not irreducible). Note that the dynamics P_i updates a current state x_{A_i} at A_i by choosing a new state y_{A_i} in Ω_{A_i} (independently of the old state x_{A_i}) according to the conditional probability $\pi(\cdot | x_{A^c})$. A "*full*" dynamics is obtained either through random [see (2.42)] or sequential [see (2.43)] sampling. In the former case, the dynamics is reversible relative to π, but in the latter it is not (although it has π as its stationary distribution). In both cases, the full dynamics is irreducible and aperiodic since $\pi(y_A | x_{A^c}) > 0$. Furthermore, the matrix of the random dynamics is positive definite and hence its spectrum lies in [0,1]. Note also that for Gibbs distribution (2.34), the partition function Z is not involved in the dynamics.

Remarks:

1) In addition to the random and sequential dynamics, one may also consider the non-homogeneous Markov chain corresponding to (2.44). The convergence of this chain to π was first proven in [35] (the proof is similar to, in fact slightly simpler than, a proof for the annealing algorithm we will present in Section 3).

2) The dynamics (random or sequential) based on (2.46) is called the *Gibbs sampler* [35]. For single-site visitation schemes and binary systems ($|\Omega_0| = 2$), the Gibbs sampler coincides with the Barker dynamics; for such systems the procedure is known in the physics literature as the *heat-bath method*. The Gibbs sampler is sometimes referred to as *stochastic relaxation*, although this term is also used for other dynamics. Throughout this article the term *Metropolis-type dynamics* (or sampling, or algorithm) refers to Metropolis-Hasting and Gibbs sampler dynamics.

3) The feasibility of the Gibbs sampler depends on the ability to sample from the conditional probabilities $\pi(y_A | x_{A^c})$; this may be difficult if A contains many sites. For this reason the As are chosen to contain a small number of sites.

4) A more general version of the Gibbs sampler, which plays a central role in *multi-grid* Monte Carlo [41] and some other *non-local* updating algorithms, is the so-called *partial resampling method* [86].

5) The sequential Gibbs sampler for Gaussian and perturbations of Gaussian distributions has been studied in [5].

2.2.3   Comparison of Metropolis-type dynamics and bounds on their relaxation times

We will restrict attention to reversible dynamics (as we mentioned in Section 2.2.1, non-reversible dynamics are much harder to analyze mathematically). For a reversible (relative to π) matrix P, we will denote by $\tau_0(P)$, $\tau_{1,f}(P)$ its relaxation constants, by $C(P)$ the limiting matrix (2.16), by $r(P)$ the spectral radius of $P - \Pi$, and by $\mu_2(P)$ the constant defined above (2.20).

The comparison of the various dynamics is based on the following simple lemma.

Lemma 2.1 *Let* $P = \{p(x,y)\}, P' = \{p'(x,y)\}$ *be irreducible and reversible relative to* π. *If*

$$p(x,y) \geq p'(x,y), \quad \text{for all } x \neq y, \ x,y \in \Omega$$

then

$$P \leq P' \text{ and } C(P) \leq C(P').$$

In particular

$$\mu_2(P) \quad \leq \mu_2(P'),$$

$$\tau_{1,f}(P) \quad \leq \tau_{1,f}(P'), \quad \text{for all } f \in \ell_2.$$

Proof: By (2.28), for all $f \in \ell_2(\pi)$,

$$< f, (I - P)f > \ \geq \ < f, (I - P')f > .$$

Hence $P \leq P'$; in particular, $\mu_2(P) \leq \mu_2(P')$. Now let $C = C(P)$ and $C' = C(P')$; then on the orthogonal complement of constant functions we have by (2.16)

$$C + I = 2(I - P)^{-1}, \ C' + I = 2(I - P')^{-1}$$

since $I - P \geq I - P' \geq 0$, by a well-known property [26, 86] we have

$$0 \leq (I - P)^{-1} \leq (I - P')^{-1}$$

which implies $C \leq C'$; in particular, $\tau_{1,f}(P) \leq \tau_{1,f}(P')$.

Remarks:

1) Notice that no aperiodicity assumption was used in Lemma 2.1.

2) Lemma 2.1 was first proven in [74]. See [10] for other comparison (properties also [29]).

3) The lemma says nothing about the relaxation times $\tau_0(P), \tau_0(P')$, unless $P \geq 0$, $P' \geq 0$, in which case $r(P) = \mu_2(P) \leq \mu_2(P') = r(P')$, and therefore $\tau_0(P) \leq \tau_0(P')$.

Next we apply Lemma 2.1 to reversible Metropolis-type dynamics. We will denote by $P^{(F)}$ the matrix corresponding to an F satisfying (2.30), and by $\tau_0(F)$, $\tau_{1,f}(F)$, $\mu_2(F)$, $r(F)$, its corresponding objects; if $F = F_M$, we will simply write $P^{(M)}$, $\tau_0(M)$, $\tau_{1,f}(M)$, $\mu_2(M)$, $r(M)$. For random updating schemes we will write $P^{(RF)}$, $\tau_0(RF)$, $\tau_{1,f}(RF)$, $\mu_2(RF)$, $r(RF)$, similarly for a random updating Gibbs sampler, $P^{(RG)}$, $\tau_0(RG)$, $\tau_{1,f}(RG)$, $\mu_2(RG)$. Note that $\mu_2(RG) = r(RG)$, since the matrix $P^{(RG)}$ is non-negative.

Since $F_M(\xi)$ is the maximal function satisfying (2.30), we have

$$p^{(M)}(x,y) \geq p^{(F)}(x,y), \text{ for all } x \neq y.$$

Hence for Fs that give irreducible dynamics

$$P^{(M)} \leq P^{(F)},$$

$$\mu_2(M) \leq \mu_2(F),$$

$$\tau_{1,f}(M) \leq \tau_{1,f}(F) \;, f \in \ell_2(\pi).$$

That is, the Metropolis dynamics has a better fluctuation constant than other irreducible Metropolis-Hasting dynamics. In particular this is true for the random updating, reversible dynamics, i.e.,

$$\mu_2(RM) \leq \mu_2(RF) \tag{2.47a}$$

$$\tau_{1,f}(RM) \leq \tau_{1,f}(RF) \;, f \in \ell_2(\pi). \tag{2.47b}$$

How does the random Gibbs sampler (RG) compare with the random Metropolis dynamics (RM)? In general, the comparison is not straightforward. But for binary systems, and single-site updating schemes, the Gibbs sampler coincides with the Barker dynamics, so in this case we have

$$\mu_2(RM) \leq \mu_2(RG) = r(RG) \tag{2.48a}$$

$$\tau_{1,f}(RM) \leq \tau_{1,f}(RG) \;, f \in \ell_2(\pi). \tag{2.48b}$$

Thus for such systems and single-site random updating dynamics, the (random) Metropolis has a better fluctuation constant than the (random) Gibbs sampler. But for the relaxation times $\tau_0(RG)$ and $\tau_0(RM)$ the situation is more complex, even for the standard nearest-neighbor, ferromagnetic Ising model that we consider next: Let S be a *periodic box (torus)* in $\mathbb{Z}^d, d \geq 1$, with "volume" $|S| = N^d$. The Gibbs distribution for the *nearest-neighbor Ising model with periodic boundary conditions* (i.e., S is a torus) reads

$$\pi_\beta(x) = \frac{1}{Z(\beta)} \exp\{\beta \sum_{<ij>} x_i x_j\}, \tag{2.49a}$$

$$Z(\beta) = \sum_{x} \exp\{\beta \sum_{<ij>} x_i x_j\}, \qquad (2.49b)$$

with $\beta \geq 0$ ("ferromagnetic"); the sum in (2.49a) is over nearest-neighbor pixels $< ij >$ (i.e., $|i - j| = 1$). Let us denote by $r(RM; \beta)$ and $r(RG; \beta)$ the spectral radii for the random Metropolis and the random Gibbs sampler, respectively. (We will use similar notation for τ_0, $\tau_{1,f}$, μ_2, and $\underline{\mu}$). We will show [30]

Proposition 2.2

a) If β is sufficiently small ($\beta \geq 0$), then

$$\tau_0(RG; \beta) < \tau_0(RM; \beta). \qquad (2.50a)$$

b) If β is sufficiently large, then

$$\tau_0(RG; \beta) > \tau_0(RM; \beta). \qquad (2.50b)$$

Proof: It is easily seen that $P^{(RG;0)}$ is irreducible and aperiodic; hence $\mu_2(RG; 0) = r(RG; 0) < 1$. On the other hand, $P^{(RM;0)}$ is irreducible and has period 2. Therefore, -1 is in the spectrum of $P^{(RM;0)}$, and hence $r(RM; \infty) = |\mu(RM; \infty)| = 1 > r(RG; 0)$. Now $r(RM; \beta)$ and $r(RG; \beta)$ are continuous functions of β; hence (2.50a) holds for sufficiently small $\beta \geq 0$.

Next we consider the matrices $P^{(RM;\infty)}$ and $P^{(RG;\infty)}$ corresponding to $\beta = +\infty$. The matrix $P^{(RG;\infty)}$ is irreducible, but it has more than one ergodic class; hence $r(RG; \infty) = \mu_2(RG; \infty) = 1$. Now $P^{(RM,\infty)}$ is not irreducible, but by suitably rearranging the elements of Ω, it can be put into a block-lower-triangular form

$$P^{(RM;\infty)} = \begin{pmatrix} A_1 & 0 & \dots & 0 & 0 \\ 0 & A_2 & \dots & 0 & 0 \\ 0 & \dots & \dots & A_m & 0 \\ D_1 & \dots & \dots & D_m & D_{m+1} \end{pmatrix}$$

where $A_1, \dots, A_m$ are ergodic matrices, and $D_1, \dots, D_{m+1}$ describe transitions of *transient states*. It is easily verified that the transient states reach the ergodic states with positive probabilities; hence the eigenvalues of D_{m+1} are strictly less than one in absolute value. Therefore $\rho(RM; +\infty) = \mu_2(RM; +\infty) = 1$. This together with (2.48a) (valid for all $\beta > 0$) implies (2.50b) for sufficiently large β.

Remarks:

1) See [30] for a more general result.

2) The proof of Proposition 2.2 shows that the validity of (2.50a) is due to the fact that for small β, $P^{(RM;\beta)}$ has spectrum near -1 and this spectrum controls $r(RM;\beta)$. On the other hand, for large β, $r(RM;\beta)$ is controlled by the spectrum near $+1$. Also, as we mentioned in sub-Section 2.1, $\tau_{1,f}$ is *always* controlled by the spectrum near $+1$, and it is for reason that we have $\tau_{1,f}(RM;\beta) \leq \tau_{1,f}(RG;\beta)$ [see (2.48b)] for all $\beta \geq 0$.

Next, we fix the dynamics P and we derive certain lower bounds of τ_0, $\tau_{1,f}$ for some systems with state spaces of the form (2.37). These bounds are especially useful for large systems, i.e., as $|S| \to +\infty$. The bounds are based on the following simple lemma [10].

Lemma 2.2. *Let $P = \{p(x,y)\}$ be irreducible and reversible relative to π. Suppose that there exists an $f \in \ell_2(\pi)$ so that $|f(x) - f(y)| \leq A$ whenever $p(x,y) > 0$, $x \neq y$, with some constant N. Then*

$$\mu_2(P) \geq 1 - \frac{A^2}{2 < f, (I - \Pi)f >}, \tag{2.51a}$$

$$\tau_{1,f}(P) \geq \frac{2 < f, (I - \Pi)f >}{A^2} - \frac{1}{2}. \tag{2.51b}$$

Proof: By (2.28)

$$< f, (I - P)f > \leq \frac{A^2}{2}.$$

Hence

$$1 - \rho_f(1) \leq \frac{A^2}{2 < f, (I - \Pi)f >}.$$

Then (2.51a) and (2.51b) are consequences of (2.27) and (2.25), resp.

Remarks:

1) If the right-hand-side of (2.51a) is positive, then

$$\tau_0 \geq [-\log(1 - \frac{A^2}{2 < f, (I - \Pi)f >})]^{-1}. \tag{2.52}$$

Furthermore, if Ω has the form (2.37), and $< f, (I - \Pi)f > \to +\infty$ as $|S| \to +\infty$, then τ_0 and $\tau_{1,f}$ go to infinity as $|S| \to +\infty$. This is the case for some important systems: Suppose that $|\Omega_0| < +\infty$, and consider single-site random updating schemes with irreducible dynamics (e.g., Metropolis or Gibbs sampler). Let

$$f(x) = \sum_{i \in S} x_i. \tag{2.53a}$$

Then $|f(x) - f(y)| \le C_0$ for some $C_0 < +\infty$ (since $|\Omega_0| < +\infty$). On the other hand

$$< f, (I - \Pi)f >= |S|\chi_S \qquad (2.53b)$$

where

$$\chi_S = \frac{1}{|S|} \sum_{i,j} [< x_i x_j > - < x_i >< x_j >]. \qquad (2.53c)$$

Except in trivial cases, $\chi_S > 0$. Thus by (2.51b) and (2.52), $\tau_{1,f}$ and τ_0 tend to infinity at least linearly in $|S|$ as $|S| \to +\infty$. But they may tend to infinity faster than $|S|$, because in some cases (see below) $\chi_S \to +\infty$ as $|S| \to +\infty$.

2) For the random Gibbs sampler (with single-site updating) and f of the form (2.53a), the condition $|\Omega_0| < +\infty$ may be relaxed, and (2.51) may be improved by a factor of 2. Notice that

$$< f, (I - P)f > \; = \frac{1}{|S|} \sum_{i,j,k} < x_i, (I - P_k)x_j >$$

$$= \frac{1}{|S|} \sum_i < x_i, (I - P_i)x_i >$$

$$\le \frac{1}{|S|} \sum_i < x_i^2 > .$$

In the second equality we have used the property that $P_k x_j = \delta_{kj} x_j$; the inequality holds because $P_i \ge 0$. Suppose that the limit

$$B = \lim_{|S| \to +\infty} \frac{1}{|S|} \sum_i < x_i^2 >$$

exists and is finite. (If $S \subset \mathbb{Z}^d$, then for "translation invariant systems" $B =< x_0^2 >$). Thus

$$1 - \rho_f(1) \le \frac{B}{|S|\chi_S}.$$

Hence

$$\tau_{1,f} \equiv \tau_{1,f}^{(S)} \ge \frac{1}{B}|S|\chi_S - \frac{1}{2} \qquad (2.54)$$

and τ_0 is bounded below by (2.52).

3) Suppose for concreteness that $S = S_N$ is a hypercube in $\mathbb{Z}^d, d \ge 1$, of size N ($|S| = N^d$). Then for large N, (2.52) yields (approximately)

$$\tau_0^{(N)} \ge C_0 N^d \chi_S,$$

where $\chi_N = \chi_{S_N}$ is given by (2.53c). Now for many statistical mechanics systems with $d \ge 2$ (such as the Ising model (2.49) and other models used in image processing problems) near a *critical point* (second-order phase transition), we have $\chi_N \sim \xi^{z_0}$ as $N \to +\infty$, where ξ is the *correlation length* of the infinite system and z_0 is an *equilibrium critical exponent* (for many systems, including the Ising model, $z_0 \simeq 2$). At a critical point $\xi = +\infty$, and hence (typically) $\chi_N \to +\infty$ as $N \to +\infty$. Thus near a critical point we have

$$\tau_0^{(N)} \geq CN^d \xi^{z_0} \tag{2.55a}$$

and at a critical point

$$\frac{1}{N^d}\tau_0^{(N)} \to +\infty, \quad \text{as } N \to +\infty. \tag{2.55b}$$

It is still an open problem, even for the Ising model (2.49) parametrized by $\beta > 0$, whether $N^{-d}\tau_0^{(N)}$ remains bounded as $N \to +\infty$, for β near but below the *critical inverse temperature* β_c. For single-site Gibbs sampler or Metropolis dynamics, and Ising-type models, it is known [3] and easily proven that $N^{-d}\tau_0^{(N)}$ remains bounded as $N \to +\infty$ for β below the Dobrushin uniqueness region [43]. One expects this to be true for all $\beta < \beta_c$, but for β near β_c it remains an open problem. The asymptotic behavior of $\tau_0^{(N)}$ as $N \to +\infty$ is one of the most interesting mathematical problems in the study of Metropolis-type simulation algorithms; see [2, 49, 50, 87, 67, 66, 68] for recent studies of the problem for various systems and dynamics.

4) Near a critical point, the relaxation time is expected to diverge, for many typical systems, as

$$\tau_0^{(N)} \sim N^d(\min(N, \xi))^z \tag{2.56}$$

where ξ is the correlation length, and z is a *dynamic critical exponent*. The divergence of $N^{-d}\tau_0^{(N)}$ at a critical point is called *critical-slowing down* — a phenomenon that severely hampers the feasibility of Metropolis-type simulation algorithms near critical points. From (2.55a), we see that $z \geq z_0 \simeq 2$, but the exact determination of z is by and large an open problem. In terms of computations, one needs about $2\tau_0^{(N)}$ iterations to generate one "effectively equilibrium" sample, which means that the computer time is of the order $N^d\xi^z$, typically a very large number.

5) The factor N^d in (2.56) is inherent in *all* Metropolis-type algorithms (each "sweep" of $S = S_N$ takes a time of order $|S_N| = N^d$). But the factor $(\min(N, \xi))^z \simeq \xi^z$ is typical of singlesite, or more generally *local updating algorithms* [including the multiple-site updating schemes (2.39)]. This is due to the fact that, in such algorithms, "information" propagates locally. Put differently, in a single algorithmic step, "information" is transmitted (in single-site updating) from one site to a neighboring site. Thus in a sense, "information" executes a random walk around the lattice S_N. In order for the system to evolve to a "new" state, the "information" has to travel a distance ξ, which means that it would take a time of the order ξ^2 to reach such a new state (this formal argument indicates that $z \geq 2$). Since the roots of the critical-slowing down factor ξ^z lie in the local nature of the algorithms, the natural way out of

this difficulty is to design *non-local updating algorithms*. At least three such algorithms have been proposed: a) Fourier acceleration [7], b) multi-grid Monte Carlo [41, 9], and c) the Swendsen-Wang algorithm [89] and its extensions [27, 28] (see [41, 86] for more information on non-local algorithms). A brief introduction to the basic Swendsen-Wang algorithm will be given in subsection 2.4.

2.3   Monte Carlo simulation via diffusion processes

In this section we will assume that $\Omega = I\!\!R^d, d \geq 1$, until the end of subsection 2.3.2, where we will consider briefly the case Ω being a bounded subset of $I\!\!R^d$, e.g., $\Omega = [0,1]^d$. The theory can be extended (at least formally) to the case when Ω is an arbitrary Riemannian manifold (compact or non-compact). We will assume that we are given a measure π on $\Omega = I\!\!R^d$ with a density, with respect to the Lebesgue measure dx, given by

$$\pi(x) = \pi_T(x) = \frac{1}{Z_T} \exp\{-\frac{1}{T}U(x)\}, \qquad (2.57a)$$

$$Z_T = \int \exp\{-\frac{1}{T}U(x)\}dx, \qquad (2.57b)$$

where $U(x)$ is a (given) real-valued function on Ω, so that $Z_T < +\infty$; T is a positive constant, the "temperature," which will play no active role in this section, but it is inserted for use in the study of simulated annealing in Section 3.2.

We seek a diffusion process $\{X(t)\}_{t \geq 0}$ on Ω having π as its unique *invariant measure*. The process will be specified via a stochastic differential equation of the form

$$dX(t) = b(X(t))dt + \sigma(X(t))dW(t) \qquad (2.58a)$$

where $b(x) = \{b_i(x)\}$, $\sigma(x) = \{\sigma_{ij}(x)\}$, $i, j = 1, 2, \cdots, d$, are the drift and diffusion matrices, respectively, and W is the standard Brownian motion on $I\!\!R^d$. Alternatively, the process is determined by its *generator*

$$\mathcal{L} = -\frac{1}{2}\sum_{i,j} a_{ij}(x)\frac{\partial^2}{\partial x_i \partial x_j} - \sum_i b_i(x)\frac{\partial}{\partial x_i} \qquad (2.58b)$$

where $a(x) = \sigma(x)\sigma^T(x)$ is the *covariance matrix*. Of course, $b(x)$, $a(x)$, or $\sigma(x)$ will be chosen so that π is the unique invariant measure of the process.

Let

$$p(t; x_0, x) = P\{X(t) = x | X(0) = x_0\}$$

be the transition probability function (we assume the existence of densities with respect to dx). If μ is an initial distribution, then the distribution at time $t > 0$ is

$$p(t, x) = (\mu e^{-t\mathcal{L}})(x) = \int dx_0 \mu(x_0) p(t; x_0, x).$$

As in the case of Markov chains (Section 2.1), there are two basic questions: the convergence of $p(t, x)$ to $\pi(x)$ as $t \rightarrow +\infty$, and the convergence of the time averages

$$\frac{1}{t} \int_0^t f(X(s))ds$$

to the equilibrium expectations

$$< f > \equiv < f >_\pi = \int f(x)\pi(dx)$$

as $t \rightarrow +\infty$. The notions of *relaxation time* and *fluctuation constant* are the same as in Section 2.1. Here we will concentrate on the former only. The deviation from equilibrium will be measured either in terms of the *total variation distance* $\|\mu e^{-t\mathcal{L}} - \pi\|_{\mathrm{var}}$ or in terms of the L_2-*distance*

$$\|e^{-t\mathcal{L}}f - < f > \|_{L_2(\pi)}$$

where $L_2(\pi)$ is the Hilbert space of complex-valued functions on Ω with inner product

$$< f, g > \equiv < f, g >_\pi = \int f^*(x)g(x)\pi(dx).$$

2.3.1   The Langevin Equation

The Langevin equation (LE), also called the Smoluchowski equation, corresponding to (2.57) reads

$$dX(t) = - \bigtriangledown U(X(t))dt + \sqrt{2T}dW(t) \tag{2.59a}$$

with generator

$$\mathcal{L} = -T\Delta + \bigtriangledown U \cdot \bigtriangledown = -T\pi^{-1} \bigtriangledown (\pi \bigtriangledown \cdot). \tag{2.59b}$$

Here we assume that $U \in C^1(\mathbb{R}^d)$; later we shall impose more conditions on U.

Equation (2.59a) was proposed by Langevin in 1908 as a generalization of Einstein's theory of Brownian motion; it describes the motion of a particle in a viscous fluid (Brownian motion assumes zero viscosity). Langevin's equation was the first mathematical equation describing non-equilibrium thermodynamics. The term $\sqrt{2T}dW(t)$ describes the microscopic fluctuations caused by the Brownian force, while $- \bigtriangledown U(X(t))dt$ (the "drag" force) is generated by the viscosity of the fluid. The Brownian fluctuations dominate during short time intervals, while the drag force controls the long-time behavior of the process.

Onsager gave a new interpretation to the LE, as an *irreversible process* by a simple correspondence: the drag force is interpreted as the "drift" of the thermodynamic system toward its equilibrium, while the velocity of the particle corresponds to the deviation of a theormodynamic quantity from its equilibrium. In Onsager's interpretation, the thermal fluctuations dominate in the short time scales, while the drift dominates in the long run; these observations are reflected in the mathematical analysis of the LE to be given below.

We will view the generator (2.59b) as an (unbounded) operator on $L_2(\pi)$; $\mathcal{L}$ is densely defined and its domain contains $C_0^2(\mathbb{R}^d)$ (C^2 functions of compact support). For $f, g \in C_0^2$, we have

$$< f, \mathcal{L}g >=< \mathcal{L}f, g >= T \int \nabla f \cdot \nabla g d\pi.$$

Hence $\mathcal{L}$ is self-adjoint and non-negative on $L_2(\pi)$. The conditions of the next theorem can be relaxed in a number of ways, but we state the theorem in a form that will be used in our study of simulated annealing in subsection 3.2.3.

Theorem 2.4 *Suppose that $U(x)$ has the following properties*

(1) $U(x) \in C^2(\mathbb{R}^d)$, $\inf\{U(x) : x \in \mathbb{R}^d\} = 0$,
$U(x) \to +\infty$ as $|x| \to +\infty$,

(2) $|\nabla U(x)| \to +\infty$ as $|x| \to +\infty$,

(3) $\frac{-\Delta U(x)}{1+|\nabla U(x)|^2}$ *is bounded below*,

(4) $Z_T = \int e^{-\frac{1}{T}U(x)} dx < +\infty$.

Then

(a) $\mathcal{L}$ *has a purely discrete spectrum*

$$\lambda_1 = 0 < \lambda_2 \leq \lambda_3 \leq \cdots$$

and $\lambda_1 = 0$ is a simple eigenvalue with normalized eigenvector $\phi_1(x) \equiv 1$.

(b)

$$\|e^{-t\mathcal{L}}f - < f >\|_{L_2(\pi)} \leq e^{-t\lambda_2}\|f - < f >\|_{L_2(\pi)} \qquad (2.60)$$

for all $f \in L_2(\pi)$. In particular

$$p(t, x) \to \pi(x) \text{ weakly as } t \to +\infty.$$

Remarks:
1) The condition $\inf\{U(x) : x \in \mathbb{R}^d\} = 0$ is for convenience only. Since $U \in C^2(\mathbb{R}^d)$ and $U(x) \to +\infty$ as $|x| \to +\infty$, $U(x)$ is bounded below; hence we can add a constant so that $U(x) \geq 0$ without changing π.
2) Condition (2) can be relaxed [85] to : For some $R > 0$, $\inf\{|\nabla U(x)| : |x| \geq R\} > 0$. In this case the spectrum of $\mathcal{L}$ is not discrete, but $\lambda_1 = 0$ is still

an isolated eigenvalue; and if λ_2 denotes the distance between $\lambda_1 = 0$ and the rest of the spectrum, then part b) of the theorem still holds.

3) Conditions (2) and (3) could be replaced by a single condition involving T [see (2.61) below]; but as we mentioned above, the theorem has been formulated in a way that will be used in the study of simulated annealing.

4) Since $U(x) \geq 0$, if $Z_{T_0} < +\infty$ for some $T_0 > 0$, then $Z_T < +\infty$ for all $T \leq T_0$.

Proof of Theorem 2.4: Define the isometry

$$\Pi^{\frac{1}{2}} : L_2(d\pi) \longrightarrow L_2(dx)$$

$$f(x) \longmapsto \sqrt{\pi(x)} f(x)$$

and its inverse

$$\Pi^{-\frac{1}{2}} : L_2(dx) \longrightarrow L_2(d\pi)$$

$$g(x) \longmapsto (\pi(x))^{-\frac{1}{2}} g(x).$$

A straightforward computation gives

$$\mathcal{H} = \Pi^{\frac{1}{2}} \mathcal{L} \Pi^{-\frac{1}{2}} = -T\Delta + \frac{T}{4T^2} [|\nabla U(x)|^2 - 2T\Delta U(x)]. \tag{2.61}$$

Clearly $\mathcal{H}$ is a densely defined, non-negative, self-adjoint operator on $L_2(dx)$. $\mathcal{H}$ is a Schrödinger-type operator on $L_2(dx)$; the conditions of the theorem guarantee [85, 77 p.249] that $\mathcal{H}$ has a purely discrete spectrum, and that $\lambda_1 = 0$ is a simple eigenvalue with (normalized) eigenvector $\psi_1(x) = \sqrt{\pi(x)}$. Since $\mathcal{L}$ is [by (2.61)] unitarily equivalent to $\mathcal{H}$, it has the same spectrum as $\mathcal{H}$. Furthermore, if $\phi_n(x)$ [resp. $\psi_n(x)$], $n = 1, 2, \ldots$, denote the normalized eigenfunctions of $\mathcal{L}$ (rep. $\mathcal{H}$), then $\psi_n(x) = \sqrt{\pi(x)}\phi_n(x)$. Part a) is readily established.

By the spectral theorem for self-adjoint operators, we have for every $f \in L_2(\pi)$

$$(e^{-t\mathcal{L}}f)(x) = \sum_{n=1}^{\infty} e^{-t\lambda_n} <f, \phi_n> \phi_n(x)$$

$$= <f> + \sum_{n\geq 2} e^{-t\lambda_n} <f, \phi_n> \phi_n(x). \tag{2.62a}$$

Hence

$$\|e^{-t\mathcal{L}}f - <f>\|^2_{L_2(\pi)} = \sum_{n\geq 2} e^{-2t\lambda_n} |<f, \phi_n>|^2 \tag{2.62b}$$

$$\leq e^{-2t\lambda_2} \|f - <f>\|_{L_2(\pi)}$$

which yield (2.60). By (2.62b), the norm of $e^{-t\mathcal{L}}$ restricted to the orthogonal complement of constant functions is bound above by $e^{-t\lambda_2}$. Hence for any $f \in L_2(\pi)$, we have

$$\left| \int p(t,x)f(x) - <f> \right| \le \|f - <f>\|_{L_2(\pi)} \cdot e^{-t\lambda_2} \left\| \frac{\mu}{\pi} - 1 \right\|_{L_2(\pi)}$$

which implies weak convergence.

Remarks:

1) From (2.62a) we see that

$$p(t; x_0, x) = (e^{-t\mathcal{L}})(x_0, x) = \sum_{n=1}^{+\infty} c^{-t\lambda_n} \phi_n(x_0) \phi_n(x) \pi(x).$$

Hence

$$\pi(x_0) p(t; x_0, x) = \pi(x) p(t; x; x_0).$$

This is the analogue of *detailed-balance* for Markov chains, and is intrinsically related to the self-adjointness of $\mathcal{L}$.

2) In the study of simulated annealing, we will need the behavior of $\lambda_2 = \lambda_2(T)$ as $T \downarrow 0$. This is a classic problem in the study [85, 48] of Schrödinger-type operators with a small parameter. It can be shown [53] that there exists a $\Delta > 0$ such that

$$\lim_{T \downarrow 0} T \log \lambda_2(T) = -\Delta. \tag{2.63}$$

3) If $\mathbb{R}^d$ is replaced by a Riemannian manifold M, then (2.59b) defines a self-adjoint operator on $L_2(M, d\pi)$, provided that $\nabla.$ and ∇ denote the gradient and divergence operations determined by the Riemannian metric on M. These types of operators are called *Dirichlet forms* [23]. In terms of local coordinates on M, these operators are similar to the ones we consider in the next subsection.

2.3.2   Generalization of the Langevin equation

Let $a(x) = \{a_{ij}(x)\}$, $i, j = 1, \ldots, d$, be a symmetric covariance matrix satisfying

$$c_1 |\xi|^2 \le \sum_{i,j} a_{ij}(x) \xi_i \xi_j \le c_2 |\xi|^2 \tag{2.64}$$

for all $x \in \mathbb{R}^d$, all vectors $\xi = (\xi_1, \ldots, \xi_n)$, and some constants $0 < c_1 \le c_2 < +\infty$. Assume that $a_{ij}(x) \in C^1(\mathbb{R}^d)$, and consider the *Dirichlet form* [π is given by (2.57)]

$$\mathcal{L} = -T\pi^{-1} \sum_{i,j} \frac{\partial}{\partial x_i} [a_{ij}(x) \pi(x) \frac{\partial}{\partial x_j}]$$

$$\tag{2.65a}$$

$$= -T \sum_{i,j} a_{ij}(x) \frac{\partial^2}{\partial x_i \partial x_j} - \sum_i b_i(x) \frac{\partial}{\partial x_i}$$

where

$$b_i(x) = \sum_j [T\frac{\partial a_{ji}}{\partial x_j} - a_{ji}(x)\frac{\partial U}{\partial x_j}]. \tag{2.65b}$$

For $f, g \in L_2(\pi)$ with compact support, we have

$$< f, \mathcal{L}g > = < \mathcal{L}f, g > = T\sum_{i,j} \int a_{ij}(x)\frac{\partial f^*}{\partial x_i}\frac{\partial g}{\partial x_j}\pi(dx).$$

Hence $\mathcal{L}$ is self-adjoint and non-negative on $L_2(\pi)$. The diffusion process $\{X(t) : t \geq 0\}$ with generator $\mathcal{L}$ satisfies the generalized Langevin equation

$$dX(t) = b(X(t))dt + \sqrt{2T}\sigma(X(t))dW(t) \tag{2.65c}$$

with $\sigma(x)$ the square root of $a(x)$. As in the LE, $\mathcal{L}$ is unitarily equivalent to the Schrödinger-type operator

$$\mathcal{H} = \Pi^{\frac{1}{2}}\mathcal{L}\Pi^{-\frac{1}{2}}$$

$$= -T\sum_{i,j}\frac{\partial}{\partial x_i}[a_{ij}(x)\frac{\partial}{\partial x_j}] + \frac{T}{4T^2}\sum_{i,j}\{a_{ij}(x)\frac{\partial U}{\partial x_i}\frac{\partial U}{\partial x_j} - 2T\frac{\partial}{\partial x_i}[a_{ij}(x)\frac{\partial}{\partial x_j}]\}.$$

If $a(x)$ satisfies (2.64), and $U(x)$ the conditions of Theorem 2.4 (both assumptions can be relaxed), then $\mathcal{H}$ (and hence $\mathcal{L}$) has a purely discrete spectrum $\lambda_1 = 0 < \lambda_2 \leq \lambda_3 \leq \ldots$, with $\lambda_1 = 0$ a simple eigenvalue. Furthermore, the other conclusions of Theorem 2.4 still hold (and their proof is the same).

A useful special case of the above, occurs when

$$a_{ij}(x) = \delta_{ij}\alpha(x_i), \quad i, j = 1, 2, \ldots, d$$

with some function $\alpha \in C^1(\mathbb{R})$ so that (2.64) holds. Then the stochastic differential equation reads

$$dX_i(t) = [T\alpha'(X_i(t)) - \alpha(X_i(t))\frac{\partial}{\partial X_i}U(X(t))]dt + \sqrt{2T\alpha(X_i(t))}dW_i(t), \tag{2.66a}$$

and the generator is

$$\mathcal{L} = -T\sum_i \alpha(x_i)\frac{\partial^2}{\partial x_i^2} - \sum_i [T\alpha'(x_i) - \alpha(x_i)\frac{\partial U}{\partial x_i}]\frac{\partial}{\partial x_i}. \tag{2.66b}$$

This special process can be used to treat the case when Ω is a bounded subset of $\mathbb{R}^d$. For simplicity assume that $\Omega = [0, 1]^d$; let $U(x)$ be a function on $[0, 1]^d$ and $\pi(x)$ the corresponding probability distribution (2.57) on $[0, 1]^d$. Choose $\alpha(\xi)$, $0 \leq \xi \leq 1$, so that $\alpha \in C^1[0, 1]$, $\alpha(0) = \alpha(1) = 0$, and $\alpha(\xi) > 0$ for all $0 < \xi < 1$ [e.g., $\alpha(\xi) = \xi(1-\xi)$]. The corresponding generator (2.65) is a non-negative self-adjoint operator on $L_2([0, 1]^d, \pi(dx))$, and has purely discrete spectrum $\lambda_1 = 0 < \lambda_2 \leq \lambda_3 \leq \ldots$ Furthermore, the process $\{X(t) : t \geq 0\}$ generated by $\mathcal{L}$ has π as its unique invariant measure, and the conclusions of Theorem 2.4 hold.

If $\Omega = [0,1]^d$, then the choice $\alpha(\xi) \equiv 1$, $\xi \in [0,1]$, in (2.65) [which gives the Langevin operator (2.59b)] does not guarantee stationarity of π. In fact, the generator (2.59b) restricted to $[0,1]^d$ is not well-defined unless we impose boundary conditions on the boundary of $\Omega = [0,1]^d$. The boundary conditions that ensure that π is the (unique) invariant distribution of the underlying process are known as *"reflecting" boundary conditions* [37].

Remarks:

1) Diffusion processes for simulated annealing were first studied in [37] for the case $\Omega = [0,1]^d$, and subsequently in [39] for $\Omega = I\!\!R^d$. A systematic study of perturbed diffusion processes with applications to simulated annealing has been carried out in [54].

2) An interesting class of *generalized diffusion processes* arises in ("scalar") quantum field theories [8, 4]. These are processes whose state space Ω is the space of *generalized functions* (Schwartz distributions) on $I\!\!R^2$. Roughly speaking, the processes correspond to the simulation of Ising type models defined not on $Z\!\!\!Z^2$ but on the continuum $I\!\!R^2$. Their study has involved Dirichlet forms [4] on infinite dimensional spaces and deep tools from the mathematical study of quantum field theories. The subject of generalized diffusion processes is often referred to as the *stochastic quantization of field theories*; it generalizes the theory of *stochastic partial differential equations* and is still in its infancy.

2.4   The Swendsen-Wang algorithm

The Swendsen-Wang (SW) algorithm was introduced in [89] for simulating the *Potts models* to be defined below. The algorithm is a dynamic Monte Carlo algorithm, but its dynamics are very different from the Metropolis-type. It is a *non-local* updating algorithm (see discussion in the last remark of Subsection 2.2.3) aimed at reducing the relaxation time. In the past few years, there have been various generalizations of the basic algorithm, with the goal of eliminating critical-slowing down and making the algorithm applicable to models other than the Potts models. Here we will describe the basic algorithm (see [41, 27, 86] and references cited there for its generalizations).

Let S be a bounded domain in $Z\!\!\!Z^d$, $d \geq 1$ (typically a hypercube), and q an integer, $q \geq 2$. The state space Ω has the form (2.37) with $\Omega_0 = \{1, 2, \ldots, q\}$. The measure π is given by (2.34) with energy function

$$H(x) = -\sum_{i,j} J_{ij}(\delta_{x_i x_j} - 1),$$

with $J_{ij} \geq 0$ for all $i, j \in S$. These models are called generalized Potts Models.

The model with $q = 2$ is equivalent to the *generalized Ising model* (i.e., to the Ising model with possibly "long range" interactions).

The basic idea of the SW algorithm is to augment the Gibbs distribution $\pi(x)$ by introducing appropriate *auxiliary* variables, and then simulate the augmented model. This is done as follows: Note that

$$\pi(x) = \tfrac{1}{Z} e^{-H(x)} \quad = \tfrac{1}{Z} \prod_{ij} \exp\{J_{ij}(\delta_{x_i x_j} - 1)\}$$

$$= \tfrac{1}{Z} \prod_{ij} \{(1 - p_{ij}) + p_{ij}\delta_{x_i x_j}\}$$

where $p_{ij} = 1 - e^{-J_{ij}}$. Using the elementary identity

$$a + b = \sum_{z=0}^{1} (a\delta_{z,0} + b\delta_{z,1}),$$

we rewrite π as

$$\pi(x) = \frac{1}{Z} \prod_{ij} \{ \sum_{z_{ij}=0}^{1} [(1 - p_{ij})\delta_{z_{ij},0} + p_{ij}\delta_{x_i x_j}\delta_{z_{ij},1}]\}.$$

A pair of sites $i, j \in s$ with $J_{ij} > 0$ is called a *bond*, and will be denoted by (i, j); let S^* be the set of bonds. The auxiliary variables z_{ij} (taking values 0 or 1) define a binary process, the *bond process*,

$$z = \{z_{ij} : (i, j) \in S^*\}$$

indexed by S^*, i.e., $z \in \{0, 1\}^{S^*}$. In terms of the bond process z we have

$$\pi(x) = \frac{1}{Z} \sum_z \prod_{z_{ij}=0} (1 - p_{ij}) \prod_{z_{ij}=1} p_{ij}\delta_{x_1 x_j} \qquad (2.67)$$

where $\prod_{z_{ij}=k}$ denotes product over bonds with $z_{ij} = k, k = 0, 1$. Now, we define a probability distribution on $\Omega_0^S \times \{0, 1\}^{S^*}$ by

$$P(x, z) = \frac{1}{Z} \prod_{z_{ij}=0} (1 - p_{ij}) \prod_{z_{ij}=1} p_{ij}\delta_{x_i x_j}, \qquad (2.68)$$

and interpret it as the joint distribution of the pair (x, z). By (2.67), the marginal of x under $P(x, z)$ is exactly $\pi(x)$.

If $z_{ij} = 1$ (resp. $z_{ij} = 0$), we say that the bond (ij) is *occupied* (resp. *unoccupied*). By (2.68), if (i, j) is occupied then $x_i = x_j$; but if (i, j) is not occupied, then there is no constraint between x_i and x_j. For a given bond configuration z, the sites of S can be grouped into *clusters*; two sites are in the same cluster if they can be connected by a path of occupied bonds (some clusters may contain only one site). Notice that *all* the x_is within a cluster have the same value, and that *distinct clusters are independent.*

For a given bond configuration z, let $C(z)$ be the number of ("connected") clusters. Then the marginal of z is

$$P(z) = \frac{1}{Z} q^{C(z)} \prod_{z_{ij}=0} (1 - p_{ij}) \prod_{z_{ij}=1} p_{ij}. \tag{2.69}$$

This defines a *generalized bond-percolation model* (the ordinary bond-percolation model corresponds to $q = 1$; also note that (2.69) is a well-defined distribution on $\{0,1\}^{S^*}$ for *any* positive q, not necessarily an integer).

The conditional probabilities

$$P(z|x) = e^{H(x)} \prod_{z_{ij}=0} (1 - p_{ij}) \prod_{z_{ij}=1} p_{ij} \delta_{x_i x_j} \tag{2.70}$$

$$P(x|z) = q^{-C(z)} \prod_{z_{ij}=1} \delta_{x_i x_j} \tag{2.71}$$

play a central role in the SW algorithm. They have the followng important properties:

(a) Given x, $P(z|x)$ says: *independently* for each bond (i, j), set $z_{ij} = 0$ if $x_i \neq x_j$, and set $z_{ij} = 1, 0$ with probability p_{ij}, $1 - p_{ij}$, respectively, if $x_i = x_j$.

(b) Given z, $P(x|z)$ says: *independently* for each (connected) cluster, set *all* the x_is in the cluster to the same value chosen from the uniform distribution on $\Omega_0 = \{1, 2, \ldots, q\}$.

The SW algorithm simulates the joint distribution (2.68) by alternating between $P(z|x)$ and $P(x|z)$ according to two steps:

Step 1   Given a configuration x, generate a bond configuration z according to $P(z|x)$.

Step 2   Using the z of step 1, generate a new configuration x according to $P(x|z)$.

The implementation of Step 1 is trivial, because of property (a). The implementation of Step 2 is also simple, except that one needs an algorithm for computing the converted cluster for a given z; this is a classic problem in computer science, for which there exist fairly efficient algorithms ([86], Section 6).

The above SW procedure induces a transition probability matrix $P = \{p(x, y) : x, y \in \Omega_0^S\}$ given by

$$p(x, y) = \sum_z p(z|x) p(y|z). \tag{2.72}$$

It is easily seen that P is reversible relative to π, irreducible, and aperiodic. This method for sampling from π is an instance of the partial resampling procedures mentioned before the last remark of subsection 2.2.2. Computer implementations [27, 28] show that the SW algorithm is quite efficient, and (for Potts models) appears to have less critical slowing-down than the Metropolis-type algorithms of Section 2.2. See [69, 70, 65] for rigorous bounds on the relaxation time (as $|S| \to +\infty$) of the SW algorithm.

3. SIMULATED ANNEALING

In Section 3.1 we state the three basic problems for which simulated annealing (SA) has been used, and lay out the basic principle of SA. The mathematical framework of SA is based on inhomogeneous Markov chains (MCs) and inhomogeneous diffusion processes; the former are treated in Section 3.2, and the latter in Section 3.4. Section 3.3 contains realizations of SA via Metropolis-type dynamics.

3.1 The problems: global optimization with and without constraints, and simulation with constraints

Let Ω be a state space and $U(x)$ a real-valued function on Ω. Simulated annealing (SA) is a general algorithm for studying the following three problems:

Problem 1 (Global optimization without constraints): Assuming that $U(x)$ is bounded below and that its infimum, $\underline{U} = \inf\{U(x) : x \in \Omega\}$, is attained, find the global minima of U, i.e., determine the set

$$\underline{\Omega} = \{x \in \Omega : U(x) = \underline{U}\}.$$

Problem 2 (Global optimization with constraints): Let $V(x)$ be another function on Ω and assume that the set

$$\Omega(V) = \{x \in \Omega : V(x) = \underline{V}\}$$

where $\underline{V} = \inf\{V(x) : x \in \Omega\}$ is non-empty. Assuming that $U(x)$ is bounded below on $\Omega(V)$ and that its infimum, $\underline{U} = \inf\{U(x) : x \in \Omega(V)\}$, is attained, find the global minima of U on $\Omega(V)$, i.e., determine the set

$$\underline{\Omega}_V = \{x \in \Omega(V) : U(x) = \underline{U}\}.$$

Problem 3 (Simulation with constraints): Let $V(x)$, $\Omega(V)$ be as in Problem 2. Simulate the measure

$$\pi(x) = \frac{1}{Z}\mathbb{1}_{\{x\in\Omega(V)\}}e^{-U(x)} \tag{3.1}$$

assuming that $\pi(x)$ is a well-defined probability density on Ω (concentrated on $\Omega(V)$) with respect to some underlying reference measure. (The general version of Problem 3 is: Suppose that $\pi(x)$ is a probability density on Ω concentrated on some subset Ω' of Ω; design a simulation algorithm for π that handles the constraint $x \in \Omega'$ in an efficient way.)

In Section 3.2 and 3.3 we assume, for simplicity, that Ω is countable (and often we specialize to finite Ω). In Section 3.4, we take Ω to be the Euclidean space $\mathbb{R}^d$ or a subset of $\mathbb{R}^d$. Much of the theory can be extended to general measurable spaces, including Riemannian manifolds [51].

The SA approach to the above three problems is based on the following observations: Consider the probability measure

$$\pi_{T,\lambda}(x) = \frac{1}{Z_{T,\lambda}} \exp\{-\frac{1}{T}[U(x) + \lambda V(x)]\} \qquad (3.2)$$

on Ω, with $T > 0$, $\lambda = 0$ for Problem 1, $T > 0$, $\lambda > 0$, for Problem 2, and $T = 1$, $\lambda > 0$ for Problem 3. The normalizing constant $Z_{T,\lambda}$, the partition function, is such that (3.2) is a probability measure on Ω with respect to some reference measure. The constants T and λ serve simply as *control parameters*; λ plays the role of a Lagrange multiplier, while T plays the role of temperature (see discussion in the Introduction). Note that at least formally (and in fact rigorously under natural assumptions on U, V)

$$\pi_{T,\lambda=0}(x) \longrightarrow \pi(x) \text{ as } T \downarrow 0, \qquad (3.3a)$$

where $\pi(x)$ concentrates on $\underline{\Omega}$ (Problem 1);

$$\pi_{T,\lambda}(x) \longrightarrow \pi(x) \text{ as } T \downarrow 0 \text{ and } \lambda \uparrow +\infty, \qquad (3.3b)$$

where $\pi(x)$ concentrates on $\underline{\Omega}_V$ (Problem 2); and

$$\pi_{1,\lambda}(x) \longrightarrow \pi(x) \text{ as } \lambda \uparrow +\infty, \qquad (3.3c)$$

where $\pi(x)$ is given by (3.1).

In SA the control parameters T and λ are replaced by monotone functions, $T(t)$ and $\lambda(t)$, of time so that as $t \to +\infty : T(t) \downarrow 0$, $\lambda(t) \equiv 0$ for Problem 1; $T(t) \downarrow 0$, $\lambda(t) \nearrow +\infty$ for Problem 2; and $T(t) \equiv 1$, $\lambda(t) \nearrow +\infty$ for Problem 3. Thus (3.2) defines

$$\pi_t(x) = \pi_{T(t),\lambda(t)}(x) \qquad (3.4)$$

which converges as $t \to +\infty$ to the limiting measures in (3.3). The idea, then, is to design (in terms of $\pi_t(x)$) a discrete or continuous time *non-homogeneous* Markov process $\{X(t) : t \geq 0\}$ on Ω, which converges (in some sense, e.g., weakly) as $t \to +\infty$ to the limiting measures in (3.3), or

$$P_\mu\{X(t) \in \underline{\Omega}\} \longrightarrow 1 \text{ as } t \to +\infty \quad \text{(Problem 1)} \qquad (3.5a)$$

$$P_\mu\{X(t) \in \underline{\Omega}_V\} \longrightarrow 1 \text{ as } t \to +\infty \quad \text{(Problem 2)} \qquad (3.5b)$$

for any initial distribution μ.

The Markov process $\{X(t)\}$ is typically chosen to be an inhomogeneous Markov chain or an inhomogeneous diffusion process. The former is constructed from $\pi_t(x)$ via the Metropolis-type dynamics of Section 2.2 (e.g., Gibbs sampler or Metropolis dynamics) or the Swendsen-Wang dynamics or any other dynamics, and the latter via the Langevin equation and its generalizations (see Section 2.3). We will see that the convergence [e.g., weakly or in the sense of (3.5)] of $\{X(t)\}$ requires that the control functions, i.e., the *annealing schedule*, $T(t)$, $\lambda(t)$ must tend to 0, $+\infty$, respectively, as $t \to +\infty$, not too fast.

3.2 General theory of simulated annealing via inhomogeneous Markov chains

An important tool in the study of inhomogeneous Markov chains (MCs) is *Dobrushin's ergodic coefficient* [58, 57, 81] defined as follows: Let Ω be an arbitrary state space and P a transition probability function on Ω with kernel $P(x, dy)$. Dobrushin's ergodic coefficient $\delta(P)$ is defined by

$$\delta(P) = \sup_{\mu \neq \nu} \frac{\|\mu P - \nu P\|}{\|\mu - \nu\|} \qquad (3.6)$$

where the sup is over probability measures μ, ν on Ω, and $\|\cdot\|$ denotes the variational norm, i.e.,

$$\|\mu - \nu\| = \sup_{\|f\|_\infty < +\infty} \frac{|\mu(f) - \nu(f)|}{\|f\|_\infty} \qquad (3.7)$$

where $\|f\|_\infty = \sup\{|f(x)| : x \in \Omega\}$. It is well-known [57] and easily proven that

$$\delta(P) = \tfrac{1}{2}\sup_{x,x'}\|\delta_x P - \delta_{x'} P\|$$

$$= \tfrac{1}{2}\sup_{x,x'}\|P(x, \cdot) - P(x', \cdot)\|$$

where δ_x denotes the Dirac measure concentrated at x. The quantity

$$\alpha(P) = 1 - \delta(P)$$

is often referred to as the *coefficient of ergodicity*. Clearly, $0 \leq \delta(P) \leq 1$ and $0 \leq \alpha(P) \leq 1$. From the definition (3.6) of $\delta(P)$ one easily derives the following two properties to be used heavily later on: For any two probability measures μ, ν, and transition probability kernels P, Q, we have

$$\|\mu P - \nu P\| \leq \delta(P)\|\mu - \nu\|, \qquad (3.8)$$

$$\delta(PQ) \le \delta(P)\delta(Q). \tag{3.9}$$

For countable Ω, (3.7) reduces to

$$\|\mu - \nu\| = \sum_x |\mu(x) - \nu(x)|$$

and (3.6) to

$$\delta(P) \equiv 1 - \alpha(P) = \frac{1}{2}\sup_{x,x'}\sum_y |p(x,y) - p(x',y)| \tag{3.10a}$$

$$= 1 - \inf_{x,x'}\sum_y \min(p(x,y), p(x',y)) \tag{3.10b}$$

where $P = \{p(x,y) : x,y \in \Omega\}$ is a transition probability matrix on Ω. The equivalence of (3.10a) and (3.10b) is verified by using the elementary identity

$$\min(a,b) = \frac{a+b}{2} - \frac{|a-b|}{2}.$$

Remarks:

1) Let μ, ν be probability measures on Ω, and f a bounded real-valued function on Ω; putting $c = \frac{1}{2}(\sup f + \inf f)$ we obtain

$$|\mu(f) - \nu(f)| = |(\mu - \nu)(f - c)| \ \le \|\mu - \nu\| \sup |f - c|$$

$$= \tfrac{1}{2}\|\mu - \nu\|(\sup f - \inf f).$$

Thus for any transition probability kernel P on Ω, we have

$$|(Pf)(x) - (Pf)(x')| \le \delta(P)(\sup f - \inf f). \tag{3.11}$$

In fact, one can show [57, p. 43] that

$$\delta(P) = \sup_f \frac{\sup Pf - \inf Pf}{\sup f - \inf f} \tag{3.12}$$

where $\sup_f$ is over all bounded real-valued functions on Ω. This yields another equivalent definition of $\delta(P)$.

2) The role of $\delta(P)$ in the study of inhomogeneous MCs is analogous to the role of the spectral radius $r(P - \pi)$ (see Section 2.1) in the study of homogeneous MCs. If Ω is finite and P is irreducible, then

$$r(P - \Pi) \le \delta(P). \tag{$*$}$$

Indeed, let λ be an eigenvalue of P with eigenfunction f; then (3.11) yields

$$|\lambda|(\sup f - \inf f) \le \delta(P)(\sup f - \inf f).$$

If $\lambda \neq 1$, then f is *not* a constant, and therefore we obtain $|\lambda| \leq \delta(P)$ for all eigenvalues $\lambda \neq 1$; this implies (*). Furthermore, for finite Ω and irreducible P, one can deduce from (3.12)

$$r(P - \Pi) = \lim_{t \to +\infty} (\delta(P^t))^{\frac{1}{t}}. \qquad (**)$$

We do not know whether (*) or (**) are true for infinite Ω.

Now we turn to a study of inhomogeneous MCs that underlies the theory of simulated annealing. From this point on, we will assume that Ω is countable. Let $\{X_t : t \geq 0\}$ be an inhomogeneous MC with one-step transition probability matrices

$$P_t = \{p_t(x, y) : x, y \in \Omega\}, \quad t \geq 1$$

$$p_t(x, y) = P\{X_t = y | X_{t-1} = x\}, \qquad t \geq 1.$$

Its $(t - (t_0 - 1))$-step transition probabilities will be denoted by

$$P_{t_0,t} = P_{t_0} P_{t_0+1} \cdots P_t = \{p_{t_0,t}(x, y) : x, y \in \Omega\}$$

$$p_{t_0,t}(x, y) = P\{X_t = y | X_{t_0-1} = x\}.$$

We assume that each P_t, $t \geq 1$, has a stationary probability distribution π_t, i.e.,

$$\pi_t P_t = \pi_t, \quad t \geq 1. \qquad (3.13)$$

Theorem 3.1 *If*

$$\sum_{t=1}^{+\infty} \|\pi_t - \pi_{t+1}\| < +\infty \qquad (3.14)$$

and

$$\lim_{t \to +\infty} \delta(P_{t_0,t}) = 0, \quad \text{for every} \ t_0 \geq 1, \qquad (3.15)$$

then there exists a probability measure π on Ω so that

$$\lim_{t \to +\infty} \|\pi_t - \pi\| = 0 \qquad (3.16)$$

and

$$\lim_{t \to +\infty} \|\mu P_{t_0,t} - \pi\| = 0, \quad \text{for every} \ t_0 \geq 1 \qquad (3.17)$$

uniformly in all initial distributions μ on Ω.

Proof: By (3.14), the sequence $\{\pi_t\}$ is Cauchy (in total variation norm); hence π satisfying (3.16) exists. Let $T_0 > t_0$ and write, using (3.8),

$$\|\mu P_{t_0,t} - \pi\| = \|(\mu P_{t_0,T_0-1} - \pi) P_{T_0,t} + \pi P_{T_0,t} - \pi\|$$

$$\leq 2\delta(P_{T_0,t}) + \|\pi P_{T_0,t} - \pi\|.$$

The first term tends to zero as $t \to +\infty$, for *every* $T_0 \geq 1$ (by (3.15)). We rewrite the second term, using (3.13) and (3.8),

$$\|\pi P_{T_0,t} - \pi\| = \left\| (\pi - \pi_{T_0}) P_{T_0,t} + \sum_{k=T_0}^{t-1} (\pi_k - \pi_{k+1}) P_{k+1,t} + \pi_t - \pi \right\|$$

$$\leq \|\pi - \pi_{T_0}\| + \|\pi_t - \pi\| + \sum_{k=T_0}^{t-1} \|\pi_k - \pi_{k+1}\|$$

$$\leq 2 \sup_{t \geq T_0} \|\pi_t - \pi\| + \sum_{t \geq T_0} \|\pi_t - \pi_{t+1}\|.$$

For T_0 sufficiently large, both terms above become arbitrarily small by (3.16) and (3.14), respectively; this yields (3.17).

Remarks:

1) For countable Ω, convergence in total variation implies pointwise convergence; hence $\pi_t(x) \to \pi(x)$ for all $x \in \Omega$. (3.17) implies that $\mu P_{t_0,t}$ converges weakly to π.

2) By Theorem V.3.2 of [58, p.151], (3.15) is equivalent to the existence of an increasing sequence of integers $\{\tau_t : t \geq 1\}$ such that

$$\sum_{t=1}^{+\infty} [1 - \delta(P_{\tau_t, \tau_{t+1}})] \equiv \sum_{t=1}^{+\infty} \alpha(P_{\tau_t, \tau_{t+1}}) = +\infty. \tag{3.18a}$$

A sufficient but *not* necessary condition for (3.15) is

$$\sum_{t=1}^{+\infty} [1 - \delta(P_t)] \equiv \sum_{t=1}^{+\infty} \alpha(P_t) = +\infty \tag{3.18b}$$

3) An inhomogeneous MC for which there exists a probability measure π such that (3.17) holds uniformly in μ and for all $t_0 \geq 1$ is called [58] *strongly ergodic*. An inhomogeneous MC for which (3.15) holds for every $t_0 \geq 1$ is called *weakly ergodic*.

The following variant [92] of Theorem 3.1 will be used in the proof of Theorem 3.3.

Theorem 3.2 *Assume that P_t, π_t satisfy (3.13), and that (3.14) holds. In addition, assume that for **some** sequence of integers $\{\tau_t : t \geq 1\}$, $\tau_t > 0$, we have*

$$\lim_{t \to +\infty} \delta(P_{t,t+\tau_t}) = 0. \tag{3.19}$$

Then there exists a π satisfying (3.16) and (3.17). Furthermore, we have

$$\|\mu P_{t_0,t} - \pi\| \longrightarrow 0 \text{ as } t_0 \to +\infty, \qquad t \geq t_0 + \tau_{t_0}, \tag{3.20}$$

uniformly in μ.

Proof: The existence of π and the validity of (3.16) are obtained as in Theorem 3.1. By (3.9) and (3.19), we have for $t \geq t_0 + \tau_{t_0}$

$$\delta(P_{t_0,t}) \leq \delta(P_{t_0,t_0+\tau_{t_0}}) \longrightarrow 0 \text{ as } t_0 \to +\infty.$$

Hence (3.17) holds by Theorem 3.1. For (3.20), we write

$$\|\mu P_{t_0,t} - \pi\| = \|(\mu - \pi_{t_0})P_{t_0,t} + \sum_{k=t_0}^{t-1}(\pi_k - \pi_{k+1})P_{k+1,t} + \pi_t - \pi\|$$

$$\leq 2\delta(P_{t_0,t}) + \sum_{t \geq t_0}\|\pi_t - \pi_{t+1}\| + \|\pi_t - \pi\|$$

which converges to zero as $t_0 \to +\infty$ (and hence $t \to +\infty$).

The following theorem was proven in [92].

Theorem 3.3 *Assume (3.13) and (3.14). Assume also that (3.19) holds for some sequence $\{\tau_t : t \geq 1\}$, $\tau_t > 0$ such that*

$$\frac{\tau_t}{t} \longrightarrow 0 \text{ as } t \to +\infty. \tag{3.21}$$

Let π be as in Theorem 3.1 and

$$< f >=< f >_\pi= \sum_x \pi(x)f(x).$$

Then for each initial distribution μ and any function f of compact support we have

$$\lim_{n\to+\infty} E_\mu\{(\frac{1}{n}\sum_{t=1}^n f(X_t)- < f >)^2\} = 0. \tag{3.22}$$

In particular, $\overline{f}^{(n)}$ defined by (2.4a) converges in probability to $< f >$.

Proof: We write

$$E_\mu\{(\overline{f}^{(n)}- < f >)^2\} = \frac{1}{n^2}\sum_{t,s=1}^n\{E_\mu(f(X_t)f(X_s)) - (< f >)^2\}$$

$$- 2 < f > \{\frac{1}{n}\sum_{t=1}^n E_\mu(f(X_t))- < f >\}.$$

The second term converges to zero as $n \to +\infty$ by Theorem 3.2 (condition (3.21) is not needed here; we only need (3.19) with $\tau_t > 0$). Let

$$\alpha_{t,s} = \begin{cases} E_\mu\{f(X_t)f(X_s)\} - (< f >)^2, & \text{if } s \geq t \\ \\ \alpha_{s,t}, & \text{if } s < t. \end{cases}$$

Then

$$E_\mu\{(\overline{f}^{(n)} - <f>)^2\} \quad = \frac{1}{n^2} \sum_{t,s=1}^{n} \alpha_{ts}$$

$$-2 <f> \{\frac{1}{n} \sum_{t=1}^{n} E_\mu(f(X_t)) - <f>\}.$$

Note that for $s \geq t + \tau_t$, by Theorem 3.2,

$$\alpha_{t,s} = \sum_x f(x) P_\mu(X_t = x) E_\mu\{f(X_s)|X_t = x\} - (<f>)^2$$

$$\longrightarrow 0 \quad \text{as} \quad t \to +\infty, s \geq t + \tau_t. \tag{3.23}$$

Again, (3.21) is not used here [only (3.19) with $\tau_t > 0$ is used]. If (3.21) holds, then (3.23) implies, by Lemma 3.1 below,

$$\frac{1}{n^2} \sum_{t,s=1}^{n} \alpha_{ts} \longrightarrow 0 \quad \text{as} \quad n \to +\infty.$$

Lemma 3.1 [92] *Let $\{\alpha_{n,m} : n, m \geq 1\}$ be a real bounded double-sequence with $\alpha_{nm} = \alpha_{mn}$, and*

$$\alpha_{n,m} \longrightarrow 0 \quad \text{as} \quad n \to +\infty, \quad m \geq n + \tau_n$$

for a sequence of integers $\{\tau_n : n \geq 1\}$ satisfying

$$\lim_{n \to +\infty} \frac{\tau_n}{n} = 0.$$

Then

$$\frac{1}{N^2} \sum_{n,m=1}^{N} \alpha_{n,m} \longrightarrow 0 \quad \text{as} \quad N \to +\infty.$$

Proof: Fix $\varepsilon > 0$, and choose M so that

$$|\alpha_{nm}| \leq \varepsilon \quad \text{for all} \quad n \geq M, \quad m \geq n + \tau_n.$$

Then using the symmetry $\alpha_{nm} = \alpha_{mn}$, write

$$\frac{1}{N^2} \sum_{n,m=1}^{N} \alpha_{nm} \quad = \frac{2}{N^2} \sum_{n=M}^{N-1} \sum_{m=n+\tau_n}^{N} \alpha_{nm} + \frac{2}{N^2} \sum_{n=M}^{N-1} \sum_{m=n+1}^{n+\tau_n-1} \alpha_{nm}$$

$$+ \frac{2}{N^2} \sum_{n=1}^{M-1} \sum_{m=n+1}^{N} \alpha_{nm} + \frac{2}{N^2} \sum_{n=1}^{N} \alpha_{nn}. \tag{*}$$

The first term in (*) is bounded above in absolute value by ε. The number of indices in the second and third terms of (*) (including the factor 2) is

$2(\tau_M + \tau_{M+1} + \cdots + \tau_{N-1} - N)$ and $(2N - M)(M - 1)$, respectively. Thus the number of indices on the left-hand side of (*) with $|\alpha_{nm}| \geq \varepsilon$ is at most $2NM + 2\sum_{n=1}^{N}\tau_n$. Thus, if $\alpha = \max|\alpha_{nm}|$, then

$$|\tfrac{1}{N^2}\sum_{n,m=1}^{N}\alpha_{nm}| \ \leq \varepsilon + 2\alpha\tfrac{M}{N} + 2\alpha\tfrac{1}{N^2}\sum_{n=1}^{N}\tau_n$$

$$\leq \varepsilon + 2\alpha\tfrac{M}{N} + 2\alpha\tfrac{1}{N}\sum_{n=1}^{N}\frac{\tau_n}{n}.$$

Since $\tau_n/n \to 0$ as $n \to +\infty$, its Cesaro means also converges to zero. This together, with the fact that ε is arbitrary, yields the lemma.

Remark: Reference [57, Section 1.2.3] contains results related to Theorem 3.3 (*the weak law of large numbers*), as well as results on the *strong law of large numbers*.

3.2.1 Continuous-Time Inhomogeneous Markov Chains

Let $\{X_t : t \geq 0\}$ be a continuous-time inhomogeneous Markov chain (MC) on a countable state space Ω, with *transition rate matrix* $\mathcal{L}(t)$ defined by

$$I - P(t, t + h) = -h\mathcal{L}(t) + o(h)$$

where

$$P(s,t) = \{p(s, x; t, y) : x, y \in \Omega\} \ , \ t \geq s$$
$$p(s, x; t, y) = P\{X_t = y | X_s = x\} \ , \ t \geq s.$$

$P(s, t)$ satisfies the *forward equation*

$$\tfrac{\partial}{\partial t}P(s, t) = -P(s, t)\mathcal{L}(t), \quad t \geq s$$

$$P(s, s) = 1,$$

(3.24a)

as well as the *backward equation*

$$\tfrac{\partial}{\partial s}P(s, t) = \mathcal{L}(s)P(s, t) \ , \ s \leq t$$

$$P(t, t) = 1.$$

(3.24b)

We will write $P(t)$ for $P(0, t)$, and $p(t; x, y)$ for $p(0, x; t, y)$. If μ is an initial distribution, then the distribution at time t is $p(t) = \mu P(t) = \{p(t, x) : x \in \Omega\}$, and satisfies

$$\frac{dp(t)}{dt} = -p(t)\mathcal{L}(t)$$

(3.24c)

$$p(0) = \mu.$$

We will assume that $\mathcal{L}(t)$ is bounded so that (3.24a) has a solution satisfying

$$\sum_y p(s, x; t, y) = 1$$

for all $t \geq s$ and all $x \in \Omega$. Then $\mathcal{L}(t) = \{\mathcal{L}(t; x, y) : x, y \in \Omega\}$ has the property

$$\sum_y \mathcal{L}(t; x, y) = 0 \qquad (3.25)$$

for all $t \geq 0$ and all $x \in \Omega$. Note that $\mathcal{L}(t; x, y) \leq 0$ for $x \neq y$, and $\mathcal{L}(t; x, x) \geq 0$.

We will assume that for each $t \geq 0$ there exists a probability distribution $\pi(t) = \{\pi(t; x) : x \in \Omega\}$ on Ω so that

$$\pi(t)\mathcal{L}(t) = 0. \qquad (3.26)$$

This is the analogue of (3.13) for continuous-time MCs. The Hilbert space $\ell_2(\pi(t))$ is defined as in Section 2.1.; its inner product will be denoted by $< \cdot, \cdot >_{\pi(t)}$, and the norm by $\| \cdot \|_{\pi(t)}$. The matrix $\mathcal{L}(t)$ acts naturally on functions according to

$$(\mathcal{L}(t)f)(x) = \sum_y \mathcal{L}(t; x, y)f(y).$$

It is easily verified that as an operator on $\ell_2(\pi(t))$, $\mathcal{L}(t)$ is self-adjoint iff it satisfies the *reversibility* or *detailed-balance* condition [see (2.19)]

$$\pi(t; x)\mathcal{L}(t; x, y) = \pi(t; y)\mathcal{L}(t; y, x), \quad \text{for all } x, y \in \Omega.$$

If $\mathcal{L}(t)$ is self-adjoint on $\ell_2(\pi(t))$, then it is easily seen that for $f, g \in \ell_2(\pi(t))$,

$$< f, \mathcal{L}(t)g >_{\pi(t)} = -\frac{1}{2}\sum_{x,y} \pi(t, x)\mathcal{L}(t; x, y)[f^*(x) - f^*(y)][g(x) - g(y)]. \quad (3.27a)$$

In particular

$$< f, \mathcal{L}(t)f >_{\pi(t)} = -\frac{1}{2}\sum_{x,y} \pi(t, x)\mathcal{L}(t; x, y)|f(x) - f(y)|^2. \qquad (3.27b)$$

Hence $\mathcal{L}(t)$ is a non-negative operator on $\ell_2(\pi(t))$. If $\mathcal{L}(t)$ is not self-adjoint on $\ell_2(\pi(t))$, then neither (3.27a) nor (3.27b) holds; but (3.27b) holds if $f \in \ell_2(\pi(t))$ is real (this property will be used below).

We will be interested in the behavior of $\mu P(t) - \pi(t)$ as $t \to +\infty$, or if $\pi(t) \longrightarrow \pi$ as $t \to +\infty$, in the convergence of $\mu P(t)$ to π as $t \to +\infty$. These will be studied either in terms of the ℓ_2-*distance* defined in (2.9a), or in terms of the *variation norm* defined in (2.10a) (in this section the variation norm is denoted by $\|\cdot\|$). One may also consider the convergence of the long-time averages

$$\overline{f}^{(t)} = \frac{1}{t} \int_0^t f(X_s)\,ds,$$

but we shall not treat this subject here.

By (3.25), zero is an eigenvalue of $\mathcal{L}(t)$ with corresponding (normalized) eigenfunction $\phi_1(x) \equiv 1$ for all $x \in \Omega$. From now on we assume that $\lambda_1 = 0$ is a *simple* eigenvalue of $\mathcal{L}(t)$ (hence $\pi(t)$ is the unique left-eigenvector with eigenvalue zero). As in Section 2.1, we define the projection operator

$$(\Pi(t)f)(x) = <f>_{\pi(t)}$$

onto the eigenspace corresponding to $\lambda_1 = 0$ (i.e., onto the constant functions). Let $\sigma(\mathcal{L}(t) - \Pi(t))$ denote the spectrum of $\mathcal{L}(t) - \Pi(t)$. If $\mathcal{L}(t)$ is self-adjoint on $\ell_2(\pi(t))$, its *mass-gap* (compare with (2.26)) is defined by

$$\lambda_2(t) = \inf \sigma(\mathcal{L}(t) - \Pi(t)). \qquad (3.28)$$

Since $\mathcal{L}(t)$ is self-adjoint and non-negative, $\lambda_2(t) \geq 0$. If $\mathcal{L}(t)$ is not self-adjoint, then $\sigma(\mathcal{L}(t) - \Pi(t))$ lies in the complex plane; it is well-known [76, p.220] that $\sigma(\mathcal{L}(t) - \Pi(t))$ is contained in the *numerical range* of $\mathcal{L}(t) - \Pi(t)$, i.e., in the set

$$\Theta(t) = \{<f, \mathcal{L}(t)f>_{\pi(t)}: \|f\|_{\pi(t)} = 1, \ <f>_{\pi(t)} = 0\}. \qquad (3.29a)$$

Let

$$\Theta_{\text{real}}(t) = \{<f, \mathcal{L}(t)f>_{\pi(t)}: \|f\|_{\pi(t)} = 1, \ <f>_{\pi(t)} = 0, \ f \ \text{real}\} \qquad (3.29b)$$

be the *real numerical range* of $\mathcal{L}(t) - \Pi(t)$ [a subset of $\Theta(t)$], and define

$$\rho(t) = \inf\{<f, [\mathcal{L}(t) - \Pi(t)]f>: \|f\|_{\pi(t)} = 1, \ f \ \text{real}\}. \qquad (3.30)$$

Since (3.27b) holds for non-self-adjoint $\mathcal{L}(t)$ provided that f is real, we have $\rho(t) \geq 0$. Note that for finite Ω we have $\lambda_2(t) > 0$, $\rho(t) > 0$ (since $\mathcal{L}(t)$ is assumed to be bounded). We have not been able to find an explicit characterization of $\rho(t)$ in terms of the spectrum of $\mathcal{L}(t)$; but we suspect that $\rho(t)$ is related to the quantity $\beta(t)$ defined at the end of this subsection.

Next we prove two theorems that will be used in Section 3.3. The first was proven in [39] and the second in [59].

Theorem 3.4 *Consider a continuous-time, inhomogeneous MC with transition rate matrix $\mathcal{L}(t)$ and initial distribution μ. Assume:*

a) *For each $t \geq 0$, there exists a probability distribution $\pi(t)$ on Ω satisfying (3.26), and*

$$|\frac{\partial}{\partial t}\pi(t, x)| \leq \gamma(t)\pi(t, x), \quad \text{for all } x \in \Omega \qquad (3.31)$$

with a positive function $\gamma(t)$ independent of x.

b) $\lambda_1 = 0$ is a *simple* eigenvalue of $\mathcal{L}(t)$.

c)

$$\int_0^{+\infty} \lambda(t)dt = +\infty, \tag{3.32}$$

$$\frac{\gamma(t)}{\lambda(t)} \to 0 \text{ as } t \to +\infty, \tag{3.33}$$

where $\lambda(t)$ denotes $\lambda_2(t)$ if $\mathcal{L}(t)$ is self-adjoint, or $\rho(t)$ defined by (3.30) if $\mathcal{L}(t)$ is not self-adjoint.

Then

$$\left\|\frac{\mu P(t)}{\pi(t)} - 1\right\|_{\pi(t)} \longrightarrow 0 \text{ as } t \to +\infty. \tag{3.34}$$

In particular $\mu P(t) - \pi(t)$ converges, as $t \to +\infty$, to zero in the ℓ_2-distance and the total variation norm. Furthermore, if there exists a probability distribution π on Ω so that $\pi(t)$ converges to π in total variation (or ℓ_2-distance), then so does $\mu P(t)$.

Proof: Let

$$F(t) = \left\|\frac{\mu P(t)}{\pi(t)} - 1\right\|_{\pi(t)}^2 = \sum_x (q(t,x) - 1)^2 \pi(t,x)$$

where $q(t) = \{q(t,x) : x \in \Omega\}$ is defined by $(\mu P(t))(x) = p(t,x) = \pi(t,x)q(t,x)$. Using (3.24c), one easily derives

$$\frac{1}{2}\frac{dF(t)}{dt} = - < q(t) - \phi_1, \mathcal{L}(t)[q(t) - \phi_1] >_{\pi(t)} -\frac{1}{2}\sum_x (q(t,x))^2 \frac{\partial \pi(t,x)}{\partial t}$$

where $\phi_1(x) \equiv 1$. By the definition of $\lambda(t)$ we obtain

$$\frac{1}{2}\frac{dF(t)}{dt} \leq -\lambda(t)F(t) + \frac{1}{2}\gamma(t)\sum_x (q(t,x))^2 \pi(t,x)$$

$$\leq -\lambda(t)F(t) + \frac{1}{2}\gamma(t)F(t) + \gamma(t).$$

Then (3.32) and (3.33) imply (3.34).

Theorem 3.5 *Consider a continuous-time, inhomogeneous MC with transition rate matrix $\mathcal{L}(t)$ and initial distribution μ. Assume:*

a) *For each $t \geq 0$ there exists a probability distribution $\pi(t)$ satisfying (3.26) and*

$$\int_0^{+\infty} \left\|\frac{d\pi(t)}{dt}\right\|dt < +\infty. \tag{3.35}$$

b) For every $t_0 \geq 1$

$$\delta(P(t_0, t)) \longrightarrow 0 \quad \text{as} \quad t \to +\infty. \tag{3.36}$$

Then

$$\|\mu P(t) - \pi(t)\| \longrightarrow 0 \quad \text{as} \quad t \to +\infty. \tag{3.37}$$

Furthermore, if there exists a π such that $\|\pi(t) - \pi\| \to 0$ as $t \to +\infty$, then $\|\mu P(t) - \pi\| \longrightarrow 0$ as $t \to +\infty$.

Proof: Let $t_0 > 0$ and write

$$\|\mu P(t) - \pi(t)\| = \|[\mu P(t_0) - \pi(t_0)]P(t_0, t) + \pi(t_0)P(t_0, t) - \pi(t)\|$$

$$\leq 2\delta(P(t_0, t)) + \|\pi(t_0)P(t_0, t) - \pi(t)\|. \tag{*}$$

Note that

$$\|\pi(t) - \pi(t_0)P(t_0, t)\| = \|\int_{t_0}^{t} \frac{\partial}{\partial s}[\pi(s)P(s, t)]ds\|$$

$$= \|\int_{t_0}^{t} [\frac{d\pi(s)}{ds}]P(s, t)ds + \int_{t_0}^{t} \pi(s)\mathcal{L}(s)P(s, t)ds\|$$

$$= \|\int_{t_0}^{t} [\frac{d\pi(s)}{ds}]P(s, t)ds\|,$$

where we have used the backward equation (3.24b) and (3.26). Since $\sum_j \frac{d}{ds}\pi_j(s) = 0$, we obtain

$$\|\pi(t_0)P(t_0, t) - \pi(t)\| \leq \int_{t_0}^{t} \|\frac{d\pi(s)}{ds}\|ds \leq \int_{t_0}^{+\infty} \|\frac{d\pi(s)}{ds}\|ds.$$

This together with (*) yields (3.37). The last part of the theorem is then easily derived.

Remarks:

1) A sufficient condition for the existence of a limiting π is [compare with (3.14)]

$$\int_{0}^{+\infty} \|\pi(t) - \pi(t+1)\|dt < +\infty. \tag{3.38}$$

2) In the applications to the problems mentioned in Section 3.1, both (3.31) and (3.35) hold (see next subsection).

3) Condition (3.36) of Theorem 3.5 [which is similar to (3.15)] corresponds to condition (3.32) of Theorem 3.2; the two conditions are similar in spirit, but we are not sure of their precise relationship [see Remark 2 below (3.12)]. It has been proven in [42] that (3.36) is implied by the following condition: Let

$$\beta(t) = \inf_{x \neq x'} \{-\mathcal{L}(t; x, x') - \mathcal{L}(t; x', x) + \sum_{y \neq x, x'} \min[-\mathcal{L}(t; x, y), -\mathcal{L}(t; x', y)]\}.$$

(3.39a)

Then (3.36) holds if

$$\lim_{t_0 \to +\infty} \int_{t_0}^{+\infty} \beta(t) dt = +\infty.$$

(3.39b)

This may be derived from the analogue of (3.18a) for continuous-time MCs.

3.3 Simulated annealing via Metropolis-type dynamics

In this section we use the general theory of inhomogeneous Markov Chains (MCs) developed in Section 3.2 to study Problems 1, 2 and 3 stated in Section 3.1. Throughout this section, we assume that Ω is finite, and specialize primarily to the case when Ω is of the form (2.37) with Ω_0 and S finite. For the sake of unifying the study of Problems 1, 2, and 3, we introduce the following general setting of [34]:

For discrete or continuous time $t \geq 1$, let $H_t(x)$ be a time-dependent energy (Hamiltonian) function on Ω satisfying

- For every $x \in \Omega$, $\{H_t(x) : t \geq 1\}$ is eventually increasing, and   (3.40a)

- There exists an $x \in \Omega$ such that $\sup_{t \geq 1} H_t(x) < +\infty$,   (3.40b)

and define the Gibbs distribution

$$\pi_t(x) = \pi(t, x) = \frac{1}{Z_t} e^{-H_t(x)}, \ Z_t = \sum_x e^{-H_t(x)}.$$

(3.41)

For Problems 1, 2, and 3, $\pi(t, x)$ is given by (3.4), and one easily shows that (3.40) holds.

Lemma 3.2 *If $\{H_t(x) : x \in \Omega, t \geq 1\}$ satisfies (3.40) then (3.14) (discrete time) and (3.38) (continuous time) hold. In particular, there exists a limiting distribution π (in the sense of total variation norm, and hence pointwise).*

Proof: Set $h_t(x) = \exp\{-H_t(x)\}$, and write

$$|\pi(t, x) - \pi(t + 1, x)| = \frac{1}{Z_t Z_{t+1}} |Z_{t+1} h_t(x) - Z_t h_{t+1}(x)|$$

$$\leq (\inf_t Z_t)^{-1}\{(\sup_t Z_t)|h_t(x) - h_{t+1}(x)| + (\sup_t h_t(x))|Z_t - Z_{t+1}|\}. \qquad (*)$$

Note that $\inf_t Z_t > 0$ by (3.40b) and $\sup_t Z_t < +\infty$ by (3.40a). Also $\{h_t(x) : t \geq 1\}$ and $\{Z_t : t \geq 1\}$ are strictly positive, and eventually decreasing by (3.40a). These properties and (*) implies (3.38).

For fixed t, $\pi_t(x)$ is a Gibbs distribution; hence we may use one of the strategies of Section 2.2 to construct a transition probability matrix P_t such that

$$\pi_t P_t = \pi_t. \qquad (3.42)$$

For *discrete time*, $\{P_t : t \geq 1\}$ defines an inhomogeneous discrete-time MC as in subsection 3.2.1. For *continuous time* t, we define a continuous-time inhomogeneous MC with generator

$$\mathcal{L}(t) = I - P_t. \qquad (3.43)$$

If Ω has no particular structure [e.g., is not of the form (2.37)], then P_t is constructed via the Metropolis-Hastings dynamics (2.29)–(2.30); the most commonly used dynamics are the Metropolis [see (2.35)]

$$p_t(x, y) = q(x, y)e^{-[H_t(y) - H_t(x)]^+}, \quad \text{for } x \neq y, \qquad (3.44)$$

and the Barker dynamics

$$p_t(x, y) = q(x, y)\frac{\exp\{-[H_t(y) - H_t(x)]\}}{1 + \exp\{-[H_t(y) - H_t(x)]\}}, \quad \text{for } x \neq y, \qquad (3.45)$$

with $p_t(x, x)$ defined by (2.29a). In both cases, $Q = \{q(x, y)\}$ is chosen to be irreducible and *symmetric*; however, from the point of view of global optimization, one may still use (3.44) and (3.45) with a Q which is *not* symmetric but satisfies (2.33). In this case, (3.42) does not hold, but it does hold if π_t is replaced by

$$\tilde{\pi}_t(x) = \frac{1}{\tilde{Z}_t}\alpha(x)e^{-H_t(x)}, \quad \tilde{Z}_t = \sum_x \alpha(x)e^{-H_t(x)}. \qquad (3.46)$$

For Problems 1 and 2, it makes no difference whether we use $\pi_t(x)$ [see (3.4)] or $\tilde{\pi}_t(x)$.

If Ω has the form (2.37), then we may use *multiple-site* (or *single-site*) *random* or *sequential* Metropolis-type dynamics [see (2.42), (2.43)]. The most commonly used dynamics are the random Metropolis or Gibbs sampler, the sequential Metropolis or Gibbs sampler, and the sequential Metropolis or Gibbs sampler corresponding to the *visitation scheme* (2.44); in the latter case, condition (2.44), which suffices for simulation (π_t independent of t), needs to be modified for t-dependent π_t (see subsection 3.3.1 below).

The study of *all* the above dynamics [including the Metropolis-Hastings generalization of (3.44) and (3.45)] is quite similar. In subsection 3.3.1 we treat (via Theorem 3.1–3.3) the discrete-time sequential Gibbs sampler corresponding to a visitation scheme of the form (2.44), while in subsection 3.3.2 we treat the continuous-time Metropolis dynamics (3.44) via Theorem 3.4.

3.3.1 Discrete-Time Sequential Gibbs Sampler

Here we assume $\Omega = \Omega_0^S$ with Ω_0 and S finite and consider a visitation scheme of the form (2.44). We require that the visitation scheme covers all of S from time to time, i.e., we assume

$$S = A_{\tau_{t-1}+1} \cup A_{\tau_{t-1}+2} \cup \cdots \cup A_{\tau_t} \quad \text{for each } t \geq 1, \tag{3.47}$$

for a sequence of "recurrence times" $0 = \tau_0 < \tau_1 < \tau_2 < \cdots$. We denote by

$$P_{\tau_{t-1}+j} = \{p_{\tau_{t-1}+j}(x,y)\}, \quad t \geq 1, \quad 1 \leq j \leq \tau_t - \tau_{t-1}$$

the matrix (2.46) corresponding to the set $A_{\tau_{t-1}+j}$. Then, the $(t-(t_0-1))$-step transition probability matrix for the sequential Gibbs sampler is

$$P_{t_0,t} = P_{t_0} P_{t_0+1} \cdots P_t, \quad t_0 \geq 1, \quad t > t_0.$$

Note that for each $t \geq 1$, there exist integers $k_t \geq 1$ and $1 \leq \ell_t \leq \tau_{k_t} - \tau_{k_t-1}$ so that $t = \tau_{k_t} + \ell_t$. The *one-full sweep* matrix

$$R_t = P_{\tau_{t-1}+1} \cdots P_{\tau_t}, \quad t \geq 1 \tag{3.48}$$

plays a useful role below.

Let $\{H_t(x) : x \in \Omega\}$, $t \geq 1$, be an energy function satisfying (3.40), and define

$$d(\tau_{t-1} + j) = \sup\{|H_{\tau_{t-1}+j}(x) - H_{\tau_{t-1}+j}(y)| : y_{A^c_{\tau_{t-1}+j}} = x_{A^c_{\tau_{t-1}+j}}\}$$

and

$$\Delta_t = \max\{d(\tau_{t-1} + j) : 1 \leq j \leq \tau_t - \tau_{t-1}\}, \quad t \geq 1.$$

Lemma 3.3 *There exists a constant $C > 0$ (independent of t) such that*

$$\delta(R_t) \equiv 1 - \alpha(R_t) \leq 1 - Ce^{-|S|\Delta_t} \tag{3.49}$$

for every $t \geq 1$.

Proof of Lemma 3.3: We will show (3.49) with $C = |\Omega|^{-|S|+1}$. From (3.10b) we see that

$$\alpha(R_t) \geq |\Omega| \min_{x,x'}(R_t(x,y), R_t(x',y)). \tag{$*$}$$

We will estimate $R_t(x, y)$, $x, y \in \Omega$, from below. First fix $k, \tau_{t-1} + 1 \leq k \leq \tau_t$ and set

$$E_k = E_k(x_{A_k^c}) = \min\{H_k(y) : y_{A_k^c} = x_{A_k^c}\}.$$

Then, using the notation introduced above (2.45), we have

$$\pi_k(y_{A_k}|x_{A_k^c}) = \frac{\exp\{-H_k(y_{A_k}x_{A_k^c})\}}{\sum_{z_{A_k}} \exp\{-H_k(z_{A_k}x_{A_k^c})\}}$$

$$= \frac{\exp\{-[H_k(y_{A_k}x_{A_k^c}) - E_k]\}}{\sum_{z_{A_k}} \exp\{-[H_k(z_{A_k}x_{A_k^c}) - E_k]\}}$$

$$\geq |\Omega_0|^{-|A_k|}e^{-d(k)} \geq |\Omega|^{-1}e^{-\Delta_t}. \tag{$**$}$$

Next, we use this to estimate $R_t(x, y)$. To this end, we construct the following sets and "recurrence times". Set

$$B_1 = A_{\tau_t}, \qquad k_1 = \tau_t,$$

and define recursively

$$k_{\ell+1} = \max\{j : \tau_{t-1} + 1 \leq j \leq k_\ell - 1, A_j \backslash \cup_{m=1}^\ell B_\ell \neq \phi\},$$

$$B_{\ell+1} = A_{k_{\ell+1}} \backslash \cup_{m=1}^\ell B_\ell.$$

This procedure generates at *most* $|S|$ non-empty, disjoint subsets $B_1, B_2, \ldots, B_\nu$, $\nu \leq |S|$, satisfying

$$S = B_1 \cup B_2 \cup \cdots \cup B_\nu.$$

Clearly, B_ℓ is the set of sites that are visited *last* at time k_ℓ. Then, denoting by $X_t^{(i)}$ the component of X_t at site $i \in S$, and using the Markov property, we get

$$R_t(x, y) = P\{X_{\tau_t} = y | X_{\tau_{t-1}} = x\}$$

$$= \sum_z P\{X_{k_\nu - 1} = z | X_{\tau_{t-1}} = x\} P\{X_{k_\ell}^{(i)} = y_i, i \in B_\ell, 1 \leq \ell \leq \nu | X_{k_\nu - 1} = z\}$$

$$= \sum_z P\{X_{k_\nu - 1} = z | X_{\tau_{t-1}} = x\} \times \prod_{\ell=1}^{\nu} P\{X_{k_\ell}^{(i)} = y_i, i \in B_\ell |$$

$$X_{k_m}^{(j)}, j \in B_m, \ell < m \leq \nu, X_{k_\nu - 1}^{(j)} = z_j, j \in B_n, 1 \leq n \leq \ell\}$$

$$= \sum_z P\{X_{k_\nu - 1} = z | X_{\tau_{t-1}} = x\} \times \prod_{\ell=1}^{\nu} P\{X_{k_\ell}^{(i)} = y_i, i \in B_\ell |$$

$$X_{k_m}^{(j)} = y_j, j \in B_m, \ell < m \leq \nu, X_{k_\ell}^{(j)} = z_j, j \in B_n, 1 \leq n \leq \ell\}$$

$$\geq \sum_z P\{X_{k_\nu - 1} = z | X_{\tau_{t-1}} = x\} |\Omega|^{-|S|} e^{-|S| \Delta t}$$

$$= |\Omega|^{-|S|} e^{-|S| \Delta t},$$

where we have used (**) in the inequality; this together with (*) yields (3.49).

Theorem 3.6 *Suppose that $\{H_t : t \geq 1\}$ satisfies (3.40), and consider the visitation scheme (3.47). Let π be the limiting probability distribution of Lemma 3.2. Then:*
  a) If

$$\sum_{t=1}^{+\infty} e^{-|S| \Delta t} = +\infty \tag{3.50}$$

then

$$\|\mu P_{1,t} - \pi\| \longrightarrow 0 \text{ as } t \to +\infty$$

uniformly in μ.

  b) If $\sup_t \{\tau_t - \tau_{t-1}\} < +\infty$ and

$$t \exp\{-|S| \max_{k \leq t} \Delta_k\} \longrightarrow +\infty \text{ as } t \to +\infty, \tag{3.51}$$

then for any initial distribution μ and any (bounded) function f on Ω,

$$\lim_{n \to +\infty} E_\mu \{ (\frac{1}{n} \sum_{t-1}^{n} f(x_t) - <f>_\pi)^2 \} = 0.$$

In particular, $\overline{f}^{(n)}$ converges to $<f>_\pi$ in probability.

Proof:

a) It suffices to verify condition (3.15) of Theorem 3.1. Note that $1 \leq t_0 < t$, there exist k_{t_0} and k_t so that

$$1 \leq t_0 \leq \tau_{k_{t_0}} < \tau_{k_t} \leq t.$$

Thus

$$\delta(P_{t_0,t}) \leq \prod_{j=k_{t_0}}^{k_t} \delta(R_j) \leq \prod_{j=k_{t_0}}^{k_t} [1 - Ce^{-|S|\Delta_j}], \tag{*}$$

which converges to zero as $t \to +\infty$, by (3.50).

b) It suffices to verify condition (3.19) of Theorem 3.2 and condition (3.21) of Theorem 3.3, with some new sequence τ_t'. Since $\tau_t - \tau_{t-1}$ is bounded, it suffices [as in (*)] to construct a $\{\tau_t'\}$ so that (3.21) and

$$\delta(R_t \cdots R_{t+\tau_t'}) \longrightarrow 0, \quad \text{as } t \to +\infty, \tag{**}$$

hold. Following [92], we define

$$\gamma_t = 1 - C \exp\{-|S| \max_{k \leq t} \Delta_k\}. \tag{3.52}$$

Clearly $0 < \gamma_t < 1$, $\{\gamma_t\}$ is increasing, $\delta(Q_t) \leq \gamma_t$, and by (3.51)

$$t(1 - \gamma_t) \longrightarrow +\infty \text{ as } t \to +\infty.$$

Let

$$\rho(t) = \inf_{s \geq t} s(1 - \gamma_s)$$

and $\tau_t' =$ the least integer greater than $t(\rho(t))^{-\frac{1}{2}}$. Then

$$\frac{\tau_t'}{t} \leq (\rho(t))^{-\frac{1}{2}} + \frac{1}{t} \longrightarrow 0 \text{ as } t \to +\infty.$$

Furthermore,

$$(\tau_t' + 1)(1 - \gamma_{t+\tau_t'}) \geq \frac{(\tau_t' + 1)\rho(t + \tau_t')}{t + \tau_t'} \geq \frac{\tau_t'\rho(t)}{t + \tau_t'} \geq \frac{t(\rho(t))^{\frac{1}{2}}}{t + \tau_t'}$$

which converges to $+\infty$ as $t \to +\infty$. Consequently

$$(\tau_t' + 1) \log \gamma_{t+\tau_t'} \longrightarrow -\infty \text{ as } t \to +\infty.$$

Hence

$$(\gamma_{t+\tau_t'})^{\tau_t'+1} \longrightarrow 0 \text{ as } t \rightarrow +\infty,$$

and by monotonicity

$$\gamma_t \cdots \gamma_{t+\tau_t'} \longrightarrow 0 \text{ as } t \rightarrow +\infty,$$

which proves (**), since $\delta(R_t \cdots R_{t+\tau_t'}) \leq \gamma_t \cdots \gamma_{t+\tau_t'}$.

Remarks:

1) Theorem 3.6 was proven in [92]; part a) is a slight generalization of a result in [34]. Gantert has observed [31] that Theorem 1.2.23 of [57, p.56] implies the strong law of large numbers for the sequential process generated by the transition matrices $\{R_t\}$ defined in (3.48), provided that

$$\sum_{t=1}^{+\infty} [t(1-\gamma_t)]^{-2} < +\infty, \tag{3.53}$$

where γ_t is defined by (3.52).

2) As we mentioned before, the procedure of this subsection can easily be adapted to other sequential or random dynamics.

Next we apply Theorem 3.6 to Problems 1, 2, and 3, of subsection 3.1:

A) Global optimization without constraints: Let $U(x), \underline{U}$, and $\underline{\Omega}$ be as in Problem 1, Section 3.1. Then

$$H_t(x) = \frac{1}{T(t)}[U(x) - \underline{U}], \tag{3.54}$$

with $T(t)$ positive and decreasing to zero, satisfies (3.40a) for all $x \in \Omega$ and (3.40b) at the minima of $U(x)$. The limiting distribution π defined by (3.3a) is the uniform distribution on $\underline{\Omega}$. Let

$$\Delta = \max\{\tilde{d}(\tau_{t-1} + j) : t \geq 1, 1 \leq j \leq \tau_t - \tau_{t-1}\},$$

$$\tilde{d}(\tau_{t-1} + j) = \max\{|U(x) - U(y)| : x_{A^c_{\tau_{t-1}+j}} = y_{A^c_{\tau_{t-1}+j}}\}.$$

Then the Δ_t of (3.49) satisfies

$$\Delta_t \leq (T(\tau_t))^{-1}\Delta.$$

Hence if

$$\sum_{t=1}^{+\infty} \exp\{-\frac{|S|\Delta}{T(\tau_t)}\} = +\infty, \tag{3.55a}$$

then part a) of Theorem 3.6 and, consequently, (3.5a) hold. Furthermore, if Y_t denotes the mean number of visits of the chain $\{X_t\}$ in $\underline{\Omega}$ up to time t, i.e.,

$$Y_t = \frac{1}{t} \sum_{n=1}^{t} \mathbb{1}_{\{X_t \in \underline{\Omega}\}},$$

then $E_\mu(Y_t) \longrightarrow 1$ as $t \to +\infty$. This shows that Y_t converges in probability to 1. Choosing an a.s. convergent subsequence, we conclude that a.s. $\{X_t\}$ cofinally stays in $\underline{\Omega}$. Clearly (3.55a) holds if

$$T(\tau_t) \geq \frac{|S|\Delta}{\log t}, \quad \text{for } t \geq t_0 \tag{3.55b}$$

with some $t_0 > 1$.

If

$$T(\tau_t) \geq \frac{(1+\varepsilon)|S|\Delta}{\log t}, \quad \text{for } t \geq t_0, \text{ some } t_0 > 1,$$

then (3.51) holds. Thus if $f(x) = \delta_{x,x_0}$ for a fixed $x_0 \in \underline{\Omega}$, then part b) of Theorem 3.6 implies that

$$E_\mu\{(\frac{1}{n} \sum_{t=1}^{n} \mathbb{1}_{\{X_t=x_0\}} - \frac{1}{|\Omega|})^2\} \longrightarrow 0 \text{ as } n \longrightarrow +\infty,$$

which means that the mean number of visits to x_0 up to time n converges in probability to $|\Omega|^{-1}$. Choosing again an a.s. convergent subsequence, we see that the process $\{X_t\}$ visits each minimum of $U(x)$ infinitely often.

Remarks:

1) Condition (3.55a), or the logarithmic schedule (3.55b), is typical of simulated annealing via any of the standard dynamics (e.g., sequential or random Metropolis or Gibbs sampler). But the constant $|S|\Delta$ can be replaced by a smaller constant. For the Metropolis dynamics (3.44) (which includes the multiple-site random Metropolis), Hajek has proven [44] the following necessary and sufficient condition for convergence in the sense of (3.5a): Assume that $Q = \{q(x,y)\}$ has the *strong irreducibility* property stated above (2.36). We say that a state $x \in \Omega$ is *reachable at height* E from state y if there is a sequence of states $x_1 = x, x_2, \ldots, x_{n-1}, x_n = y$ such that

$$q(x_\ell, x_{\ell+1}) > 0, \quad 1 \leq \ell \leq n-1$$

and

$$U(x_\ell) \leq E \text{ for } 1 \leq \ell \leq n.$$

Assume that Q is *weakly reversible* in the sense that for any real E and any two states x and y, x is reachable at height E from y iff y is reachable at height E from x [this condition is weaker than the reversibility condition (2.33)]. We say that $x \in \Omega$ is a *local minimum* if no state x' with $U(x') < U(x)$ is reachable from x at height $U(x)$. The *depth* $d(x)$ of a local minimum that is not a global minimum is defined to be the smallest $E > 0$ such that there exists a $y \in \Omega$ with $U(y) < U(x)$, and y can be reached from x at height $U(x) + E$. Let

$$d^* = \max\{d(x) : x \text{ is a local but not global minimum}\}.$$

Assume that $T(t) \downarrow 0$ as $t \to +\infty$. Then (3.5a) holds iff

$$\sum_{t=1}^{+\infty} \exp\{-\frac{d^*}{T(t)}\} = +\infty. \tag{3.56}$$

This condition is *not* sufficient, in general, for weak convergence (or convergence in the total variation norm) of $\mu P_{1,t}$ to π if $U(x)$ has more than one global minima. In subsection 3.3.2, we will derive (via Theorem 3.4) a condition similar to (3.56) for weak convergence.

2) It is concievable that more careful estimates in the proof of Threorem 3.6 or in the asymptotic behavior of Dobrushin's ergodic coefficients $\delta(P_{t_0,t})$ as $t \to +\infty$ [e.g., along lines mentioned in the last remark after (3.12)] may yield a result similar to (3.56) for weak convergence (with d^* replaced by another constant; see next subsection).

3) References [90] and [21, 22] contain studies of certain classes of inhomogeneous MCs and applications to the global optimization problem without constraints, via very different techniques than those we have presented here.

4) Interesting results on the convergence rate and cooling schedules for the annealing algorithm, via the Wentzell-Freidlin theory of the long-time behavior of dynamical systems, have been obtained in [11, 12, 13].

5) For a study of *parallel simulated annealing*, see [6]; and for a study on the finite behavior of simulated annealing via Metropolis dynamics, see [32].

6) In [40], simulated annealing was combined with *renormalization group* ideas for speeding up the algorithm and for establishing a *multiresolution* method for image processing tasks.

B) Global Optimization with Constraints: Let $U(x)$, $V(x)$, $\underline{U}, \underline{V}, \underline{\Omega}_V$ be as in Problem 2, Section 3.1. Let

$$H_t(x) = \frac{1}{T(t)}\{U(x) - \underline{U} + \lambda[V(x) - \underline{V}]\}, \tag{3.57}$$

with $T(t)$ decreasing to zero and $\lambda(t)$ increasing to infinity, as $t \to +\infty$. Clearly (3.40a) holds, and (3.40b) is valid at the minima on $\underline{\Omega}_V$. Let Δ be as in **A)**, and Γ defined the same way but with $U(x)$ replaced by $V(x)$. Then

$$\Delta_t \leq \frac{1}{T(\tau_t)}[\Delta + \lambda(\tau_t)\Gamma].$$

Thus (3.50) holds if

$$\sum_{t=1}^{+\infty} \exp\{-\frac{|S|[\Delta + \lambda(\tau_t)\Gamma]}{T(\tau_t)}\} = +\infty, \tag{3.58a}$$

and therefore in particular if

$$\frac{1}{T(\tau_t)}[\Delta + \lambda(\tau_t)\Gamma] \le \frac{1}{|S|}\log t \tag{3.58b}$$

for $t \ge t_0$, some $t_0 > 1$. Similarly (3.51) holds if

$$\frac{1}{T(\tau_t)}[\Delta + \lambda(\tau_t)\Gamma] \le \frac{1-\varepsilon}{|S|}\log t \tag{3.58c}$$

with some $\varepsilon > 0$, and $t \ge t_0$, some $t_0 > 1$. All the conclusion in **A)** are valid if $\underline{\Omega}$ is replaced by $\underline{\Omega}_V$.

C) Simulation with Constraints: Let $\pi(x)$ be as in (3.1), and set

$$H_t(x) = U(x) + \lambda(t)[V(x) - \underline{V}]. \tag{3.59}$$

Let Γ be as in **B)**. Then (3.50) holds if

$$\sum_{t=1}^{+\infty} \exp\{-|S|\Gamma\lambda(\tau_t)\} = +\infty, \tag{3.60a}$$

and in particular, if

$$\lambda(\tau_t) \le \frac{1}{|S|\Gamma}\log t, \qquad t \ge t_0, \text{ some } t_0 > 0. \tag{3.60b}$$

Similarly, (3.51) holds if

$$\lambda(\tau_t) \le \frac{1-\varepsilon}{|S|\Gamma}\log t \tag{3.60c}$$

with some $\varepsilon > 0$.

3.3.2 Continuous-Time Metropolis Dynamics

Here we shall study the continuous-time Metropolis dynamics (3.44). The study can easily be extended to other dynamics. We assume that $Q = \{q(x,y)\}$ is irreducible and symmetric. Hence the generator $\mathcal{L}(t)$ defined by (3.43) is self-adjoint on $\ell_2(\pi(t))$. Let $\lambda_2(t)$ be the *mass-gap* of $\mathcal{L}(t)$, defined by (3.28). From (3.41) we obtain

$$\frac{\partial}{\partial t}\pi(t,x) = -\{\frac{\partial H_t(x)}{\partial t} - <\frac{\partial H_t}{\partial t}>_{\pi(t)}\}\pi(t,x).$$

If $\lambda_2(t)$ satisfies (3.32), and if there exists a $\gamma(t)$ satisfying (3.33) and

$$|\frac{\partial H_t(x)}{\partial t} - < \frac{\partial H_t}{\partial t} >_{\pi(t)} | \leq \gamma(t), \text{ for all } x \in \Omega,$$

then Theorem 3.4 applies, and we obtain weak convergence of $\mu P_{1,t}$ to π (where π is the limiting vector in Lemma 3.2). Now we apply these remarks to Problems 1, 2, and 3, of Section 3.1.

A) Global optimization without constraints: From (3.54) we see that

$$|\frac{\partial \pi(t,x)}{\partial t}| \leq \gamma(t)\pi(t,x)$$

with

$$\gamma(t) = \frac{|T'(t)|}{T^2(t)}\{\max_x U(x) - \underline{U}\}. \tag{3.61}$$

Let $\lambda_2(T)$ be the mass-gap corresponding to (3.2) with $\lambda = 0$. It is well-known [91, 52] that there exists a constant $d > 0$ (depending on U) and constants $0 < c_1 \leq c_2 < +\infty$ such that

$$c_1 e^{-\frac{d}{T}} \leq \lambda_2(T) \leq c_2 e^{-\frac{d}{T}}, \text{ for all } T > 0. \tag{3.62a}$$

In particular

$$\lim_{T \downarrow 0} T \log \lambda_2(T) = -d \tag{3.62b}$$

[compare with (2.63)]. The constant d has the following representation in terms of the definitions above (3.56) (recall that here we assume Q to be symmetric): If $U(x)$ has a single global minimum, then $d = d^*$. If $U(x)$ has more than one global minima, then for $x, y \in \underline{\Omega}, x \neq y$, let $h(x,y)$ be the smallest $E > 0$ so that x and y are reachable from one another at height $E + U(x)$. Define

$$\overline{d} = \sup\{h(x,y) : x, y \in \underline{\Omega}, x \neq y\}.$$

Then

$$d = \max\{\overline{d}, d^*\}$$

where d^* is the constant in (3.56).

It is easily seen that d has the following intuitive and geometric interpretation: Let $x, y \in \Omega$, and consider a sequence of points (a *path*) $x_1 = x, x_2 \ldots, x_n = y$, from x to y so that $q(x_\ell, x_{\ell+1}) > 0, \ell = 1, \ldots, n-1$. Denote such a path by $s = \{s_\ell\}_{\ell=1}^n$, $s_\ell = x_\ell$, and the set of paths from x to y by $\mathcal{S}_{x,y}$. The highest elevation along a path $s \in \mathcal{S}_{x,y}$ is

$$E_s = \max\{U(s_\ell) - \underline{U} : s_\ell \in s\}$$

and the lowest possible elevation along any path from x to y is

$$g(x, y) = \min\{E_s : s \in S_{x,y}\}.$$

Then
$$d = \max\{g(x, y) - [U(x) - \underline{U}] - [U(y) - \underline{U}] : x, y \in \Omega\}.$$

Notice that $g(x, y) = g(y, x)$ and $g(x, y) \leq \max(g(z, x), g(z, y))$ for all $x, y, z \in \Omega$. Hence, if x, y are such that

$$d = g(x, y) - [U(x) - \underline{U}] - [U(y) - \underline{U}],$$

then either $U(x) = \underline{U}$ or $U(y) = \underline{U}$. Therefore,

$$d = \max\{g(x, y) - [U(y) - \underline{U}] : x \in \underline{\Omega}, y \in \Omega\}.$$

From (3.62), we see that (3.32) is similar to Hajek's condition (3.56). Clearly (3.32) holds if

$$T(t) \geq \frac{d}{\log t}, \qquad t \geq t_0, \text{ some } t_0 > 0. \tag{3.63a}$$

On the other hand, by (3.61), (3.33) holds if

$$T(t) \geq \frac{c_0}{\log t}, \text{ with } c_0 > d \tag{3.63b}$$

for $t \geq t_0$, some $t_0 > 0$. Note that the validity of Theorem 3.4 requires that $c_0 > d$ even if $U(x)$ has a single global minimum (in which case $d = d^*$). However, a more careful analysis of the forward equation underlying the proof of Theorem 3.4 shows [53] that if $U(x)$ has a single global minimum then the result holds even with $c_0 = d$.

The constant d gives the optimal annealing schedule for weak convergence, in the sense that weak convergence fails if $c_0 < d$.

Remark: Some general convergence results for $c_0 \geq d$, $c_0 > d^*$ have been established in [53, 55, 16, 17, 18].

B) Global Optimization with constraints: From (3.57), we see that (3.31) holds with

$$\gamma(t) = \frac{|T'(t)|}{T^2(t)}\{\max_x U(x) - \underline{U}\} + \frac{d}{dt}\left(\frac{\lambda(t)}{T(t)}\right)\{\max_x V(x) - \underline{V}\}.$$

As in (3.62), one can show that there exists a constant $d_0 > 0$ such that

$$-\frac{T}{\lambda} \log \lambda_2(T, \lambda) \longrightarrow d_0, \text{ as } T \downarrow 0, \ \lambda \uparrow +\infty.$$

Then for

$$\frac{\lambda(t)}{T(t)} \leq \frac{1}{c_0} \log t, \quad t \geq t_0, \text{ some } t_0 > 1,$$

with $c_0 > d_0$, (3.32) and (3.33) hold. Again, it can be shown that weak convergence fails, if $c_0 < d_0$.

C) Simulation with constraints: The analysis here is the same as in **B)** with $T(t) \equiv 1$.

3.4 Simulated annealing via inhomogeneous diffusion Processes

Here we assume that Ω and U are as in Section 2.3. The parameter T (the "temperature") appearing in Section 2.3 is now taken to be a monotone function $T(t)$ of the time and is assumed to tend to zero as $t \to +\infty$. For simplicity we consider only the logarithmic schedule

$$T(t) = \frac{c}{\log(2 + t)}, \quad t \geq 0, \tag{3.64}$$

with c a positive constant. The probability density (2.57) becomes time-dependent and we set $\pi(t, x) = \pi_{T(t)}(x)$; the generators (2.59) and (2.65) also become time-dependent, and will be denoted by $\mathcal{L}_t$. Then the corresponding diffusion processes, (2.59a) or (2.65c), are inhomogeneous processes.

The following theorem was proven in [78] using a variant of the procedure of Theorem 3.4 [39], hypercontractive estimates of [23], and results from [19].

Theorem 3.7 *Assume that $U(x)$ satisfies conditions (1)-(4) of Theorem 2.4 (condition (4) need only be satisfied for $T \leq T_0$, some $T_0 > 0$). Let $\{X(t) : t \geq 0\}$ be the inhomogeneous process determined by (2.59a) with $T = T(t)$ given by (3.64), and let $p(0, x_0; t, x)$ be its transition probability function. Also, let Δ be the constant defined in (2.63).*

If $c > \Delta$, then

$$\|p(0, x_0; t, \cdot) - \pi(t)\| \longrightarrow 0 \text{ as } t \to +\infty, \tag{3.65}$$

uniformly for x_0 in compact subsets of $\mathbb{R}^d$; here $\| \cdot \|$ denotes total variation norm.

Remarks:

1) Under reasonable conditions, the measure $\pi(t, x)$ converges as $t \to +\infty$ to a distribution π concentrated at the global minima of $U(x)$. Then Theorem 3.7 implies that $p(0, x_0; t, \cdot)$ converges weakly to π.

2) See [63] for a study of the Langevin equation as a global optimization algorithm, via large deviation techniques.

3) The proof in [78] of Theorem 3.7 is a mixture of probabilistic arguments and analytic tools. We believe that the following formal arguments and resulting differential equations (3.68) and (3.69) can be turned into an alternative PDE proof of Theorem 3.7.

Next, we will establish a result for the process on $\Omega = [0,1]^d$ determined by (2.66) with $T = T(t)$. But first we consider the general generator $\mathcal{L}$ given by (2.65a,b). Its formal adjoint reads

$$\mathcal{L}^* f = -T \sum_{i,j} \frac{\partial}{\partial x_i} \{a_{ij}(x)\pi_T(x)\frac{\partial}{\partial x_j}(\pi_T^{-1}f)\}$$

$$= -T \sum_{i,j} \frac{\partial}{\partial x_i}(a_{ij}\frac{\partial f}{\partial x_j}) - \sum_{ij} \frac{\partial}{\partial x_i}(a_{ij}\frac{\partial U}{\partial x_j}f)$$

$$= -T \sum_{ij} \frac{\partial^2}{\partial x_i \partial x_j}(a_{ij}f) - \sum_{i,j} \frac{\partial}{\partial x_i}[a_{ij}\frac{\partial U}{\partial x_j}f - T\frac{\partial a_{ij}}{\partial x_j}f].$$

For time-dependent $T = T(t)$, we write $\mathcal{L}_t$, $\mathcal{L}_t^*$ [and as before $\pi(t,x) = \pi_{T(t)}(x)$]. Let $\mu(x)$ be an initial distribution density. Then the distribution density $p(t,x)$ at time t satisfies the forward equation

$$\frac{\partial p}{\partial t} = -\mathcal{L}^* p, \tag{3.66}$$

$$p(0,x) = \mu(x).$$

Let

$$p(t,x) = \pi(t,x)q(t,x).$$

Then

$$\frac{\partial p}{\partial t} = \pi\frac{\partial q}{\partial t} + \frac{\partial \pi}{\partial t}q$$

$$= \pi\frac{\partial q}{\partial t} + \pi\frac{T'}{T^2}[U(x) - \overline{U}(t)]q, \tag{3.67}$$

where $T' = \frac{dT(t)}{dt}$ and

$$\overline{U}(t) = <U>_{\pi(t)} = \int U(x)\pi(t,x)dx.$$

An easy calculation shows that $\mathcal{L}^*\pi f = (\mathcal{L}f)\pi$. Then (3.66) and (3.67) yield

$$\frac{\partial q}{\partial t} = -\mathcal{L}_t q - \frac{T'}{T^2}[U(x) - \overline{U}(t)]q. \tag{3.68}$$

Let

$$F(t) = \int[q(t,x) - 1]^2\pi(t,x)dx$$

$$= \int[q(t,x)]^2\pi(t,x)dx - 1$$

(compare with the $F(t)$ defined in the proof of Theorem 3.4). Then using (3.68)

$$\frac{1}{2}\frac{dF(t)}{dt} = \int q\frac{\partial q}{\partial t}\pi dx + \frac{1}{2}\int q^2\frac{\partial \pi}{\partial t}$$

$$= -\int(q\mathcal{L}_t q)\pi dx - \frac{1}{2}\frac{T'}{T^2}\int[U(x) - \overline{U}(t)]q^2\pi dx$$

$$= -\int(q\mathcal{L}_t q)\pi dx \qquad (3.69)$$

$$-\frac{1}{2}\frac{T'}{T^2}\int[U(x) - \overline{U}(t)](q-1)^2\pi dx$$

$$-\frac{T'}{T^2}\int[U(x) - \overline{U}(t)](q-1)\pi dx$$

(compare with the corresponding equation derived in the proof of Theorem 3.4). Recall that $\mathcal{L}_t$ is a self-adjoint non-negative operator on $L_2(\Omega, \pi(t, dx))$, and, under reasonable conditions (see Section 2.3.2), it has purely discrete spectrum $\lambda_1 = 0 < \lambda_2(t) \le \lambda_3(t) \le \cdots$, with $\lambda_1 = 0$ as a simple eigenvalue (under weaker conditions $\lambda_1 = 0$ is an isolated simple eigenvalue).

We believe that (3.68), (3.69), and the spectral properties of $\mathcal{L}_t$ can be used to provide a PDE proof of Theorem 3.7 [even in the general case with a_{ij}s satisfying (2.64)]. Next, we carry out this strategy for the process on $\Omega = [0,1]^d$ determined by (2.66) with $\alpha \in C^1[0,1]$, $\alpha(0) = \alpha(1) = 0$, and $\alpha(\xi) > 0$ for all $0 < \xi < 1$. (The procedure applies also to the case when Ω is a compact Riemannian manifold with $\mathcal{L}$ generated by π and the Riemannian metric on Ω as stated in the last remark of Section 2.3.1; see [51].) Note that $q(t,x) - 1$ is orthogonal (in $L_2([0,1]^d, \pi(t, dx))$) to the constant functions, i.e., to the eigenspace corresponding to the eigenvalue $\lambda_1 = 0$. The above assumptions on $\alpha(\cdot)$, and Poincare's inequality, can be used to bound the first term on the right-hand-side of (3.69), i.e.,

$$\int(q\mathcal{L}_t q)\pi dx \ge \lambda_2(t)F(t), \qquad (3.70)$$

where $\lambda_2(t)$ is the second eigenvalue of $\mathcal{L}_t$. Assuming (without loss of generality) $\inf\{U(x) : x \in [0,1]^d\} = 0$, let

$$M = \sup\{U(x) : x \in [0,1]^d\}.$$

Then using (3.70), and bounding the second and third terms in (3.69) in terms of M, we obtain

$$\frac{1}{2}\frac{dF(t)}{dt} \le -\lambda_2(t)F(t) + \frac{1}{2}\frac{|T'|}{T^2}MF(t) + \frac{|T'|}{T^2}M(F(t))^{\frac{1}{2}}. \qquad (3.71)$$

This is the analogue of the final inequality in the proof of Theorem 3.4. Let

$$\gamma(t) = \frac{|T'|}{T^2}M = -\frac{T'}{T^2}M.$$

If [compare with (3.32) and (3.33)]

$$\int_0^{+\infty} \lambda_2(t)dt = +\infty \tag{3.72}$$

and

$$\frac{\gamma(t)}{\lambda_2(t)} \longrightarrow 0 \text{ as } t \to +\infty, \tag{3.73}$$

then $F(t) \longrightarrow 0$ as $t \to +\infty$, and hence we obtain convergence in total variation norm (and weakly).

It can be shown, as in [51], that $\lambda_2(t)$ satisfies

$$T(t) \log \lambda_2(t) \longrightarrow -\Delta, \text{ as } t \to +\infty.$$

Hence, for the logarithmic schedule (3.64), (3.72) is satisfied for $c > \Delta$. It can be shown, as in [51], that convergence fails if $c < \Delta$. In this sense, Δ determines the optimal annealing schedule.

References

[1] **Aarts, E. and Korst, J.,** *Simulated Annealing and Boltzmann Machines*, John Wiley & Sons, 1989.

[2] **Adachi, S. H.,** *Convergence to Equilibrium and Critical Slowing Down of Dynamical Models in Statistical Mechanics*, Ph.D. Thesis, Division of Applied Mathematics, Brown University, 1987.

[3] **Aizenman, M. and Holley, R.,** Rapid Convergence to Equilibrium of Stochastic Ising Model in the Dobrushin-Shlosman Regime, in *Proceedings of the IMA Workshop on Percolation and Ergodic Theory of Infinite Particle Systems*, Springer-Verlag, 1987, ed. H. Kesten.

[4] **Albererio, S. and Röckner, M.,** Classical Dirichlet Forms on Topological Vector Spaces: The Construction of the Associated Diffusion Process, *Prob. Theory and Related Fields*, **83**, 405–434, 1989.

[5] **Amit, Y.,** On Rates of Convergence of Stochastic Relaxation for Gaussian and Non-Gaussian Distributions, preprint 1990, Div. of Applied Math., Brown University.

[6] **Azencott, R.,** Synchronous Boltzman Machines and Gibbs Fields: Learning Algorithms, in *Neurocomputing, Algorithms, Architectures, and Applications*, NATO ASI series, F **68**, Springer-Verlag, 51–63, eds. F. Fogelman-Soulie and J. Herault.

[7] **Batrouni, G. G., Katz, G. R., Knonfeld, A. S., Lapage, G. P., Svetitsky, B. and Wilson, K. G.,** Langevin Simulations of Lattice Field Theories, *Phys. Rev.*, D **32**, 2736–2747, 1985.

[8] **Borkar, V. S., Chari, R. T. and Mitter, S. K.,** Stochastic Quantization of Field Theory in Finite and Infinite Volume, *J. Funct. Analysis*, **81**, 184–206, 1988.

[9] **Brandt, A., Ron, D. and Amit, D. J.,** Multilevel Approaches to Discrete State and Stochastic Problems, in *Multigrid Methods II*, Springer-Verlag, 66–99, 1986, eds. W. Hackbusch and V. Trottenberg.

[10] **Caracciolo, S., Pelissetto, A. and Sokal, A. D.,** Nonlocal Monte Carlo Algorithms for Self - Avoiding Walks with Fixed Endpoints, *J. Stat. Phys.*, **60**, 7–53, 1990.

[11] **Catoni, O.,** Rough Large Deviations Estimates for Simulated Annealing: Application to Exponential Schedules, *Ann. Prob.*, **20**, 109–146, 1992.

[12] **Catoni, O.,** Applications of Sharp Large Deviations Estimates to Optimal Cooling Schedules, Annales de l' Institut Henri Poincaré, **27**, 291–383, 1991.

[13] **Catoni, O.,** Sharp Large Deviations Estimates for Simulated Annealing Algorithms, to appear in *Annales de l' Institut Henri Poincaré Probab. Stat.*

[14] **Cerny, V.,** Thermodynamical Approach to the Traveling Salesman Problem: an Efficient Simulation Algorithm, *J. Opt. Theory Appl.*, **45**, 41–51, 1985.

[15] **Cheeger, J.,** A Lower Bound for the Lowest Eigenvalue of the Laplacian, Problems in Analysis: A Symposium in Honor of S. Bochner, Princeton Univ. Press, Princeton, 195–199, 1970, ed. R.C. Gunning.

[16] **Chiang, T.-S. and Chow, Y.,** A Limit Theorem for a Class of Inhomogeneous Markov Processes, *Ann. Prob.*, **17**, 1483–1502, 1989.

[17] **Chiang, T.-S. and Chow, Y.,** The Asymptotic Behavior of Simulated Annealing Processes with Absorption, Technical Report, Inst. of Mark., Academia Sinica, 1989.

[18] **Chiang, T.-S. and Chow, Y.,** On the Convergence Rate of Annealing Processes, SIAM *J. Control Optim.*, **26**, 1455–1470, 1988.

[19] **Chiang, T.-S., Hwang, C.-R. and Sheu, S.-J.,** Diffusion for Global Optimization in $I\!R^n$, SIAM *J. Control Optim.*, **25**, 737–753, 1987.

[20] **Chung, K. L.**, *Markov Chains with Stationary Transition Probabilities*, second edition, Springer-Verlag, 1967.

[21] **Connors, D. P. and Kumar, P. R.**, Balance of Recurrence Order in Time-Inhomogeneous Markov Chains with Applications to Simulated Annealing, *Prob. in the Engineering and Informatio Sciences*, **2**, 157–184, 1988.

[22] **Connors, D. P. and Kumar, P. R.**, Simulated Annealing Type Markov Chains and Their Order Balance Equations, SIAM *J. Control and Optimization*, **27**, 1440–1462, 1989.

[23] **Davies, E. B. and Simon, B.**, Ultracontractivity and the Heat Kernel for Schrödinger Operators and Dirichlet Laplacians, *J. Functional Analysis*, **59**, 335–395, 1984.

[24] **Diaconis, P. and Stroock, D.**, Geometric Bounds for Eigenvalues of Markov Chains, *The Annals of Applied Prob.*, **1**, 36–61, 1991.

[25] **Dixon, L. C. W. and Szegö, G. P. (eds.)**, *Towards Global Optimization*, **2**, North-Holland, 1978.

[26] **Donoghue, Jr., W. F.**, *Monotone Matrix Functions and Analytic Continuation*, Springer-Verlag, 1974.

[27] **Edwards, R. G. and Sokal, A. D.**, Generalization of the Fortuin-Kasteleyn-Swendsen-Wang Representation and Monte Carlo Algorithm, *Phys. Rev.*, D **38**, 2009–2012, 1988.

[28] **Edwards, R. G. and Sokal, A. D.**, Dynamic Critical Behavior of Wolff's Collective-Mode Monte Carlo Algorithm for the Two-Dimensional $O(n)$ Nonlinear σ-model, *Phys. Rev.*, D **40**, 1374–1377, 1989.

[29] **Frigessi, A., Hwang, C. R. and Younes, L.**, Optimal Spectral Structure of Reversible Stochastic Matrices, Monte Carlo Methods and the Simulation of Markov Random Fields, *Ann. Appl. Prob.*, **2**, 610–628, 1992.

[30] **Frigessi, A., Hwang, C. R., Sheu, S. J. and DiStefano, P.**, Convergence Rates of the Gibbs Sampler, the Metropolis Algorithm, and Other Single-Site Updating Dynamics, *J. Royal Stat. Soc.*, Series B **55**, 205–219, 1993.

[31] **Gantert, N.**, Laws of Large Numbers for the Annealing Algorithm, *Stoch. Processes and Their Appl.*, **35**, 309–313, 1990.

[32] **Gelfand, S. B. and Mitter, S. K.,** Analysis of Simulated Annealing for Optimization, Proceedings of the 24th IEEE Conference on Decision and Control, Ft. Lauderdale, FL, 779–786, 1985.

[33] **Geman, D.,** *Random Fields and Inverse Problems In Imaging, Lecture Notes in Mathematics,* **1427**, Springer-Verlag, 1990.

[34] **Geman, D. and Geman, S.,** Relaxation and Annealing with Constraints, Complex Systems Technical Report No. 35, Division of Applied Mthematics, Brown University, 1987.

[35] **Geman, S. and Geman, D.,** Stochastic Relaxation, Gibbs Distributions, and the Bayesian Restoration of Images, IEEE Trans. PAMI-6, 721–741, 1984.

[36] **Geman, D. and Gidas, B.,** Image Analysis and Computer Vision, in *Spatial Statistics and Image Processing,* National Research Council report 1991.

[37] **Geman, S. and Hwang, C.-R.,** Diffusions for Global Optimization, SIAM *J. Control Optim.,* **24**, 1031–1043, 1986.

[38] **Gidas, B.,** Nonstationary Markov Chains and Convergence of the Annealing Algorithm, *J. Stat. Phys.,* **39**, 73–131, 1985.

[39] **Gidas, B.,** Global Optimization via the Laugevin Equation, Proceedings of 24th Conference on Decision and Control, Ft. Lauderdale, FL, December 1985.

[40] **Gidas, B.,** Renormalization Group Approach to Image Processing Problems, IEEE Trans. PAMI-11, 164–180, 1989.

[41] **Goodman, J. and Sokal, A. D.,** Multigrid Monte Carlo Method. Conceptual Foundations, *Phys. Rev.,* D **40**, 2035–2071, 1989.

[42] **Griffeath, D.,** Uniform Coupling of Non-homogeneous Markov Chains, *J. Appl. Prob.,* **12**, 753–762, 1975.

[43] **Gross, L.,** Absence of Second-Order Phase Transitions in the Dobrushin Uniqueness Region, *J. Stat. Phys.,* **25**, 57–72, 1981.

[44] **Hajek, B.,** Cooling Schedules for Optimal Annealing, *Math. Oper. Research,* **13**, 311–329, 1988.

[45] **Hajek, B.,** A Tutorial Survey of Theory and Applications of Simulated Annealing, in Proceed of 24th Conference on Decision and Control, Ft. Lauderdale, FL, December 1985.

[46] **Hammersley, J. M. and Handscomb, D. C.**, *Monte Carlo Methods*, Methuen, London, 1965.

[47] **Hastings, W. K.**, Monte Carlo Sampling Methods Using Markov Chains and Their Applications, *Biometrilea*, **57**, 72–89, 1970.

[48] **Helffer, B. and Sjöstrand, I.**, Multiple Wells in the Semiclassical Limit, I, *Comm. Part. Diff. Equ.*, **9**, 337–408, 1984.

[49] **Holley, R.**, Rapid Convergence to Equilibrium in One Dimensional Stochastic Ising Models, *Ann. Prob.*, **13**, 72–89, 1985.

[50] **Holley, R.**, Possible Rates of Convergence in Finite Range, Attractive Spin Systems, *Contemporary Mathematics*, **41**, 215–234, 1985.

[51] **Holley, R. A., Kusuoka, S. and Stroock, D. W.**, Asymptotics of the spectral Gap with Applications to the Theory of Simulated Annealing, *J. Funct. Anal.*, **83**, 333–347, 1989.

[52] **Holley, R. and Stroock, D.**, Simulated Annealing via Sobolev Inequalities, *Comm. Math. Phys.*, **115**, 553–569, 1988.

[53] **Hwang, C.-R. and Sheu, S.-J.**, Large-Time Behavior of Perturbed Diffusion Markov Processes with Applications to the Second Eigenvalue Problem for Fokker-Planck Operators and Simulated Annealing, *Acta Applicandae Mathematicae*, **19**, 253–295, 1990.

[54] **Hwang, C.-R. and Sheu, S.-J.**, Large-Time Behavior for Perturbed Diffusion Markov Processes with Applications, I, II, III, Technical Reports, Institute of Math., Academia Simica, 1986.

[55] **Hwang, C.-R. and Sheu, S.-J.**, Singular Perturbed Markov Chains and the Exact Behaviors of Simulated Annealing Processes, to appear in *J. of Theoretical Probability*.

[56] **Iosifescu, M.**, *Finite Markov Processes & Their Applications*, John Wiley & Sons, 1980.

[57] **Iosifescu, M. and Theodorescu, R.**, *Random Processes and Learning*, Springer-Verlag, 1969.

[58] **Isaacson, D. L. and Madsen, R. W.**, *Markov Chains Theory and Applications*, John Wiley & Sons, 1976.

[59] **Johnson, J. and Issacson, D.**, Conditions for Strong Ergodicity Using Intensity Matrices, preprint 1986, Department of Statistics, Iowa State University.

[60] **Johnson, D. S., Papadimitriou, C. H. and Yannakakis, A. M.,** How easy is local search? in *Proc. Annual Symp. on Foundations of Computer Science*, Los Angeles, 39–42, 1985.

[61] **Kemeny, J. G. and Snell, J. L.,** *Finite Markov Chains*, Springer-Verlag, 1983.

[62] **Kirkpatrick, S., Gellatt, C. D. Jr. and Vecchi, M. P.,** Optimization by Simulated Annealing, *Science*, **220**, 671–680, 1983.

[63] **Kushner, H. J.,** Asymptotic Global Behavior for Stochastic Approximation and Diffusions with slowly Decreasing Noise Effects: Global Minimization via Monte Carlo, SIAM *J. Appl. Math.*, **47**, 169–185, 1987.

[64] **Lawler, G. F. and Sokal, A. D.,** Bounds on the L^2 Spectrum for Markov Chains and Markov Processes: A Generalization of Cheeger's Inequality, *Trans. Amer. Math. Soc.*, **309**, 557–580, 1988.

[65] **Li, X.-J., Sokal, A. D.,** Rigorous Lower Bound on the Dynamic Critical Exponent of the Swendsen-Wang Algorithms, *Phys. Rev. Letters*, **63**, 827–830, 1989.

[66] **Liggett, T. M.,** *Interacting Particle Systems*, Springer-Verlag, 1985.

[67] **Madras, N. and Sokal, A. D.,** The Pivot Algorithm: A Highly Efficient Monte Carlo Method for Self-Avoiding Walk, *J. Stat. Phys.*, **50**, 109–186, 1988.

[68] **Martinelli, F., Olivieri, E. and Scoppola, E.,** Metastability and Exponential Approach to Equilibrium for Low Temperature Stochastic Ising Models, *J. Stat. Physics*, **61**, 1105–1119, 1990.

[69] **Martinelli, F., Olivieri, E. and Scoppola, E.,** On the Swendsen-Wang Dynamics I: Exponential Convergence to Equilibrium, *J. Stat. Physics*, **62**, 117–133, 1991.

[70] **Martinelli, F., Olivieri, E. and Scoppola, E.,** On the Swendsen-Wang Dynamics II: Critical Droplets and Homogeneous Nucleation at Low Temperatures for the Two-Dimensional Ising Model, *J. Stat. Physics*, **62**, 135–159, 1991.

[71] **Metropolis, N., Rosenbluth, A. W., Rosenbluth, M. N., Teller A. H. and Teller, E.,** Equations of State Calculations by Fast Computing Machines, *J. Chemical Phys.*, **21**, 1087–1091, 1953.

[72] **Nummelin, E.,** *General Irreducible Markov Chains and Non-Negative Operators*, Cambridge Univ. Press, Cambridge, 1984.

[73] **Padimitriou, C. H. and Steinglitz, K.**, *Combinatorial Optimization: Algorithms and Complexity*, Prentice-Hall, Inc. 1982.

[74] **Peskun, P. H.**, Optimal Monte-Carlo Sampling using Markov Chains, *Biometrika*, **60**, 607–612, 1973.

[75] **Pincus, M.**, A Monte-Carlo Method for the Approximate Solution of Certain Types of Constrained Optimization Problems, *Oper. Research*, **18**, 1225–1228, 1970.

[76] **Reed, M. and Simon, B.**, *Methods of Modern Mathematical Physics, I: Functional Analysis*, Academic Press, 1980 (second edition).

[77] **Reed, M. and Simon, B.**, *Methods of Modern Mathematical Physics IV: Analysis of Operators*, Academic Press, 1978.

[78] **Royer, G.**, A Remark on Simulated Annealing of Diffusion Processes, SIAM *J. Control Optim.*, **27**, 1403–1408, 1989.

[79] **Rubinstein, R. Y.**, *Monte Carlo Optimization, Simulation and Sensitivity of Queueing Networks*, John Willey & Sons, 1986.

[80] **Rudin, W.**, *Real and Complex Analysis*, Third Edition, McGraw-Hill Book Company, 1987.

[81] **Seneta, E.**, *Non-Negative Matrices and Markov Chains*, Springer-Verlag, 1981.

[82] **Shiryayev, A. N.**, *Probability*, Springer-Verlag, 1984.

[83] **Sidak, Z.**, Eigenvalues of Operators in ℓ_p-Spaces in Denumerable Markov Chains, *Czechoslovak Math. J.*, **14**, 438–443, 1964.

[84] **Simon, B.**, Fifty Years of Eigenvalue Perturbation Theory, *Amer. Math. Soc. Bulletin*, **24**, 303–319, 1991.

[85] **Simon, B.**, Semiclassical Analysis of Low Lying Eigenvalues I. Nondegenerate Minima: Asymptotic Expansions, *Ann. Inst. Henri Poincaré*, **38**, 295–307, 1983.

[86] **Sokal, A. D.**, Monte Carlo Methods in Statistical Mechanics: Foundations and New Algorithms, Cours de Troisième Cycle de la Physique en Suisse Romande, Lausanne, June 1989.

[87] **Sokal, A. D. and Thomas, L. E.**, Exponential Convergence to Equilibrium for a Class of Random-Walk Models, *J. Stat. Phys.*, **54**, 797–807, 1989.

[88] **Sokal, A. D. and Thomas, L. E.**, Absence of Mass Gap for a Class of Stochastic Countour Models, *J. Stat. Phys.*, **51**, 907–947, 1988.

[89] **Swendsen, R. H. and Wang, J.-S.,** Nonuniversal Critical Dynamics in Monte Carlo Simulations, *Phys. Rev. Letters*, **58**, 86–88, 1987.

[90] **Tsitsiklis, J. N.,** Markov Chains with Rare Transitions and Simulated Annealing, *Math. Op. Research*, **14**, 1–12, 1989.

[91] **Ventcel, A. D.,** On the Asymptotics of Eigenvalues of Matrices with Elements of Order $\exp\{-V_{ij}/2\varepsilon^2\}$, *Dokl. Acad. Nauk SSR*, **202**, 65–68, 1972.

[92] **Winkler, G.,** An Ergodic L^2-Theorem for Simulated Annealing in Bayesian Image Reconstruction, *J. Appl. Prob.*, December 1990.

Chapter 8

RANDOM GRAPHS IN ECOLOGY

Joel E. Cohen
Rockefeller University

ABSTRACT

Random graphs have described relations among biological species at least since
Charles Darwin's *Origin of Species* (1859). In ecology, random directed graphs
called food webs describe which species eat which species. When large numbers of real food webs are viewed as an ensemble, empirical patterns appear.
Stochastic models of food webs can explain and unify some of these empirical
patterns. This chapter gives an elementary account of how stochastic models
of food webs are constructed, analyzed and interpreted. Connections are described with models of parallel computing and the qualitative of large systems
of nonlinear ordinary differential equations.

1. HISTORICAL BACKGROUND

The first edition of Charles Darwin's book *On the Origin of Species by Means
of Natural Selection, or the Preservation of Favoured Races in the Struggle
for Life*, contains only one picture. The picture is a fold-out inserted between
pages 116 and 117. Because the first edition has been reissued in a facsimile
edition (Darwin 1859 [1964], [14]), it is not necessary to collect rare books to
enjoy the picture (Figure 1).

The picture is a diagrammatic sketch of a small portion of what Darwin
later (p. 130) calls "the great Tree of Life," showing how (p. 123) "two or more
genera are produced by descent, with modification, from two or more species
of the same genus." One node of the tree is labelled (A). Darwin explains:
"Let (A) be a common, widely-diffused, and varying species, belonging to
a genus large in its own country. The little fan of diverging dotted lines
of unequal lengths proceeding from (A), may represent its varying offspring.
The variations are supposed to be extremely slight, but of the most diversified
nature; they are not supposed all to appear simultaneously, but often after
long intervals of time; nor are they all supposed to endure for equal periods.
Only those variations which are in some way profitable will be preserved or
naturally selected." Some of the dotted lines terminate in extinction; others

0-8493-8073-1/95/$0.00+$.50

give rise to new varieties or species. Lines emanating from these new varieties or species give rise to other new varieties or species, which in turn eventually give rise to new genera.

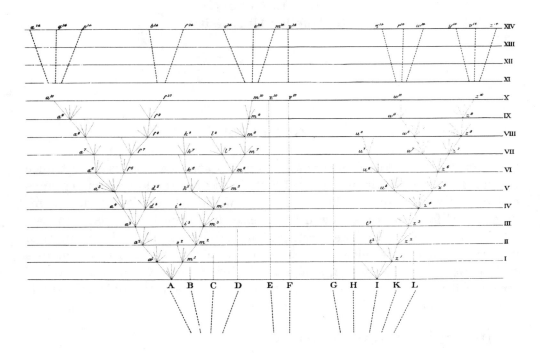

FIGURE 1. Evolution of new genera from old, according to the sole illustration of Charles Darwin's *Origin of Species*. Source: Darwin (1859 [1964], between p. 116 and 117).

The picture appeared in 1859, when the theory of probability was relatively young (Pierre Simon Laplace had first published his *Théorie des Probabilités* in 1812, but Isaac Todhunter was not to publish his *History of the Mathematical Theory of Probability* until 1865), the theory of graphs (apart from Leonhard Euler's great initiatives) was little developed, and the theory of random graphs lay a century in the future.

Yet to modern eyes, the picture is clearly a realization of a random directed graph. It is in fact a random directed tree. Charles Darwin made a simplified graph-theoretic model of nature, though I do not know whether he was the first to do so. Darwin symbolized all the complexity of a species or variety by a node of a graph, and all the complexity of a relation between two species or varieties (in Darwin's case, the relation of descent) by a directed edge, or arc, of the graph.

Darwin's picture has been as influential scientifically as the book it illustrated. His picture provides the prototype of today's phylogenetic trees, which are drawn based on results of molecular biology and numerical taxonomy that Darwin could not even imagine. Beyond summarizing evolutionary relationships, the picture's way of representing relationships among varieties or species influenced ecologists, though the words "ecology" and "ecologist" had not been invented when *Origin* appeared. We shall pursue the ecological thread of the story.

In the 1870s, in America, according to Stephen Alfred Forbes, "under the influence of Darwin and Agassiz and Huxley, a transforming wave of progress was sweeping through college and school, a wave whose strong upward swing was a joy to those fortunate enough to ride on its crest, but which smothered miserably many an unfortunate whose feet were mired in marsh mud" (Forbes 1977, [16], p. 10), in a retrospective speech he gave in 1907). Forbes, then at the Illinois State Normal University, published detailed empirical studies of the food of birds (1877) and fishes (1878) of Illinois and the Great Lakes. Forbes's major papers from 1878 onward have recently been reissued in facsimile (Forbes 1977, [16]).

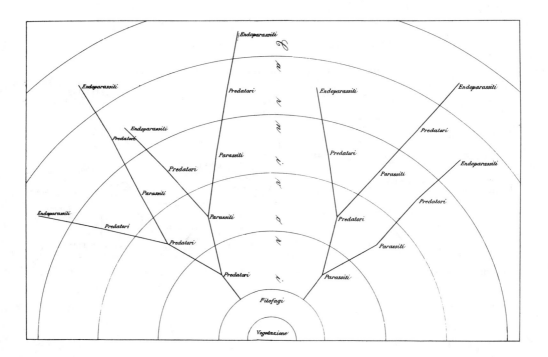

FIGURE 2. Camerano's schema of feeding relations. *Vegetazione* = vegetation, *Fitofagi* = phytophagous animals, *Predatori* = predators, *Parassiti* = parasites, *Carnivori* = carnivores, *Endoparassiti* = endoparasites. Source: Camerano (1880, [3]), plate IX.

In 1880, Forbes summarized long lists of the food of Illinois fishes in tabular form. He listed categories of food eaten by fishes as the headings of columns, and the names of particular species of fishes as the labels of rows. As the entries in each row, he reported how many of each kind of fish he examined and how frequently he found each kind of fish eating each kind of food. In later papers, he exchanged the roles of rows and columns, listing predators as the column headings, prey as the row labels, and the frequencies of predators consuming prey items as entries of the tables. This is still the format of modern so-called "predation matrices".

Also in 1880, a 24-year-old Italian entomologist named Lorenzo Camerano, working at the Museum of Zoology of the University of Turin, used a relabeled form of Darwin's random directed tree to describe the feeding relations in a community of plants and animals (Figure 2).

Camerano did not refer explicitly to Darwin, so I do not know whether he borrowed Darwin's graphical idea or came upon it independently. The nodes of Camerano's tree represent groups of species, and an arc goes from one group to another if members of the second consume members of the first. For example, "predators" eat "phytophages", "parasites" eat "predators", and so on. To my knowledge, this is the first picture of a food web as a directed graph.

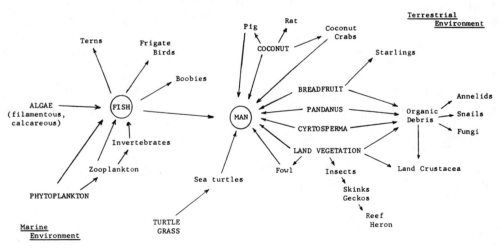

FIGURE 3. A contemporary food web: Niering's web of the Kapingamarangi Atoll. Source: Niering (1963, [29],p. 157), reprinted with permission.

These brief historical comments are not intended as exhaustive scholarship, and certainly not as establishing priorities, but rather as evidence that the mathematical models of food webs below are rooted in biological experience and thought.

In the century and more since Forbes described food webs in empirical detail and Camerano described generalized, abstract food webs, other ecologists have reported many hundreds, perhaps thousands, of food webs. A minority of reported webs is quantitative, like Forbes's tables. Most show only the feeding links among species (e.g., Figure 3). Henceforth, except in section 7, we shall be concerned exclusively with qualitative food webs that show who eats whom, but not how much or how often.

If a collection of species living in one place is like a city, a web is like a street map of the city; it shows where road traffic can and does go. A street map usually omits many important details, e.g., the flow of pedestrian and bicycle traffic, how much traffic flows along the available streets, what kind of vehicular traffic it is, the reasons for the traffic, the laws governing traffic flow, rush hours, and the origin of the vehicles. By analogy, a web often omits small flows of food or predation on minor species, the quantities of food or energy consumed, the chemical composition of food flows, the behavioral and physical constraints on predation, variations over time, whether periodic or random, in eating, and the population dynamics of the species involved. Thus a web gives at best very sketchy information about the functioning of a community. But just as a map provides a helpful framework for organizing more detailed information, a web helps biologists picture how a community works.

Until the second half of this century, each web was treated much as a species of butterfly was treated before Darwin — as unique, complex, and beautiful, perhaps sharing some similarities with other webs from similar habitats, but essentially individual and lawless. In recent decades, in addition to relishing the features that set each individual web apart, ecologists have given more attention to the patterns and order to be found in collections of a large number of food webs. Some recent accounts of the results obtained from this relatively new perspective are by Pimm (1982, [31]), Yodzis (1989, Chap. 8, [37]), Lawton (1989, [24]), Schoener (1989, [36]), Cohen, Briand, and Newman (1990, [9]), and Pimm, Lawton, and Cohen (1991, [33]).

This paper will review some theoretical ideas about random directed graphs that have been used to interpret data on ensembles of food webs. Little attention will be given to comparing the results of theoretical derivations with data, though such comparisons are the vital scientific rationale for theoretical work.

The mathematical custom, which will be followed here, of showing only theoretical derivations without the detailed empirical research that leads a scientist to require them is a traditional but unfortunate fraud. If data analysis is thought of as one level of research, and mathematical analysis is thought of as another level of research (let us take the level of mathematical analysis as higher, to flatter mathematicians), then the trajectory of my own research in this area is something like the letter M, starting from data and moving up to theory, then back down to data, then again up to theory, and so on. During a long-term effort to collect, analyze, and summarize real food webs (starting in 1968, and still continuing), I noticed along the way some remarkably simple patterns; for example, the ratio of the number of predators to the number of prey, and the ratio of the total number of links to the total number of species, are roughly independent of the total number of species in different webs. These and several other data-analytic patterns required theoretical explanation. I constructed and, with mathematical colleagues, analyzed dozens of models to see which ones could explain the patterns quantitatively and qualitatively. The few surviving models made novel predictions not tested previously, so I returned to data analysis. When the data suggested modifications in the models, more mathematics was required. Most of my part of the story so far appears in gruesome detail in two books (Cohen 1978, [4], Cohen, Briand, Newman 1990, [9]), but the story is far from complete.

Before turning to some mathematical ideas, I would like to urge budding mathematicians and mathematical scientists to study at least one area of empirical science. I do not mean other people's mathematical theories about empirical science. Rather I urge contact with some domain of nature, contact that is as direct as possible. If that contact turns out to be fun, then a strenuous effort to formulate, analyze and interpret it by using mathematical language is almost guaranteed to lead to novel mathematical questions, and may well lead to new scientific insights.

2. CHARACTERISTICS OF DIRECTED GRAPHS USEFUL FOR MEASUREMENT AND THEORY

Some characteristics of directed graphs will now be defined. The purpose of these definitions is to specify properties of food webs, which will be treated as directed graphs, that can be measured in the real world and compared with predictions calculated from probabilistic food web models. This section will give definitions; the next, some models; and, the following sections, some calculated properties of the models leading to testable predictions.

A directed graph (henceforth abbreviated to "digraph") consists of a finite set of vertices together with a set of arcs, or directed edges (Robinson and Foulds 1980, [35]). Each vertex corresponds to a species, or sometimes to a stage in the life-cycle of a species (for example, the larvae of a certain insect or the eggs of a certain fish), or sometimes to a group of species whose diet and predators are not distinguished (for example, a size-class of phytoplankton or zooplankton species). For simplicity we shall just say that each vertex represents a species. Let the number of vertices (species) be $S > 1$. We label the vertices (and species) $1, 2, \ldots, S$ and call the set of vertices $[S] = \{1, 2, \ldots, S\}$.

If $i \in [S]$, $j \in [S]$, we interpret the arc (i, j) to mean that species i is eaten by species j. Graphically, as in Figure 3, we draw an arrow from i to j, showing the direction that the food flows. An arc of a food web is called a trophic link, or a link, in ecology. Denote by A the set of arcs or links or directed edges, $A \subseteq [S] \times [S]$. Then a digraph D_S on S vertices is the ordered pair $([S], A)$.

The adjacency matrix $W = W(D_S)$ of D_S is an $S \times S$ matrix such that $w_{ij} = 1$ if and only if $(i, j) \in A$ and $w_{ij} = 0$ otherwise (Robinson and Foulds 1980, [35] p. 169). Ecologists call this matrix the predation matrix. Its column j has an element equal to 1 in each row that represents a species in the diet of species j, and 0 elsewhere. Its row i has an element equal to 1 in each column that represents a predator or consumer of species i, and 0 elsewhere. Sometimes we shall call W the web. The digraph and the adjacency matrix contain equivalent information. Sometimes it is more convenient to think in terms of a graphical representation of a web, other times in terms of a matrix representation.

A basal species is a species that eats no other species. A basal species is called a source in digraph theory, and is identified by a column of W in which all elements are equal to zero, or by a vertex of the digraph that has no incoming arcs. A top species is a species that is eaten by no other species. A top species is called a sink in digraph theory, and is identified by a row of W in which all elements are equal to zero, or by a vertex of the digraph that has no outgoing arcs. A species that is simultaneously basal and top is said to be isolated. A basal species that is not isolated is said to be a proper basal species. A proper top species is a top species that is not isolated. An intermediate species is one that eats and is eaten by other species.

A walk in a digraph is a sequence of alternating vertices and edges, starting and ending with vertices. For example, if $(i, j) \in A$, $(j, k) \in A$, then $i, (i, j), j, (j, k), k$ is a walk from vertex i to vertex k. Cannibalism is represented by a walk of the form $i, (i, i), i$, and is present if W has any nonzero diagonal element. The length of a walk is the number of edges in it. An n-walk is a walk of length n. A digraph is acyclic if and only if no vertex (or species) appears as a vertex more than once in any walk in the digraph. Thus an acyclic web has no cannibalism, no 2-loops such as $i, (i, j), j, (j, i), i$, where $i \neq j$, nor any longer loops.

A chain is a walk from a basal species to a top species. A chain in this sense is identical to a "maximal food chain" as defined by Cohen (1978, [4] p. 56). An n-chain is a chain of length n, i.e., a chain with n links. The height of a web is the length of the longest chain.

A real web can be described by its number of species, links, top species, intermediate species, basal species, number of chains of each length $n = 1, 2, \ldots, S - 1$, and other characteristics. These observations can then be compared with predictions of stochastic models of webs, to which we now turn.

3. MODELS

In modeling an ensemble of food webs, the elements w_{ij} of W, which represent the presence or absence of feeding on species i by species j, are treated as random variables. This step in modeling food webs is very troubling for some ecologists. After all, much of ecology is devoted to documenting the exquisite adaptation of organisms to their diet and to their predators. The teeth of the tiger adapt it to carnivory. The teeth of the antelope adapt it to herbivory. Some plants provide refuge and food to certain species of ants that protect the plants against more deleterious consumers. What then is the sense of modeling the presence or absence of feeding between two species as a random variable? While the interactions between species may be very complex in many feeding relations, it may be possible to summarize the pattern resulting from these interactions in simple ways. By analogy, each driver on the congested highways of New York has his or her own special personal, cultural and ethnic history, his or her own reasons for urgency or patience, but the traffic jams of morning and evening rush hour are substantially predictable. Traffic engineers have long found simplified models of driver behavior useful approximations for understanding repeatable phenomena of traffic flow.

The probability that $w_{ij} = 1$, i.e., that species j eats species i, is written $P\{w_{ij} = 1\}$. The heterogeneous cascade model (Cohen 1990, [7]) assumes that $P\{w_{ij} = 1\} = p_{ij}$, $0 \leq p_{ij} \leq 1$, where

$$p_{ij} = 0 \quad \text{if} \quad i \geq j, \tag{3.1}$$

$$p_{ij} > 0 \quad \text{if} \quad i < j, \text{ and} \tag{3.2}$$

the events $\{w_{ij} = 1\}$ are mutually independent for all $1 \leq i < j \leq S$. (3.3)

The matrix P with elements p_{ij} is called the predation probability matrix. Assumptions (3.1) and (3.2) guarantee that a heterogeneous cascade digraph is acyclic with probability one, because food is permitted to flow only from a species with a lower label i to a species with a higher label j. There is thus a cascade, or hierarchy, of feeding relations. The cascade is heterogeneous because it is not yet assumed that all nonzero elements p_{ij} are equal to one another.

More general cascade models could weaken the assumption (3.3) of independence among trophic links or relax (3.2) to $p_{ij} \geq 0$ for $i < j$. Such extensions will not be considered here.

It seems difficult to estimate separately all the elements of the predation probability matrix P. Indeed, even if one had a large number of webs with the same S, it might not be obvious how to label species, because, if links were relatively sparse, there could be multiple orderings of the species compatible with the assumption of a cascade. Therefore, it is convenient to assume that certain elements of P are equal in order to estimate their common value.

Many special cases of the heterogeneous cascade model are of biological interest (Cohen 1990, [7]). For example, suppose that each consumer is equally likely to consume any of the prey species available to it, but, because of different behavioral capacities and morphology, different predators have different probabilities of preying on the species available to them. Then the model is said to be predator-dominant, and $p_{ij} = b_j > 0$, for $j = 2, \ldots, S$ and $i < j$. For a second example, suppose that the chances of predation are determined by the relative abundances or defensive abilities of the available prey species, so that different prey species have different probabilities of being preyed on but, as a first approximation, each prey is equally like to be preyed on by any of its possible consumers. Then the model is said to be prey-dominant, and $p_{ij} = a_i > 0$, for $j = 2, \ldots, S$ and $i < j$.

The simplest model assumes $p_{ij} = p > 0$, for $1 \leq i < j \leq S$. This homogeneous cascade model $C(S, p)$ on S species (vertices) with link (arc) probability p is the only case that will be considered further here.

Because real webs have widely varying numbers of species, it has been of interest to describe empirically and explain theoretically how the properties of webs vary as the number S of species varies. It is found empirically that the average number of links (averaging over different webs with roughly the same number of species) increases with the number of species. The increase in the expected number of links is at least proportional to the number of species. Clearly, the expected number of links cannot increase faster than S^2 asymptotically for large S, because W has only S^2 elements. There is still some uncertainty among ecologists whether the expected number of links increases linearly with S or superlinearly, that is, as some power, greater than one, of S. The homogeneous cascade model can encompass both of these possibilities, as follows.

Fix the positive real number c and the nonpositive real number d, $0 \geq d \geq -1$. For $1 \leq i < j \leq S$, let the probability of any arc be $p = cS^d$. Then the expected number of arcs is

$$E(L) = \binom{S}{2} p \sim \frac{c}{2} S^{2+d} \qquad (3.4)$$

asymptotically. When $d = -1$, $E(L)$ is linear in S, hence this special case is called the linear cascade model; when $d > -1$, $E(L)$ rises faster than linearly in S, hence this special case is called the superlinear cascade model (Cohen 1990, [7]).

4. CONNECTIONS

The homogeneous cascade model $C(S, p)$ on S species (vertices) with link (arc) probability $p_{ij} = p > 0$, for $1 \leq i < j \leq S$, is identical to a model that was proposed independently for parallel computation. The homogeneous cascade model is also closely connected with a basic model of random (undirected) graphs.

The homogeneous cascade model may be interpreted as a model of parallel computation by supposing that each vertex represents a task that must be processed, and that an arc goes from task i to task j if task i must be completed before task j; thus the arcs represent precedence (Gelenbe, Nelson, Philips, Tantawi, 1986, [18]). If each task requires unit time, then the duration of the entire computation is just one plus the length (number of arcs) in the longest chain, or one plus the height of the digraph. If each task requires an amount of time given by some random variable, then the duration of the entire computation is the maximum of the sum of the random task durations, where the sum is taken along each chain and the maximum is taken over all chains.

A basic model $G(S, p)$ in the theory of random (undirected) graphs (Erdös and Rényi 1960, [15]; see also e.g., Bollobás 1985, [1], p. 32; Palmer 1985, [30], p. 6) constructs a random graph on the vertex set $[S]$ by placing an undirected edge $\{i, j\}$ between vertices i and j with probability p and no edge between vertices i and j with probability $1 - p$, independently for all i and j with $1 \leq i < j \leq S$. To go from the random digraphs specified by $C(S, p)$ to the random graphs specified by $G(S, p)$, simply drop the orientation of arcs, i.e., replace the arc (i, j) from i to j by the edge $\{i, j\}$ between i and j. To go from $G(S, p)$ to $C(S, p)$, do the reverse, i.e., orient each edge from the vertex with the lower number to the vertex with the higher number.

Suppose the edge probability $p = p(S)$ in $G(S, p)$ depends on S as the number S of vertices increases without bound. The marvellous accomplishment of Erdös and Rényi (1960) was to discover that the structure of $G(S, p)$ changes quite suddenly as p changes smoothly from 0 to 1, and to invent methods for calculating these changes in structure (see Bollobás [1985], [1]). Without going into the methods, it is worthwhile to give a simple example of the powerful results. The complete graph K_n is defined as the graph on n vertices in which all possible $n(n-1)/2$ edges are present. Erdös and Rényi (1960) proved that, in the limit as $S \to \infty$, if $pS^{2/(n-1)} \to 0$, then the probability that the random graph $G(S, p)$ contains K_n approaches zero, while, if $pS^{2/(n-1)} \to \infty$, then the probability that the random graph $G(S, p)$ contains K_n approaches one. Thus, loosely speaking, complete graphs on $n = 4$ vertices are virtually absent from large random graphs if $p(S) < S^{-2/3}$ for large S, but suddenly appear as soon as $p(S) > S^{-2/3}$.

Define the competition graph $G(D) = ([S], E)$ of a digraph $D = ([S], A)$ as an undirected simple graph on the vertex set $[S]$ with edge set E, where $\{i, j\} \in E$ if and only if there exists a vertex k in $[S]$ such that both $(k, i) \in A$, $(k, j) \in A$. Thus consumers i and j are joined by an edge in the competition graph if and only if there is at least one prey species that both consumers eat. [Ecologists also call $G(D)$ the trophic niche overlap graph of D.] When the random digraph is specified by the homogeneous cascade model $C(S, p)$, the competition graph $G(C(S,p))$ is a random graph model that differs from the classical model $G(S,p)$. Like $G(S,p)$, $G(C(S,p))$ displays abrupt changes in structure, but the changes occur for different values of $p(S)$ (Cohen and Palka 1990, [13]). For example, in the limit as $S \to \infty$, if $pS^{1+1/n} \to 0$, then the probability that the random competition graph $G(C(S,p))$ contains K_n approaches zero, while if $pS^{1+1/n} \to \infty$, then the probability that the random competition graph $G(C(S,p))$ contains K_n approaches one. Thus, loosely speaking, complete graphs on $n = 4$ vertices are virtually absent from large random competitions graphs derived from the homogeneous cascade model if $p(S) < S^{-5/4}$ for large S, but suddenly appear as soon as $p(S) > S^{-5/4}$. Some structural properties of the competition graph of the homogeneous cascade model explain facts about observed competition graphs (those derived from real food webs) that have previously lacked a quantitative explanation (Cohen and Palka 1990, [13]).

5. BASIC PROPERTIES OF THE HETEROGENEOUS CASCADE MODEL FOR FINITES

Define

$$q_{ij} = 1 - p_{ij} = P\{w_{ij} = 0\}, \qquad \text{for } i,j = 1,\ldots,S, \tag{5.1}$$

$$r_S = 1, \ r_i = \prod_{j=i+1}^{S} q_{ij}, \qquad \text{for } i = 1,\ldots,S-1, \tag{5.2}$$

$$c_1 = 1, \ c_j = \prod_{i=1}^{j-1} q_{ij}, \qquad \text{for } j = 2,\ldots,S. \tag{5.3}$$

The probability that row i of W is entirely 0 is given by r_i and the probability that column j of W is entirely 0 is given by c_j. For $i = 1,\ldots,S$, it follows that

P{species i is not isolated $\} = 1 - r_i c_i$,
P{species i is top$\} = r_i$,
P{species i is basal$\} = c_i$,
P{species i is intermediate$\} = (1 - r_i)(1 - c_i) = 1 - r_i - c_i + r_i c_i$,
P{species i is proper top$\} = r_i(1 - c_i)$,
P{species i is proper basal$\} = c_i(1 - r_i)$.

Consequently, if N, T, B, I, T_P, and B_P respectively denote the numbers of not isolated, top, basal, intermediate, proper top, and proper basal species (these numbers are random variables), then

$$E(N) = S - \sum_{i=1}^{S} r_i c_i, \tag{5.4}$$

$$E(T) = \sum_{i=1}^{S} r_i, \qquad \text{var}(T) = \sum_{i=1}^{S} r_i(1 - r_i), \tag{5.5}$$

$$E(B) = \sum_{i=1}^{S} c_i, \qquad \mathrm{var}(B) = \sum_{i=1}^{S} c_i(1 - c_i), \qquad (5.6)$$

$$E(I) = 2S - E(T) - E(B) - E(N), \qquad (5.7)$$

$$E(T_P) = E(T) + E(N) - S, \qquad (5.8)$$

$$E(B_P) = E(B) + E(N) - S. \qquad (5.9)$$

A weak component is a maximal set of species linked to each other, directly or indirectly, by arcs regardless of their orientation. A web has more than one weak component with very low probability in the models to be developed and in the data, so this possibility receives no special treatment here.

Let the number of links, i.e., the number of elements of W equal to 1, be

$$L = |A| = \sum_{1 \le i < j \le S} w_{ij}, \qquad (5.10)$$

L_{BT} the number of basal-top links (links from some basal species to some top species), L_{BI} the number of basal-intermediate links, L_{IT} the number of intermediate-top links, and L_{II} the number of intermediate-intermediate links. For $i < j$, there is a basal-top link from i to j if and only if there is a link from i to j (with probability p_{ij}) and j is top (with probability r_j) and i is basal (with probability c_i). By such arguments, it is obvious that

$$E(L) = \sum_{j=2}^{S} \sum_{i=1}^{j-1} p_{ij}, \qquad (5.11)$$

$$\mathrm{var}(L) = \sum_{j=2}^{S} \sum_{i=1}^{j-1} p_{ij} q_{ij}, \qquad (5.12)$$

$$E(L_{BT}) = \sum_{j=2}^{S} \sum_{i=1}^{j-1} p_{ij} c_i r_j, \qquad (5.13)$$

$$E(L_{BI}) = \sum_{j=2}^{S} \sum_{i=1}^{j-1} p_{ij} c_i (1 - r_j), \qquad (5.14)$$

$$E(L_{IT}) = \sum_{j=2}^{S} \sum_{i=1}^{j-1} p_{ij} (1 - c_i) r_j, \qquad (5.15)$$

$$E(L_{II}) = \sum_{j=2}^{S} \sum_{i=1}^{j-1} p_{ij} (1 - c_i)(1 - r_j). \qquad (5.16)$$

There is a chain of length n, where n counts the number of links, that involves the $n + 1$ species $i_0, i_1, \ldots, i_n$ if and only if the following conditions all hold: $1 \leq i_0 < i_1 < \ldots < i_n \leq S$; and there is a link from i_h to i_{h+1} for all $h = 0, \ldots, n - 1$; and i_0 is basal; and i_n is top. Let C_n denote the number of chains of length n. Then

$$E(C_n) = \sum_{i_0=1}^{S-n} \sum_{i_1=i_0+1}^{S-n+1} \cdots \sum_{i_h=i_{h-1}+1}^{S-n+h} \cdots \sum_{i_n=i_{n-1}+1}^{S} \prod_{h=0}^{n-1} p_{i_h i_{h+1}} c_{i_0} r_{i_n} \qquad (5.17)$$

or

$$E(C_n) = \sum_{i_0=1}^{S-n} c_{i_0} \sum_{i_1=i_0+1}^{S-n+1} p_{i_0 i_1} \cdots \sum_{i_h=i_{h-1}+1}^{S-n+h} p_{i_{h-1} i_h} \cdots \sum_{i_n=i_{n-1}+1}^{S} p_{i_{n-1} i_n} r_{i_n}. \qquad (5.18)$$

For different values of n, different numbers of summations are required to evaluate $E(C_n)$. For numerical and symbolic computation, a recursive function is convenient.

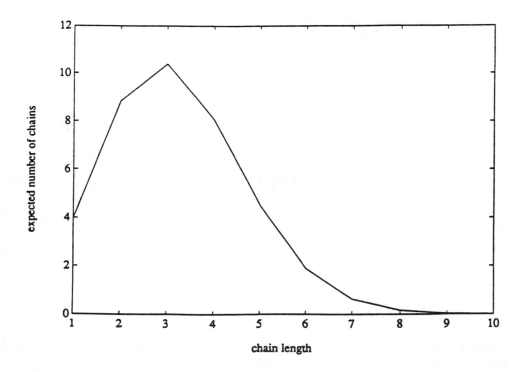

FIGURE 4. Expected number of chains of each length according to the linear cascade model of a web with $S = 17$ species and link probability $4/17$, based on $c = 4$. The expected total number of chains is 38.458. The expected numbers of chains of length 11 through 16 are not shown because their sum is less than 0.001.

Exercise. Simplify the above formulas (5.1) to (5.18) when $p_{ij} = p > 0$, for $1 \leq i < j \leq S$. In particular, prove (Cohen, Briand and Newman 1990, [9], p. 119) that (5.18) can be transformed to

$$E(C_n) = p^n q^{S-1} \sum_{k=n}^{S-1} (S - k) \binom{k-1}{n-1} q^{-k}, \quad n = 1, 2, \ldots, S - 1. \tag{5.19}$$

Figure 4 plots (5.19), the expected number of chains of each length according to the linear cascade model, for a web with $S = 17$ species and link probability 4/17, based on $c = 4.0$. The choice of this value of c is explained at (6.8).

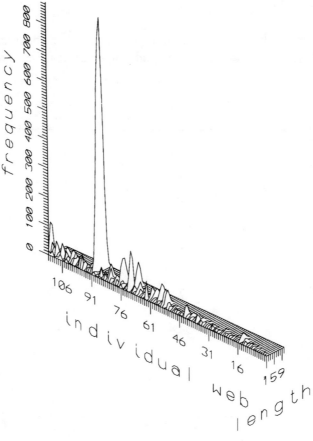

FIGURE 5. Observed numbers of chains with one to nine links in 113 community food webs. No chains longer than nine links were observed. Each slice parallel to the axis labeled "length" is the observed frequency distribution for one web. The height of a slice at each value of length is the observed number of chains of that length. The webs are ranked from the smallest (web 1) to the largest (web 113) number of species. The mean number of species was close to 17. Source: Cohen (1990, [7], p. 59), reprinted with permission.

For qualitative comparison, Figure 5 shows the observed number of chains of each length in 113 real webs. The raw data are given in Cohen, Briand, Newman (1990, Chap. 4, [9]). A detailed quantitative comparison between the observations and the predictions of the linear cascade model (Cohen, Briand, Newman 1990, [9], Chap. 3.4) shows that the predictions are not bad for all but 16 or 17 of the webs.

6. LIMIT THEORY OF THE LINEAR CASCADE MODEL FOR LARGE S

One reason for developing the linear cascade model was to explain a simple empirical pattern — as the number of species in a web increases, the fraction of all species that are top species seems neither to increase nor to decrease systematically, but fluctuates around a fixed proportion. (Caveat! This finding was based on webs with at most 48 species. The finding is under challenge by the very few webs reported so far with more than 48 species. In a science as young as this part of ecology, models and theories often aim at a shifting target.) How can the linear cascade model account for this (tentative) finding?

Giving away part of the answer to the previous exercise, the expected number of top species and the expected number of basal species in the linear cascade model are

$$E(T) = E(B) = [1 - q^S]/p , \qquad (6.1)$$

where $q = 1 - p$. Hence, since $p = c/S$, asymptotically

$$\lim_{S \to \infty} E(T)/S = \lim_{S \to \infty} E(B)/S = (1/c)(1 - e^{-c}) , \qquad (6.2)$$

Exercise. Prove that

$$\mathrm{var}(T) = \mathrm{var}(B) = (1 - q^S)/p - (1 - q^{2S})/(1 - q^2) \qquad (6.3)$$

and hence

$$\lim_{S \to \infty} \mathrm{var}(T/S) = \lim_{S \to \infty} \mathrm{var}(B/S) = 0 . \qquad (6.4)$$

The right side of (6.2) is a constant, independent of S. The model predicts that the proportion of species that are top species will asymptotically be independent of S and will [according to (6.4)] have vanishing variability, relative to S.

For quantitative comparison with actual data, it is necessary to examine the number of proper top species as a fraction of nonisolated species, because isolated species have been removed from the data. The right quantity to calculate would be $E(T_P/N)$, but that appears to be too difficult!

Exercise. Show that

$$E(T_P) = E(B_P) = S[(1 - q^S)/c - q^{S-1}] , \quad E(N) = S(1 - q^{S-1}) , \qquad (6.5)$$

and asymptotically

$$\lim_{S \to \infty} \frac{E(T_P)}{E(N)} = \lim_{S \to \infty} \frac{E(B_P)}{E(N)} = \frac{(1 - e^{-c})/c - e^{-c}}{1 - e^{-c}}. \tag{6.6}$$

Table 1 compares the linear cascade model's predicted asymptotic (as $S \to \infty$) fractions of nonisolated species that are proper basal, proper top, and intermediate with the observed fractions. It also compares the predicted asymptotic fractions of links of each possible kind with the observed fractions. To calculate these numerical predictions requires not only the formulas above but also a specific numerical value for the parameter c. The asymptotic formula (3.4) with $d = -1$ gives a rough-and-ready estimate

$$\hat{c} = 2E(L)/S. \tag{6.7}$$

The predictions in Table 1 take the ratio of links to species as exactly 2, though the actual ratio is 1.99, hence $\hat{c} = 4.0$.

113 Community food webs	Observed number	Observed fraction	Linear cascade model predicted fraction
proper basal species	353	0.186	0.231
intermediate species	1038	0.546	0.537
proper top species	511	0.269	0.231
all species	1902	1.000	1.000
basal-intermediate links	1029	0.272	0.264
basal-top links	230	0.061	0.114
intermediate -intermediate links	1194	0.316	0.359
intermediate-top links	1327	0.351	0.264
all links	3780	1.000	1.000

TABLE 1. Observed fractions of nonisolated species that are proper top, intermediate and proper basal, and observed fractions of links of each kind, in 113 community food webs. Source: Cohen (1990, [7], p. 56).

The fact that the right side of (6.7) is independent of S illustrates a property of the linear cascade model, and possibly a property of the real data, called scale-invariance — the form of the web (as measured by the proportion of its nonisolated species that are proper top species) is asymptotically independent of its scale, or size, or number of species. In the superlinear cascade model, by contrast, the proportion of nonisolated species that are proper top species slowly but markedly declines as the number of species increases (Cohen 1990, [7], p. 70). Whether this is a virtue or a defect of the superlinear cascade model will depend on large food webs now being collected.

Chain lengths may also be analyzed in detail, but the analysis is substantially more difficult (Newman and Cohen 1986, [28]; reprinted in Cohen, Briand, Newman 1990, [9], pp. 149-173). Let

$$C = \sum_{n=1}^{S-1} C_n \qquad (6.8)$$

be the total number of chains of all possible lengths. It may be shown that

$$\lim_{S \to \infty} E(C_n/C) = \lim_{S \to \infty} E(C_n)/E(C), \quad n = 1, 2, \ldots, S - 1, \qquad (6.9)$$

and both limits may be calculated from the coefficients of a power-series expansion of an explicitly stated generating function. Because, in the linear cascade model,

$$\lim_{S \to \infty} \mathrm{var}(C_n/S) = 0, \qquad (6.10)$$

the random variable C_n/C converges in probability as $S \to \infty$ to the limit in (6.9). For large c, say $c = 10$, the asymptotic relative frequency of chains of each length is normally distributed with mean equal to c and variance equal to $c - 1/2$ (Cohen, Briand and Newman 1990, [9], p. 155). For realistic values of c around 4.0, the asymptotic relative frequency of chains of each length is skewed to the right, with modal value close to c. Apparently $S = 17$ is close to infinity, because the asymptotic (large S) distribution of chain lengths closely resembles that shown for $S = 17$ in Figure 4.

The linear cascade model explains qualitatively an observation made independently by many biologists, that the longest chain length is short relative to the number of species in a web, even for webs with very large numbers of species. Let M be the maximum chain length, or height, in a realization of the linear cascade model; M is a random variable. Then for extremely large S, M grows like $\ln(S)/\ln(\ln(S))$, even more slowly than $\ln(S)$. The exact results (Newman and Cohen 1986, [28]) are a bit more surprising than this.

Define m^* to be the smallest positive integer m such that $c^{m+1}S/(m+2)! \leq (m+2)^{-1/2}$. Then, for large enough S, m^* is a nondecreasing sequence such that

$$\lim_{S \to \infty} \frac{m^*}{\ln S / \ln \ln S} = 1, \qquad (6.11)$$

and

$$\lim_{S \to \infty} P[M = m^* \text{ or } M = m^* - 1] = 1. \qquad (6.12)$$

That is, the distribution of the height is concentrated on just two numbers (which depend on S), m^* and $m^* - 1$. For very, very large numbers S of species, m^* grows at a rate that is essentially independent of c or $p = c/S$ (provided $c > 0$) and depends only on S. For extremely large S, m^* is approximately $\ln S / \ln \ln S$ in the sense that their ratio approaches 1. This does not imply that the difference between M and $\ln(S)/\ln(\ln(S))$ is arbitrarily small with high probability, because M and $\ln(S)/\ln(\ln(S))$ could differ by an amount that goes to infinity more slowly than $\ln(S)/\ln(\ln(S))$. Higher order correction terms to the asymptotic expansion (6.11) are given by Newman and Cohen (1986, [28]).

The estimated rate of convergence of $P[M = m^*$ or $M = m^* - 1]$ to 1 is very slow, namely,

$$1 - P[M = m^* \text{ or } M = m^* - 1] = O(m^{*-1/2}). \tag{6.13}$$

A qualitatively similar phenomenon has been observed elsewhere in the theory of random graphs. Bollobás and Erdös (1976, [2]) proved that the size of the maximal complete subgraph (clique) in a random graph takes one of at most two values (that depend on the size of the random graph) with a probability that approaches 1 as the random graph gets large, when the edge probability is held fixed, independent of the number of vertices. However, the asymptotic behavior of the at most two possible values for the size of the largest clique does depend on the fixed probability that there is an edge between any two given vertices, according to Bollobás and Erdös (1976, [2]).

According to the superlinear cascade model (Newman 1991, [27]; Cohen and Newman 1991 , [12], when S becomes very large, the ratio between M and epS is arbitrarily close to one, where $p = cS^d$, with a probability that approaches one. So M grows like ecS^{1+d}, much faster than in the linear cascade model.

The average chain length $\mu(S)$ in the homogeneous cascade model web with S species and probability p of a link is the same for both the linear and the superlinear cascade models (Newman and Cohen 1986, [28], p. 358):

$$\mu(S) = Sp\frac{(1+p)^{S-1} - (1-p)^{S-1}}{(1+p)^S - (1-p)^S - 2Sp(1-p)^{S-1}} . \tag{6.14}$$

In the linear cascade model, the average chain length approaches a finite limit for large S:

$$\lim_{S\to\infty} \mu(S) = \frac{c(1 - e^{-2c})}{1 - e^{-2c}(1 + 2c)} , \tag{6.15}$$

which is approximately 4.01 if $c = 4.0$. This limit is approached fairly rapidly. For example, the average chain length is predicted to exceed 3.5 as soon as $S \geq 30$; if $S = 100, M(100) = 3.86$.

Comparing (6.11) and (6.15) shows that the ratio of the height to the average chain length in the linear cascade model increases without limit as the number of species gets large. Here is an intuitive explanation for this phenomenon. First, as mentioned above, there is a fixed limiting distribution of chain lengths as S gets large. The limiting average chain length given by (6.15) is the mean of this fixed distribution. Second, as S gets large, more and more chains are sampled from this fixed distribution of chain lengths, so that very long chains, with a fixed low probability, are more likely to be observed and the height increases without limit. The particular form of the increase given by (6.11) is expected with a sample size proportional to S taken from a fixed distribution whose tail probabilities behave approximately like those of a Poisson distribution. Such a tail behavior can be proved by using the exact formulas for the generating function of the distribution. The expected number of chains of all lengths, which represents the sample size, is proportional to S (Newman and Cohen 1986, [28]).

In the superlinear cascade model, as S increases without limit, the ratio of the average chain length to Sp approaches 1 (Cohen and Newman 1991, [12]). The ratio of the height to the average chain length in the superlinear cascade model approaches the limit e as the number of species gets large. Thus there is a qualitative difference in the ratio of the height to the average chain length between the cascade and the superlinear homogeneous models.

7. DYNAMICS OF FOOD WEBS

When food webs were still young, few people had hopes of comprehending their structural and dynamic complexity in simple models. Instead, early explorers focused on a single feeding link, and on the dynamics of the two populations interacting through that link. Kingsland (1985, [22]) tells the history in an interesting way. Alfred J. Lotka, trained as a physical chemist, borrowed the style of the kinetic equations of chemistry to model a single predatory population eating a single prey population (his great 1924 book [25] summarizes his work). If $N(t)$ is the number or biomass (quantity of living material) of the resource population and $P(t)$ is the number or biomass of the predatory population, the Lotka-Volterra equations for predator-prey interactions are (May 1981, Chap. 5, [26]):

$$\frac{dN}{dt} = aN(t) - \alpha N(t)P(t), \qquad (7.1)$$

$$\frac{dP}{dt} = -bP(t) + \beta N(t)P(t), \qquad (7.2)$$

where the coefficients a, b, α and β are all assumed positive. Among other simplifications, these equations ignore genetic variations within the predator and prey populations, differences in ages within a population with respect to probabilities or rates of predation, reproduction and mortality, the spatial distribution of predators and prey, the possible influences of environmental changes (weather, diurnal cycles, volcanoes, tides) on both species and their interactions, all other species that might interact with either or both species, and any physical and chemical constraints (such as limited nutrients or limited space) that affect either or both species.

Dynamic equations of the same embarrassing simplicity as (7.1) and (7.2) are often presented as more realistic than the structural models described in the previous sections, and are often accepted as such by both ecologists and mathematicians. The reason for this acceptance is simply familiarity, not evidence. In general, the empirical support for dynamic equations such as the Lotka-Volterra predator-prey equations, though non-zero (Gause 1935, [17]; see Krebs 1972, [23], pp. 250-272 for a review), is slimmer than that for the cascade models.

Realistic or not, the Lotka-Volterra predator-prey equations (hereafter abbreviated to LV's) can obviously be generalized to S interacting species. If $u_i(t)$ is the abundance or biomass of the ith species at time t, then it is assumed that there exist a real $S \times S$ matrix $B = (b_{ij})$ and a real $S \times 1$ vector $e = (e_i)$, with both B and e independent of time t, such that, for all $t \geq 0$,

$$\frac{du_i}{dt} = u_i\left(e_i + \sum_{j=1}^{S} b_{ij}u_j\right), \qquad u_i(0) > 0, \qquad i = 1, \ldots, S. \qquad (7.3)$$

The coefficient b_{ij} measures the effect of species j on the growth rate of species i. It is assumed that (7.3) has a stationary solution in the first orthant, i.e., that there exists a constant $S \times 1$ vector $q = (q_i)$ such that

$$0 = e_i + \sum_{j=1}^{S} b_{ij}q_j, \qquad q_i > 0, \qquad i = 1, \ldots, S. \qquad (7.4)$$

LVs make several further assumptions that could be shortcomings empirically. The LVs assume that there are no mutualistic interactions between species, that all interactions among species are strictly pairwise, and that the pairwise interactions follow a simple mass-action law specified by the product of abundances or biomasses.

Nevertheless, LVs have been extensively used for food web theory (Pimm 1982, [31]; Cohen, Luczak, Newman, Zhou 1990, [10]; Pimm, Lawton, Cohen 1991, [33]). A major problem is that the LVs have $S^2 + S$ parameters (the b_{ij} and the e_i), not counting the initial conditions. Yet, in the real world, there are only roughly twice as many links as species. It seems unlikely that all the possible interactions really matter dynamically.

Since the values of the interaction coefficients in the matrix B in (7.3) can never be known exactly, but the signs of the coefficients can be estimated more reliably, it is natural to consider the behavior of (7.3) when the value of any interaction coefficient b_{ij} is changed to some other number with the same sign, while an interaction coefficient that is zero is left at zero. Under certain conditions, some of which will be described in a moment, the LVs bequeath their stability to the whole family of equations obtained by replacing the interaction coefficients by others with the same signs. Under such conditions, (7.3) is said to be qualitatively stable. Qualitative stability is a natural concept for linking dynamic models with structural models based on random graphs, because structural models deal with the presence or absence of certain interactions, and not with the magnitudes of those interactions. What is surprising is that it is possible to say things about qualitative stability that connect usefully with structural food web models.

A recently proposed hybrid of the LVs and the linear cascade model, called the Lotka-Volterra cascade model (LVCM), assumes the population dynamics of the LVs while letting a refinement of the cascade model determine the interactions between species (Cohen, Luczak, Newman, and Zhou 1990, [10]). Mathematical analysis of the LVCM combines the theory of large random digraphs with the qualitative theory of nonlinear differential equations to characterize the global asymptotic stability of ecological communities in the double limit of large time and large numbers of species.

The LVCM links the cascade model to the Lotka-Volterra model by discriminating among the possible population dynamical effects caused by each feeding link. In principle, when species j eats species i, there could be a positive, a negative, or no effect on the population growth rate of species j, and a positive, a negative, or no effect on the population growth rate of species i, for a total of nine possible pairs of effects. The LVCM ignores the five possible pairs of effects where j eating i hurts the population growth rate of species j or helps the population growth rate of species i. Thus if species j eats species i, the LVCM supposes that one of four biological effects occurs:

(i) the feeding has no effect on the growth of species j but hurts the growth of species i; or

(ii) the feeding helps the growth of species j but has no effect on the growth of species i; or

(iii) the feeding helps species j and hurts species i; or

(iv) the feeding has no effect on the growth of either j or i.

Corresponding to each biological effect, assume that:

(i) $b_{ji} = 0$ and $b_{ij} < 0$; or

(ii) $b_{ji} > 0$ and $b_{ij} = 0$; or

(iii) $b_{ji} > 0$ and $b_{ij} < 0$; or

(iv) $b_{ji} = b_{ij} = 0$.

Because of (iv), the event:

(iv') there is no dynamic interaction of any kind between species i and j represents two biologically distinct situations: predation without dynamic effects [described by (iv)] (e.g., the old lady who accidentally swallowed a fly), and the absence of predation [i.e., the absence of an edge (i, j) in W].

The LVCM assumes that events (i), (ii), (iii), and (iv') occur independently for each pair i, $j = 1$, $\ldots$, S such that $i < j$, with probabilities, respectively, r/S, s/S, t/S and $1 - (r + s + t)/S$, where r, s, t are nonnegative constants that do not depend on S. [Predation without dynamic effects (iv) occurs with probability $c/S - (r + s + t)/S$.]

More formally, for $S = 1$, 2, $\ldots$, let N_S be the system (7.3) with randomly chosen coefficients where, with probability 1, $b_{ii} < 0$ for $i = 1, \ldots, S$ and the pairs $\{b_{ji}, b_{ij}\}$ for each $i, j = 1, \ldots, S$ with $i < j$ are chosen independently with probabilities

$$
\begin{aligned}
\text{P}\,\{b_{ji} = 0 \text{ and } b_{ij} < 0\} &= r/S, & (i) \\
\text{P}\,\{b_{ji} > 0 \text{ and } b_{ij} = 0\} &= s/S, & (ii) \\
\text{P}\,\{b_{ji} > 0 \text{ and } b_{ij} < 0\} &= t/S, & (iii) \\
\text{P}\,\{b_{ji} = 0 \text{ and } b_{ij} = 0\} &= 1 - (r + s + t)/S. & (iv')
\end{aligned}
$$

and the vector $e_S = (e_i)_{i=1}^S$ is chosen (depending on $b_S = (b_{ij})_{i,j=1}^S$) so that for some vector $q_S > 0$ (also depending on b_S), $0 = e_S + b_S q_S$. The sequence of systems $\{N_S\}_{S=1}^\infty$ defines the LVCM.

To state the interesting facts about the LVCM requires some further definitions. For any real finite scalar s, define $\text{sign}(s) = +1$ if $s > 0$, $\text{sign}(s) = -1$ if $s < 0$ and $\text{sign}(s) = 0$ if $s = 0$. Define $\bar{B} \sim B$ (read: $\bar{B}$ is sign-equivalent to B) if and only if, for all i, j, $\text{sign}(b_{ij}) = \text{sign}(\bar{b}_{ij})$. Let $\bar{N}$ refer to the family of equations (7.3) when (i) B is replaced by any $S \times S$ matrix $\bar{B} \sim B$, and (ii) e is replaced by any $S \times 1$ vector $\bar{e}$ such that $0 = \bar{e} + \bar{B}\bar{q}$ has a positive solution $\bar{q} > 0$. Positive initial conditions $u(0) > 0$ are assumed throughout. A result is considered "qualitative" if it refers to all of $\bar{N}$.

Define LVs (7.3) to be qualitatively globally asymptotically stable (q.g.a.s.) if and only if every solution of every system in the family $\bar{N}$ is bounded and has a limit as $t \to \infty$ and that limit is independent of the initial conditions. This property may formalize what some ecologists mean by ecological stability (Pimm 1984, [32]) for some ecological communities, but no claim is made that it formalizes ecologists' concept of ecological stability for all communities.

The LVCM has what physicists call a phase transition. As the parameters of the LVCM cross a certain critical surface, the probability of being qualitatively globally asymptotically stable changes from positive to zero. Cohen, Łuczak, Newman, and Zhou (1990, [10]) show where the critical surface is and how the LVCM behaves on either side of the surface as well as on it.

Let $x: [1, \infty) \to (0, 1]$ be the smallest root of $x(z)e^{-x(z)} = ze^{-z}$.

(i) If

$$r + t < 1 \text{ and } s + t < 1, \quad or \tag{7.5}$$

$$r + t \geq 1 \text{ but } s + t < x(r + t), \quad or \tag{7.6}$$

$$s + t \geq 1 \text{ but } r + t < x(s + t) \tag{7.7}$$

then

$$\lim_{S \to \infty} \text{P}\,\{N_S \text{ is q.g.a.s.}\} = \rho > 0, \quad where \tag{7.8}$$

$$
\rho = \begin{cases}
e^{(r+t)(s+t)/2} \dfrac{(r + t)e^{s+t} - (s + t)e^{r+t}}{r - s}, & r \neq s, \\
e^{(r+t)^2/2} e^{r+t}(1 - r - t), & r = s.
\end{cases} \tag{7.9}
$$

When $(r + t)(s + t) > 0$, then $\rho < 1$.

(ii) If

$$r + t \geq 1 \quad \text{and} \quad s + t \geq x(r + t), \quad or \tag{7.10}$$

$$s + t \geq 1 \quad \text{and} \quad r + t \geq x(s + t) \tag{7.11}$$

then

$$\lim_{S \to \infty} \mathrm{P} \{N_S \text{ is q.g.a.s.}\} = 0. \tag{7.12}$$

The proof of this theorem is difficult and will not be attempted here. The proof depends crucially on facts, first developed by Quirk and Ruppert (1965, [34]), about the qualitative stability of linear systems.

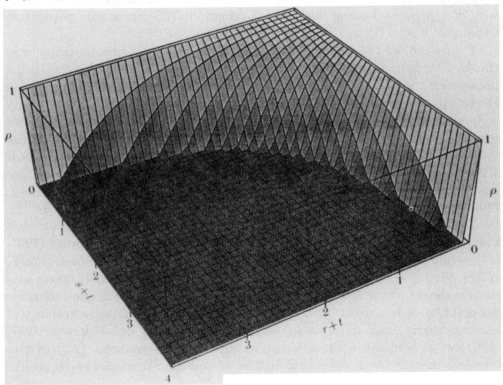

FIGURE 6. Perspective view of the probability of qualitative global asymptotic stability in the Lotka-Volterra cascade model (LVCM), in the limit as the number of species S approaches infinity, as a function of the parameters $r + t$ and $s + t$, which are defined in the text. The limiting probability is 0 in the flat region in the foreground. Small values of $r + t$ and $s + t$ assure a high limiting probability of qualitative global asymptotic stability. Source: Cohen, Łuczak, Newman, Zhou (1990, [10], p. 616), reprinted with permission.

Informally speaking, a critical surface divides the three-dimensional parameter space $\{(r, s, t) | r \geq 0, s \geq 0, t \geq 0\}$ of the LVCM into two regions. In region (i), where (7.5) or (7.6) or (7.7) holds, as the number of species becomes large, the probability that the LVCM is qualitatively globally asymptotically stable (q.g.a.s.) approaches a positive limit. This limiting probability is given explicitly by (7.9). In region (ii), where (7.10) or (7.11) holds, as the number of species becomes large, the probability that the LVCM is q.g.a.s. approaches zero. The critical surface for the phase transition in the stability of the LVCM is exactly the same as the critical surface for a phase transition when a giant strongly connected component suddenly appears in a certain digraph $D(B)$, called the interaction digraph, associated with the matrix B of interaction coefficients. $D(B)$ has S vertices, an arrow (j, i) from vertex j to vertex i if $b_{ij} \neq 0$ and $b_{ji} = 0$, and a bidirectional arrow $\{i, j\}$ between i and j (equivalent to a pair of unidirectional arrows (j, i) and (i, j)) if $b_{ij} b_{ji} < 0$. (The case $b_{ij} b_{ji} > 0$ with $i \neq j$ was excluded in constructing the LVCM.) Whereas the web W represents feeding relations, the digraph $D(B)$ represents population dynamical interactions.

Figure 6 plots the probability of qualitative global asymptotic stability in the LVCM, in the limit of large numbers of species S, as a function of the parameters $r+t$ and $s+t$. The limiting probability is zero in the flat region in the foreground. Small values of $r+t$ and $s+t$ assure a high limiting probability of qualitative global asymptotic stability. The transition from the region of positive probability to the region of zero probability is abrupt in the sense that the derivative of ρ changes discontinuously as the frontier of stability is crossed. For example, along the diagonal cross-section through the surface defined by $r + t = s + t$, the derivative of ρ is evidently zero when $r + t > 1$, but, as $r + t$ approaches 1 from below, the derivative of ρ approaches $-e^{3/2}$.

The theory of the qualitative stability of linear and nonlinear systems has made great progress since the pioneering paper of Quirk and Ruppert (1965, [34]). Jeffries, Klee, and van den Driessche (1987, [19]) and Jeffries (1988b, [21]) review and extend this progress. The qualitative stability of linear systems is closely connected to the qualitative stability of LVs. An open scientific opportunity is to exploit recent discoveries about the qualitative stability of linear systems (such as those in Jeffries, Klee and van den Driessche 1987, [19]) and of nonlinear systems (such as those in Jeffries 1988a, [20]) to give new information about stochastic families of nonlinear dynamical systems like the LVCM.

Information about qualitative global asymptotic stability could assist the design of managed ecological systems such as closed ecological life support systems for space travel, nature reserves, and complex chemostats, microcosms, and mesocosms. Qualitatively globally asymptotically stable systems may be desirable for practical and aesthetic reasons, because perturbations that do not change the signs of the interactions between species will not alter the existence of a long-run globally stable equilibrium. If future empirical studies confirm its usefulness, the LVCM would suggest designs that maximize (subject to some constraints) the probability of being q.g.a.s., asymptotically for large numbers of species. They would be those designs that satisfy the hypotheses (7.5) to (7.7) with large values of ρ. The LVCM would suggest avoiding ecological designs that satisfy (7.10) or (7.11), which have little chance of being q.g.a.s., asymptotically for large numbers of species.

Apart from its potential uses in ecological design, the LVCM warns of the possibility that gradual, smooth changes in the probabilities $r, s,$ and t of various kinds of dynamic interactions related to feeding can have abrupt effects on the long-run probability of qualitative global stability or instability of ecological communities. Such phase transitions in stability are driven by phase transitions in the structure of underlying random graphs or digraphs, and are found in many large, random structures (see Cohen [1988], [5] for an expository account).

ACKNOWLEDGMENTS

J.E.C. acknowledges helpful comments from Charles M. Newman, the support of U.S. National Science Foundation grant BSR87-05047, and the hospitality of Mr. and Mrs. William T. Golden.

References

[1] **Bollobás, B.,** *Random Graphs.* Academic Press, London, 1985.

[2] **Bollobás B. and Erdös, P.,** Cliques in random graphs, *Math. Proc. Camb. Phil. Soc.*, **80**, 419–427, 1976.

[3] **Camerano, L.,** Dell'equilibrio dei viventi mercè la reciproca distruzione, *Atti della Reale Accademia delle Scienze di Torino*, **15**, 393–414, 1880.

[4] **Cohen, J. E.,** *Food Webs and Niche Space.* Princeton University Press, Princeton, NJ. Monographs in Population Biology, Vol. 11, 1978, 190.

[5] **Cohen, J. E.,** Threshold phenomena in random structures, *Discrete Applied Mathematics*, 113–128, 1988. Reprinted in *Applications of Graphs in Chemistry and Physics*, Ed. J. W. Kennedy and L. V. Quintas, Amsterdam, North-Holland, 113–128, 1988.

[6] **Cohen, J. E.,** Food webs and community structure. in *Perspectives on Ecological Theory*, Eds. J. Roughgarden, R. M. May, S. Levin, Princeton University Press, Princeton, 1989, 181–202.

[7] **Cohen, J. E.,** A stochastic theory of community food webs. VI. Heterogeneous alternatives to the cascade model, *Theoretical Population Biology*, **37**, 55–90, 1990.

[8] **Cohen, J. E., Briand, F. and Newman, C. M.,** A stochastic theory of community food webs. III. Predicted and observed lengths of food chains, *Proc. Roy. Soc.*, (London), **B. 228**, 317–353, 1986.

[9] **Cohen, J. E., Briand, F. and Newman, C. M.,** *Community Food Webs: Data and Theory*, Biomathematics, **20**, Springer-Verlag, Heidelberg, Berlin, New York, 1990.

[10] **Cohen, J. E., Łuczak, T., Newman, C. M. and Zhou, Z. M.,** Stochastic structure and nonlinear dynamics of food webs: qualitative stability in a Lotka-Volterra cascade model, *Proc. Roy. Soc.*, (London), series **B. 240**, 607–627, 1990.

[11] **Cohen, J. E., and Newman, C. M.,** A stochastic theory of community food webs. I. Models and aggregated data, *Proc. Roy. Soc.*, (London), **B. 224**, 421–448, 1985.

[12] **Cohen, J. E. and Newman, C. M.,** 1991 Community area and food chain length: theoretical predictions, *American Naturalist*, **138**, 1542–1554, 1991.

[13] **Cohen, J. E. and Palka, Z. J.,** A stochastic theory of community food webs: V. Intervality and triangulation in the trophic niche overlap graph, *American Naturalist*, **135(3)**, 435–463, 1990.

[14] **Darwin, Charles,** 1859 *On the Origin of Species*. Facsimile of 1st ed. with Intro. by Ernst Mayr. Harvard University Press, Cambridge, MA, 1964.

[15] **Erdös, P. and Rényi, A.,** On the evolution of random graphs, *Publ. Math. Inst. Hung. Acad. Sci.*, **5**, 17–61, 1960.

[16] **Forbes, S. A.,** *Ecological Investigations of Stephen Alfred Forbes*, [Reprints.] Arno Press, New York, 1977.

[17] **Gause, G. F.,** 1934 *The Struggle for Existence*. Reprinted 1964. Hafner, New York.

[18] **Gelenbe, E., Nelson, R., Philips, T. and Tantawi, A.,** An approximation of the processing time for a random graph model of parallel computation, in *Proceedings of the Fall Joint Computer Conference*, H. S. Stone, Ed., IEEE Computer Society Press, Washington, DC, 691–697, 1986.

[19] **Jeffries, C., Klee, V. and van den Driessche, P.,** Qualitative stability of linear systems, *Linear Algebra and Its Applications*, **87**, 1–48, 1987.

[20] **Jeffries, C.,** Ecosystem modelling: qualitative stability. in M. G. Singh, *Systems and Control Encyclopedia; Theory, Technology, Applications*, Pergamon, Oxford, 1988, 1348–1351.

[21] **Jeffries, C.**, Eigenvalues, stability, and color tests, *Linear Algebra and Its Applications*, **107**, 1988, 65–76.

[22] **Kingsland, S. E.**, *Modeling Nature: Episodes in the History of Population Ecology*. University of Chicago Press, Chicago, 1985.

[23] **Krebs, Charles J.**, *Ecology: The Experimental Analysis of Distribution and Abundance*. Harper and Row, New York, 1972.

[24] **Lawton, J. H.**, Food webs. In J. M. Cherrett, Ed., Ecological Concepts: *The Contribution of Ecology to an Understanding of the Natural World*, Blackwell Scientific, Oxford, 1989, 43–78.

[25] **Lotka, A. J.**, 1924 *Elements of Physical Biology*. Williams and Wilkins, Baltimore. Rep. 1956: Elements of Mathematical Biology. Dover, New York.

[26] **May, R. M., ed.** *Theoretical Ecology: Principles and Applications*. 2d ed. Blackwell Scientific, Oxford, 1981.

[27] **Newman, C. M.**, Chain lengths in certain random directed graphs, *Random Structures and Algorithms*, 1991. In Press.

[28] **Newman, C. M. and Cohen, J. E.**, A stochastic theory of community food webs. IV. Theory of food chain lengths in large webs, *Proc. Roy. Soc.*, (London), **B. 228**, 355–377, 1986.

[29] **Niering, W. A.**, Terrestrial ecology of Kapingamarangi Atoll, Caroline Islands, *Ecological Monographs*, **33**, 131–160, 1963.

[30] **Palmer, E. M.**, *Graphical Evolution: An Introduction to the Theory of Random Graphs*. John Wiley & Sons, New York, 1985.

[31] **Pimm, S. L.**, *Food Webs*. Chapman and Hall, London, 1982.

[32] **Pimm, S. L.**, The complexity and stability of ecosystems, *Nature*, **307**, 321–326, 1984.

[33] **Pimm, S., Lawton, J., Cohen, J. E.**, Food web patterns and their consequences, *Nature*, **350**, 669–674, 1991.

[34] **Quirk, J. and Ruppert, R.**, Qualitative economics and the stability of equilibrium, *Review of Economic Studies*, **32**, 311–326, 1965.

[35] **Robinson, D. F. and Foulds, L. R.**, *Digraphs: Theory and Techniques*. Gordon and Breach, New York, 1980.

[36] **Schoener, T. W.**, Food webs from the small to the large, *Ecology*, **70**, 1559–1589, 1989.

[37] **Yodzis, P.,** *Introduction to Theoretical Ecology.* Harper and Row, New York, 1989.

Chapter 9

HOW MANY TIMES SHOULD YOU SHUFFLE A DECK OF CARDS?[*]

Brad Mann
Department of Mathematics
Harvard University

ABSTRACT

In this paper a mathematical model of card shuffling is constructed and used to determine how much shuffling is necessary to randomize a deck of cards. The crucial aspect of this model is rising sequences of permutations, or equivalently descents in their inverses. The probability of an arrangement of cards occurring under shuffling is a function only of the number of rising sequences in the permutation. This fact makes computation of variation distance, a measure of randomness, feasible; for in an n card deck there are at most n rising sequences but $n!$ possible arrangements. This computation is done exactly for $n = 52$, and other approximation methods are considered.

1. INTRODUCTION

How many times do you have to shuffle a deck of cards in order to mix them reasonably well? The answer is about seven for a deck of fifty-two cards, or so claims Persi Diaconis. This somewhat surprising result made the *New York Times* [5] a few years ago. It can be seen by an intriguing and yet understandable analysis of the process of shuffling. This paper is an exposition of such an analysis in Bayer and Diaconis [2], though many people have done work on shuffling. These have included E. Gilbert and Claude Shannon at Bell Labs in the 1950s, and more recently Jim Reeds and David Aldous.

---

[*]This article was written for the Chance Project at Dartmouth College supported by the National Science Foundation and The New England Consortium for Undergraduate Education.

2. WHAT IS A SHUFFLE, REALLY?

2.1 Permutations

Let us suppose we have a deck of n cards, labeled by the integers from 1 to n. We will write the deck with the order of the cards going from left to right, so that a virgin unshuffled deck would be written $123\cdots n$. Hereafter we will call this the natural order. The deck after complete reversal would look like $n\cdots321$.

A concise mathematical way to think about changing orderings of the deck is given by permutations. A permutation of n things is just a one-to-one map from the set of integers, between 1 and n inclusive, to itself. Let S_n stand for the set of all such permutations. We will write the permutations in S_n by lower case Greek letters, such as π, and can associate with each permutation a way of rearranging the deck. This will be done so that the card in position i after the deck is rearranged was in position $\pi(i)$ before the deck was rearranged. For instance, consider the rearrangement of a 5 card deck by moving the first card to the end of the deck and every other card up one position. The corresponding permutation π_1 would be written

i	1	2	3	4	5
$\pi_1(i)$	2	3	4	5	1

Or consider the so-called "perfect shuffle" rearrangement of an 8 card deck, which is accomplished by cutting the deck exactly in half and then alternating cards from each half, such that the top card comes from the top half and the bottom card from the bottom half. The corresponding permutation π_2 is

i	1	2	3	4	5	6	7	8
$\pi_2(i)$	1	5	2	6	3	7	4	8

Now we don't always want to give a small table to specify permutations. So we may condense notation and just write the second line of the table, assuming the first line was the positions 1 through n in order. We will use brackets when we do this to indicate that we are talking about permutations and not orders of the deck. So in the above examples we can write $\pi_1 = [23451]$ and $\pi_2 = [15263748]$.

It is important to remember the distinction between orderings of the deck and permutations. An ordering is the specific order in which the cards lie in the deck. A permutation, on the other hand, does not say anything about the specific order of a deck. It only specifies some rearrangement, i.e., how one ordering changes to another, regardless of what the first ordering is. For example, the permutation $\pi_1 = [23451]$ changes the ordering 12345 to 23451, as well as rearranging 41325 to 13254, and 25431 to 54312. (What will be true, however, is that the numbers we write down for a permutation will always be the same as the numbers for the ordering that results when the rearrangement corresponding to this permutation is done to the naturally ordered deck.) Mathematicians say this convention gives an *action* of the group of permutations S_n on the set of orderings of the deck. (In fact, the action is a simply transitive one, which just means there is always a unique permutation that rearranges the deck from any given order to any other given order.)

Now we want to consider what happens when we perform a rearrangement corresponding to some permutation π, and then follow it by a rearrangement corresponding to some other permutation τ. This will be important later when we wish to condense several rearrangements into one, as in shuffling a deck of cards repeatedly. The card in position i after both rearrangements are done was in position $\tau(i)$ when the first but not the second rearrangement was done. But the card in position j after the first but not the second rearrangement was in position $\pi(j)$ before any rearrangements. So set $j = \tau(i)$ and get that the card in position i after both rearrangements was in position $\pi(\tau(i))$ before any rearrangements. For this reason we define the *composition* $\pi \circ \tau$ of π and τ to be the map which takes i to $\pi(\tau(i))$, and we see that doing the rearrangement corresponding to π and then the one corresponding to τ is equivalent to a single rearrangement given by $\pi \circ \tau$. (Note that we have $\pi \circ \tau$ and not $\tau \circ \pi$ when π is done first and τ second. In short, the order matters greatly when composing permutations, and mathematicians say that S_n is noncommutative.) For example, we see the complete reversal of a 5 card deck is given by $\pi_3 = [54321]$, and we can compute the composition $\pi_1 \circ \pi_3$.

i	1	2	3	4	5
$\pi_3(i)$	5	4	3	2	1
$\pi_1 \circ \pi_3(i)$	1	5	4	3	2

2.2 Shuffles

Now we must define what a shuffle, or method of shuffling, is. It's just a probability density on S_n, considering each permutation as a way of rearranging the deck. This means that each permutation is given a certain fixed probability of occurring, and that all such probabilities add up to one. A well-known example is the top-in shuffle. This is accomplished by taking the top card off the deck and reinserting it in any of the n positions between the $n-1$ cards in the remainder of the deck, doing so randomly according to a uniform choice. This means the density on S_n is given by $1/n$ for each of the cyclic permutations $[234 \cdots (k-1)k1(k+1)(k+2) \cdots (n-1)n]$ for $1 \leq k \leq n$, and 0 for all other permutations. This is given for a deck of size $n = 3$ in the following example:

permutation	[123]	[213]	[231]	[132]	[321]	[312]
probability under top-in	1/3	1/3	1/3	0	0	0

What this definition of shuffle leads to, when the deck is repeatedly shuffled, is a random walk on the group of permutations S_n. Suppose you are given a method of shuffling Q, meaning each permutation π is given a certain probability $Q(\pi)$ of occurring. Start at the identity of S_n, i.e., the trivial rearrangement of the deck that does not change its order at all. Now take a step in the random walk, which means choose a permutation π_1 randomly, according to the probabilities specified by the density Q. (So π_1 is really a random variable.) Rearrange the deck as directed by π_1, so that the card now in position i was in position $\pi_1(i)$ before the rearrangement. The probability of each of these various rearrangings of the deck is obviously just the density of π_1, given by Q. Now repeat the procedure for a second step in the random walk, choosing another permutation π_2, again randomly according to the density Q (i.e., π_2 is a second, independent random variable with the same density as π_1). Rearrange the deck according to π_2. We saw in the last section on permutations that the effective rearrangement of the deck including both permutations is given by $\pi_1 \circ \pi_2$.

What is the probabiltiy of any particular permutation now, i.e., what is the density for $\pi_1 \circ \pi_2$? Call this density $Q^{(2)}$. To compute it, note the probability of π_1 being chosen, and then π_2, is given by $Q(\pi_1) \cdot Q(\pi_2)$, since the choices are independent of each other. So for any particular permutation π, $Q^{(2)}(\pi)$ is given by the sum of $Q(\pi_1) \cdot Q(\pi_2)$ for all pairs π_1, π_2 such that $\pi = \pi_1 \circ \pi_2$, since in general there may be many different ways of choosing π_1 and then π_2 to get the same $\pi = \pi_1 \circ \pi_2$. (For instance, completely reversing the deck and then switching the first two cards gives the same overall rearrangement as first switching the last two cards and then reversing the deck.) This way of combining Q with itself is called a *convolution* and written $Q * Q$:

$$Q^{(2)}(\pi) = Q * Q(\pi) = \sum_{\pi_1 \circ \pi_2 = \pi} Q(\pi_1) Q(\pi_2) = \sum_{\pi_1} Q(\pi_1) Q(\pi_1^{-1} \circ \pi).$$

Here π_1^{-1} denotes the inverse of π_1, which is the permutation that "undoes" π_1, in the sense that $\pi_1 \circ \pi_1^{-1}$ and $\pi_1^{-1} \circ \pi_1$ are both equal to the identity permutation which leaves the deck unchanged. For instance, the inverse of [253641] is [613524].

So we now have a shorthand way of expressing the overall probability density on S_n after two steps of the random walk, each step determined by the same density Q. More generally, we may let each step be specified by a different density, say Q_1 and then Q_2. Then the resulting density is given by the convolution

$$Q_1 * Q_2(\pi) = \sum_{\pi_1 \circ \pi_2 = \pi} Q_1(\pi_1) Q_2(\pi_2) = \sum_{\pi_1} Q_1(\pi_1) Q_2(\pi_1^{-1} \circ \pi).$$

Further, we may run the random walk for an arbitrary number, say k, of steps, the density on S_n being given at each step i by some Q_i. Then the resulting density on S_n after these k steps will be given by $Q_1 * Q_2 * \cdots * Q_k$. Equivalently, doing the shuffle specified by Q_1, and then the shuffle specified by Q_2, and so on, up through the shuffle given by Q_k, is the same as doing the single shuffle specified by $Q_1 * Q_2 * \cdots * Q_k$. In short, repeated shuffling corresponds to convoluting densities. This method of convolutions is complicated, however, and we will see later that, for a realistic type of shuffle, there is a much easier way to compute the probability of any particular permutation after any particular number of shuffles.

3. THE RIFFLE SHUFFLE

We would now like to choose a realistic model of how actual cards are physically shuffled by people. A particular one with nice mathematical properties is given by the "riffle shuffle". (Sometimes called the GSR shuffle, it was developed by Gilbert and Shannon, and independently by Reeds.) It goes as follows. First cut the deck into two packets, the first containing k cards, and the other the remaining $n - k$ cards. Choose k, the number of cards cut, according to the binomial density, meaning the probability of the cut occurring exactly after k cards is given by $\binom{n}{k}/2^n$.

Once the deck has been cut into two packets, interleave the cards from each packet in any possible way, such that the cards of each packet maintain their own relative order. This means that the cards originally in positions $1, 2, 3, \ldots k$ must still be in the same order in the deck after it is shuffled, even if there are other cards in-between; the same goes for the cards originally in positions $k + 1, k + 2, \ldots n$. This requirement is quite natural when you think of how a person shuffles two packets of cards, one in each hand. The cards in the left hand must still be in the same relative order in the shuffled deck, no matter how they are interleaved with the cards from the other packet, because the cards in the left hand are dropped in order when shuffling; the same goes for the cards in the right hand.

Choose among all such interleavings uniformly, meaning each is equally likely. Since there are $\binom{n}{k}$ possible interleavings (as we only need choose k spots among n places for the first packet, the spots for the cards of the other packet then being determined), this means any particular interleaving has probability $1/\binom{n}{k}$ of occurring. Hence the probability of any particular cut followed by a particular interleaving, with k the size of the cut, is $\binom{n}{k}/2^n \cdot 1/\binom{n}{k} = 1/2^n$. Note that this probability $1/2^n$ contains no information about the cut or the interleaving! In other words, the density of cuts and interleavings is uniform — every pair of a cut and a possible resulting interleaving has the same probability.

This uniform density on the set of cuts and interleavings now induces in a natural way a density on the set of permutations, i.e., a shuffle, according to our definition. We will call this the riffle shuffle and denote it by R. It is defined for π in S_n by $R(\pi)$ = the sum of the probabilities of each cut and interleaving that gives the rearrangement of the deck corresponding to π, which is $1/2^n$ times the number of ways of cutting and interleaving that give the rearrangement of the deck corresponding to π. In short, the chance of any arrangement of cards occurring under riffle shuffling is simply the proportion of ways of riffling that give that arrangement.

Here is a particular example of the riffle shuffle in the case $n = 3$, with the deck starting in natural order 123.

k = cut position	cut deck	probability of this cut	possible interleavings
0	|123	1/8	123
1	1|23	3/8	123,213,231
2	12|3	3/8	123,132,312
3	123|	1/8	123

Note that 0 or all 3 cards may be cut, in which case one packet is empty and the other is the whole deck. Now let us compute the probability of each particular ordering occurring in the above example. First, look for 213. It occurs only in the cut $k=1$, which has probability 3/8. There it is one of three possibilities, and hence has the conditional probability 1/3, given $k = 1$. So the overall probability for 213 is $\frac{1}{3} \cdot \frac{3}{8} = \frac{1}{8}$, where of course $\frac{1}{8} = \frac{1}{2^3}$ is the probability of any particular cut and interleaving pair. Similar analyses hold for 312, 132, and 231, since they all occur only through a single cut and interleaving. For 123, it is different; there are four cuts and interleavings that give rise to it. It occurs for $k = 0, 1, 2$, and 3, these situations having probabilities 1/8, 3/8, 3/8, and 1/8, respectively. In these cases, the conditional probability of 123, given the cut, is 1, 1/3, 1/3, and 1. So the overall probability of the ordering is $\frac{1}{8} \cdot 1 + \frac{3}{8} \cdot \frac{1}{3} + \frac{3}{8} \cdot \frac{1}{3} + \frac{1}{8} \cdot 1 = \frac{1}{2}$, which also equals $4 \cdot \frac{1}{2^3}$, the number of ways of cutting and interleaving that give rise to the ordering times the probability of any particular cut and interleaving. We may write down the entire density, now dropping the assumption that the deck started in the natural order, which means we must use permutations instead of orderings.

permutation π	[123]	[213]	[231]	[132]	[312]	[321]
probability $R(\pi)$ under riffle	1/2	1/8	1/8	1/8	1/8	0

It is worth making obvious a point that should be apparent. The information specified by a cut and an interleaving is richer than the information specified by the resulting permutation. In other words, there may be several different ways of cutting and interleaving that give rise to the same permutation, but different permutations necessarily arise from distinct cut/interleaving pairs. (An exercise for the reader is to show that, for the riffle shuffle, this distinction is nontrivial only when the permutation is the identity, i.e., the only time distinct cut/interleaving pairs give rise to the same permutation is when the permutation is the identity.)

There is a second, equivalent way of describing the riffle shuffle. Start the same way, by cutting the deck according to the binomial density into two packets of size k and $n - k$. Now we are going to drop a card from the bottom of one of the two packets onto a table, face down. Choose between the packets with probability proportional to packet size, meaning, if the two packets are of size p_1 and p_2, then the probability of the card dropping from the first is $\frac{p_1}{p_1+p_2}$, and $\frac{p_2}{p_1+p_2}$ from the second. So this first time, the probabilities would be $\frac{k}{n}$ and $\frac{n-k}{n}$. Now repeat the process, with the numbers p_1 and p_2 being updated to reflect the actual packet sizes by subtracting one from the size of whichever packet had the card dropped last time. For instance, if the first card was dropped from the first packet, then the probabilities for the next drop would be $\frac{k-1}{n-1}$ and $\frac{n-k}{n-1}$. Keep going until all cards are dropped. This method is equivalent to the first description of the riffle in that this process also assigns uniform probability $1/\binom{n}{k}$ to each possible resulting interleaving of the cards.

To see this, let us figure out the probability for some particular way of dropping the cards, say, for the sake of definiteness, from the first packet and then from the first, second, second, second, first, and so on. The probability of the drops occurring this way is

$$\frac{k}{n} \cdot \frac{k-1}{n-1} \cdot \frac{n-k}{n-2} \cdot \frac{n-k-1}{n-3} \cdot \frac{n-k-2}{n-4} \cdot \frac{k-2}{n-5} \cdots,$$

where we have multiplied probabilities since each drop decision is independent of the others once the packet sizes have been readjusted. Now the product of the denominators of these fractions is $n!$, since it is just the product of the total number of cards left in both packets before each drop, and this number decreases by one each time. What is the product of the numerators? Well, we get one factor every time a card is dropped from one of the packets, this factor being the size of the packet at that time. But then we get all the numbers $k, k-1, \ldots, 1$ and $n-k, n-k-1, \ldots, 1$ as factors in some order, since each packet passes through all of the sizes in its respective list as the cards are dropped from the two packets. So the numerator is $k!(n-k)!$, which makes the overall probability $k!(n-k)!/n! = 1/\binom{n}{k}$, which is obviously valid for any particular sequence of drops, and not just the above example. So we have now shown the two descriptions of the riffle shuffle are equivalent, as they have the same uniform probability of interleaving after a binomial cut.

Now let $R^{(k)}$ stand for convoluting R with itself k times. This corresponds to the density after k riffle shuffles. For which k does $R^{(k)}$ produce a randomized deck? The next section begins to answer this question.

4. HOW FAR AWAY FROM RANDOMNESS?

Before we consider the question of how many times we need to shuffle, we must decide what we want to achieve by shuffling. The answer should be randomness of some sort. What does randomness mean? Simply put, any arrangement of cards is equaly likely; no one ordering should be favored over another. This means the uniform density U on S_n, each permutation having probability $U(\pi) = 1/|S_n| = 1/n!$.

Now it turns out that for any fixed number of shuffles, no matter how large, riffle shuffling does not produce complete randomness in this sense. (We will, in fact, give an explicit formula that shows that, after any number of riffle shuffles, the identity permutation is always more likely than any other to occur.) So, when we ask how many times we need to shuffle, we are not asking how far to go in order to achieve randomness, but rather to get close to randomness. So we must define what we mean by close, or far, i.e., we need a distance between densities.

The concept we will use is called variation distance (which is essentially the L^1 metric on the space of densities). Suppose we are given two probability densities, Q_1 and Q_2, on S_n. Then the variation distance between Q_1 and Q_2 is defined to be

$$\|Q_1 - Q_2\| = \frac{1}{2} \sum_{\pi \in S_n} |Q_1(\pi) - Q_2(\pi)|.$$

The $\frac{1}{2}$ normalizes the result to always be between 0 and 1.

Here is an example. Let $Q_1 = R$ be the density calculated abcve for the three card riffle shuffle. Let Q_2 be the complete reversal — the density that gives probability 1 for [321], i.e., certainty, and 0 for all other permutations, i.e., nonoccurrence.

π	$Q_1(\pi)$	$Q_2(\pi)$	$|Q_1(\pi) - Q_2(\pi)|$
[123]	1/2	0	1/2
[213]	1/8	0	1/8
[312]	1/8	0	1/8
[132]	1/8	0	1/8
[231]	1/8	0	1/8
[321]	0	1	1
		Total	2

So here $\|Q_1 - Q_2\| = 2/2 = 1$, and the densities are as far apart as possible.

Now the question we really want to ask is: how big must we take k to make the variation distance $\|R^{(k)} - U\|$ between the riffle and uniform small? This can be best answered by a graph of $||R^{(k)} - U||$ versus k. The following theory is directed toward constructing this graph.

5. RISING SEQUENCES

To begin to determine what the density $R^{(k)}$ is, we need to consider a fundamental concept, that of a rising sequence. A rising sequence of a permutation is a maximal consecutively increasing subsequence. What does this really mean for cards? Well, perform the rearrangement corresponding to the permutation on a naturally ordered deck. Pick any card, labeled x say, and look after it in the deck for the card labeled $x + 1$. If you find it, repeat the procedure, now looking after the $x + 1$ card for the $x + 2$ card. Keep going in this manner until you have to stop because you can't find the next card after a given card. Now go back to your original card x and reverse the procedure, looking before the original card for the $x - 1$ card, and so on. When you are done, you have a rising sequence. It turns out that a deck breaks down as a disjoint union of its rising sequences, since the union of any two consecutively increasing subsequences containing a given element is also a consecutively increasing subsequence that contains that element.

Let's look at an example. Suppose we know that the order of an eight card deck after shuffling the natural order is 45162378. Start with any card, say 3. We look for the next card in value after it, 4, and do not find it. So we stop looking after and look before the 3. We find 2, and then we look for 1 before 2 and find it. So one of the rising sequences is given by 123. Now start again with 6. We find 7 and then 8 after it, and 5 and then 4 before it. So another rising sequence is 45678. We have accounted for all the cards, and are therefore done. Thus this deck has only two rising sequences. This is immediately clear if we write the order of the deck this way, $45_16_23_78$, offsetting the two rising sequences.

It is clear that a trained eye may pick out rising sequences immediately, and this forms the basis for some card tricks. Suppose a brand new deck of cards is riffle shuffled three times by a spectator, who then takes the top card, looks at it without showing it to a magician, and places it back in the deck at random. The magician then tries to identify the reinserted card. He is often able to do so because the reinserted card will often form a singleton rising sequence, consisting of just itself. Most likely, all the other cards will fall into $2^3 = 8$ rising sequences of length 6 to 7, since repeated riffle shuffling, at least the first few times, roughly tends to double the number of the rising sequences and halve the length of each one each time. Diaconis, himself a magician, and Bayer [2] describe variants of this trick that magicians have actually used.

It is interesting to note that the order of the deck in our example, $45_16_23_78$, is a possible result of a riffle shuffle with a cut after 3 cards. In fact, any ordering with just two rising sequences is a possible result of a riffle shuffle. Here the cut must divide the deck into two packets such that the length of each is the same as the length of the corresponding rising sequence. So if we started in the natural order 12345678 and cut the deck into 123 and 45678, we would interleave by taking 4, then 5, then 1, then 6, then 2, then 3, then 7, then 8, thus obtaining the given order through riffling. The converse of this result is that the riffle shuffle always gives decks with either one or two rising sequences.

6. BIGGER AND BETTER: a-SHUFFLES

The result that a permutation has nonzero probability under the riffle shuffle if and only if it has exactly one or two rising sequences is true, but it only holds for a single riffle shuffle. We would like similar results on what happens after multiple riffle suffles. This can ingeniously be accomplished by considering a-shuffles, a generalization of the riffle shuffle. An a-shuffle is another probability density on S_n, achieved as follows. Let a stand for any positive integer. Cut the deck into a packets, of nonnegative sizes $p_1, p_2, \ldots, p_a$, with the probability of this particular packet structure given by the multinomial density: $\binom{n}{p_1, p_2, \ldots, p_a} / a^n$. Note we must have $p_1 + \cdots + p_a = n$, but some of the p_i may be zero. Now interleave the cards from each packet in any way, so long as the cards from each packet maintain their relative order among themselves. With a fixed packet structure, consider all interleavings equally likely. Let us count the number of such interleavings. We simply want the number of different ways of choosing, among n positions in the deck, p_1 places for things of one type, p_2 places for things of another type, etc. This is given by the multinomial coefficient $\binom{n}{p_1, p_2, \ldots, p_a}$. Hence the probability of a particular rearrangement, i.e., a cut of the deck and an interleaving, is

$$\binom{n}{p_1, p_2, \ldots, p_a} / a^n \cdot \binom{n}{p_1, p_2, \ldots, p_a} = \frac{1}{a^n}.$$

So it turns out that each combination of a particular cut into a packets and a particular interleaving is equally likely, just as in the riffle shuffle. The induced density on the permutations corresponding to the cuts and interleavings is then called the a-shuffle. We will denote it by R_a. It is apparent that the riffle is just the 2-shuffle, so $R = R_2$.

An equivalent description of the a-shuffle begins the same way, by cutting the deck into packets multinomially. But then drop cards from the bottom of the packets, one at a time, such that the probability of choosing a particular packet to drop from is proportional to the relative size of that packet compared to the number of cards left in all the packets. The proof that this description is indeed equivalent is exactly analogous to the $a = 2$ case. A third equivalent description is given by cutting multinomially into $p_1, p_2, \ldots, p_a$ and riffling p_1 and p_2 together (meaning choose uniformly among all interleavings which maintain the relative order of each packet), then riffling the resulting pile with p_3, then riffling that resulting pile with p_4, and so on.

There is a useful code that we can construct to specify how a particular a-shuffle is done. (Note that we are abusing terminology slightly and using shuffle here to indicate a particular way of rearranging the deck, and not the density on all such rearrangements.) This is done through n digit base a numbers. Let A be any one of these n digit numbers. Count the number of 0s in A. This will be the size of the first packet in the a-shuffle, p_1. Then p_2 is the number of 1s in A, and so on, up through $p_a =$ the number of $(a-1)$s. This cuts the deck cut into a packets. Now take the beginning packet of cards, of size p_1. Envision placing these cards on top of all the 0 digits of A, maintaining their relative order as a rising sequence. Do the same for the next packet, p_2, except placing them on the 1s. Again, continue up through the $(a-1)$s. This particular way of rearranging the cards will then be the particular cut and interleaving corresponding to A.

Here is an example, with the deck starting in natural order. Let $A = 23004103$ be the code for a particular 5-shuffle of the 8 card deck. There are three 0s, one 1, one 2, two 3s, and one 4. Thus $p_1 = 3$, $p_2 = 1$, $p_3 = 1$, $p_4 = 2$, and $p_5 = 1$. So the deck is cut into $123 \mid 4 \mid 5 \mid 67 \mid 8$. So we place 123 where the 0s are in A, 4 where the 1 is, 5 where the 2 is, 67 where the 3s are, and 8 where the 4 is. We then get a shuffled deck of 56128437 when A is applied to the natural order.

Reflection shows that this code gives a bijective correspondence between n digit base a numbers and the set of all ways of cutting and interleaving an n card deck according to the a-shuffle. In fact, if we put the uniform density on the set of n digit base a numbers, this transfers to the correct uniform probability for cutting and interleaving in an a-shuffle, which means the correct density is induced on S_n, i.e., we get the right probabilities for an a-shuffle. This code will prove useful later on.

7. VIRTUES OF THE a-SHUFFLE

7.1 Relation to rising sequences

There is a great advantage to considering a-shuffles. It turns out that, when you perform a single a-shuffle, the probability of achieving a particular permutation π does not depend upon all the information contained in π, but only on the number of rising sequence that π has. In other words, we immediately know that the permutations [12534], [34512], [51234], and [23451] all have the same probability under any a-shuffle, since they all have exactly two rising sequences. Here is the exact result:

The probablity of achieving a permutation π when doing an a-shuffle is given by $\binom{n+a-r}{n}/a^n$, where r is the number of rising sequences in π.

Proof: First note that, if we establish and fix where the $a-1$ cuts occur in an a-shuffle, then whatever permutations can actually be achieved by interleaving the cards from this cut/packet structure are achieved in exactly one way; namely, just drop the cards in exactly the order of the permutation. Thus the probability of achieving a particular permutation is the number of possible ways of making cuts that could actually give rise to that permutation, divided by the total number of ways of making cuts and interleaving for an a-shuffle.

So let us count the ways of making cuts in the naturally ordered deck that could give the ordering that results when π is applied. If we have r rising sequences in π, we know exactly where $r-1$ of the cuts have to have been; they must have occurred between pairs of consecutive cards in the naturally ordered deck such that the first card ends one rising sequence of π and the second begins another rising sequence of π. This means we have $a-1-(r-1) = a-r$ unspecified, or free, cuts. These are free in the sense that they can in fact go anywhere. So we must count the number of ways of putting $a-r$ cuts among n cards. This can easily be done by considering a sequence of $(a-r)+n$ blank spots that must be filled by $(a-r)$ things of one type (cuts) and n things of another type (cards). There are $\binom{(a-r)+n}{n}$ ways to do this, i.e., choosing n places among $(a-r)+n$.

This is the numerator for our probability expressed as a fraction; the denominator is the number of possible ways to cut and interleave for an a-shuffle. By considering the encoding of shuffles we see there are a^n ways to do this, as there are this many n digit base a numbers. Hence our result is true.

This allows us to envision the probability density associated with an a-shufle in a nice way. Order all the permutation in S_n in any way such that the number of rising sequences is non-decreasing. If we label these permutations as points on a horizontal axis, we may take the vertical axis to be the numbers between 0 and 1, and at each permutation place a point whose vertical coordinate is the probability of the permutation. Obviously, the above result means we will have sets of points of the same height. Here is an example for a 7-shuffle of the five card deck (solid line), along with the uniform density $U \equiv 1/5! = 1/120$ (dashed line).

Notice the probability $\binom{n+a-r}{n}/a^n$ is a monotone decreasing function of r. This means, if $1 \le r_1 < r_2 \le n$, then a particular permutation with r_1 rising sequences is always more likely than a permutation with r_2 rising sequences under any a-shuffle. Hence the graph of the density for an a-shuffle, if the permutations are ordered as above, will always be nonincreasing. In par-

ticular, the probability starts above uniform for the identity, the only permutation with $r = 1$. [In our example $R_7(\text{identity}) = \begin{pmatrix} 5 + 7 - 1 \\ 5 \end{pmatrix}/7^5 = .0275.$]
It then decreases for increasing r, at some point crossing below uniform (from $r = 2$ to 3 in the example). The greatest r value such that the probability is above uniform is called the *crossover point*. Eventually at $r = n$, which occurs only for the permutation corresponding to complete reversal of the deck, the probability is at its lowest value. [In the example $\begin{pmatrix} 5 + 7 - 5 \\ 5 \end{pmatrix}/7^5 = .0012.$]
All this explains the earlier statement that, after an a-shuffle, the identity is always more likely than it would be under a truly random density, and is always more likely than any other particular permutation after the same a-shuffle.

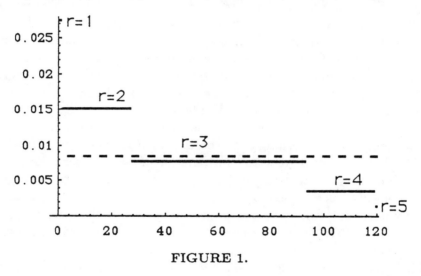

FIGURE 1.

For a fixed deck size n, it is interesting to note the behavior of the crossover point as a increases. By analyzing the inequality

$$\begin{pmatrix} n + a - r \\ n \end{pmatrix}/a^n \geq \frac{1}{n!},$$

the reader may prove that the crossover point never moves to the left, i.e., it is a nondecreasing function of a, and that it eventually moves to the right, up to $n/2$ for n even and $(n-1)/2$ for n odd, but never beyond. Furthermore, it will reach this halfway point for a approximately the size of $n^2/12$. Combining with the results of the next section, this means roughly $2\log_2 n$ riffle shuffles are needed to bring the crossover point to halfway.

7.2 The multiplication theorem

Why bother with an a-shuffle? In spite of the nice formula for a density dependent only on the number of rising sequences, a-shuffles seem of little practical use to any creature that is not a-handed. This turns out to be false. After we establish another major result that addresses this question, we will be in business to construct our variation distance graph.

This result concerns multiple shuffles. Suppose you do a riffle shuffle twice. Is there any simple way to describe what happens, all in one step, other than the convolution of densities described in section 2.2? Or more generally, if you do an a-shuffle and then do a b-shuffle, how can you describe the result? The answer is the following:

An a-shuffle followed by a b-shuffle is equivalent to a single ab-shuffle, in the sense that both processes give exactly the same resulting probability density on the set of permutations.

Proof: Let us use the previously described code for shuffles. Suppose that A is an n digit base a number, and B is an n digit base b number. Then first doing the cut and interleaving encoded by A and then doing the cut and interleaving encoded by B gives the same permutation as the one resulting from the cut and interleaving encoded by the n digit base ab number given by $A^B \& B$, as John Finn figured out. (The proof for this formula will be deferred until section 9.4, where the inverse shuffle is discussed.) This formula needs some explanation. A^B is defined to be the code that has the same base a digits as A, but rearranged according to the permutation specified by B. The symbol $\&$ in $A^B \& B$ stands for digit-wise concatenation of two numbers, meaning treat the base a digit A_i^B in the ith place of A^B together with the base b digit B_i in the ith place of B as the base ab digit given by $A_i^B \cdot b + B_i$. In other words, treat the combination $A_i^B \& B_i$ as a two digit number, the right-most place having value 1, and the left-most place having value b, and then treat the result as a one digit base ab number.

Why this formula holds is better shown by an example than by general formulas. Suppose $A = 012210$ is the code for a particular 3-shuffle, and $B = 310100$ is the code for a particular 4-shuffle. (Again we are abusing terminology slightly.) Let π_A and π_B be the respective permutations. Then in the tables below note that $\pi_A \circ \pi_B$, the result of a particular 3-shuffle followed by a particular 4-shuffle, and $\pi_{AB\&B}$, the result of a particular 12-shuffle, are the same permutation.

i	1	2	3	4	5	6
$\pi_A(i)$	1	3	5	6	4	2
$\pi_B(i)$	6	4	1	5	2	3
$\pi_A \circ \pi_B(i)$	2	6	1	4	3	5

A	0	1	2	2	1	0
B	3	1	0	1	0	0
A^B	0	2	0	1	1	2
B	3	1	0	1	0	0
$A^B \& B$	3	9	0	5	4	8

i	1	2	3	4	5	6
$\pi_{A^B\&B}(i)$	2	6	1	4	3	5

We now have a formula $A^B\&B$ that is really a one-to-one correspondence between the set of pairs, consisting of one n digit base a number and one n digit base b number, and the set of n digit base ab numbers; further, this formula has the property that the cut and interleaving specified by A, followed by the cut and interleaving specified by B, result in the same permutation of the deck as that resulting from the cut and interleaving specified by $A^B\&B$. Since the probability densities for a, b, and ab-shuffles are induced by the uniform densities on the sets of n digit base a, b, or ab codes, respectively, the properties of the one-to-one correspondence imply the induced densities on S_n of an a-shuffle followed by a b-shuffle and an ab-shuffle are the same. Hence our result is true.

7.3 Expected happenings after an a-shuffle

It is of theoretical interest to measure the expected value of various quantities after an a-shuffle of the deck. For instance, we may ask what is the expected number of rising sequences after an a-shuffle? I've found an approach to this question which has too much computation to be presented here, but gives the answer as

$$a - \frac{n+1}{a^n} \sum_{r=0}^{a-1} r^n.$$

As $a \to \infty$, this expression tends to $\frac{n+1}{2}$, which is the expected number of rising sequences for a random permutation. When $n \to \infty$, the expression goes to a. This makes sense, since, when the number of packets is much less than the size of the deck, the expected number of rising sequences is the same as the number of packets.

The expected number of fixed points of a permutation after an a-shuffle is given by $\sum_{i=0}^{n-1} a^{-i}$, as mentioned in [2]. As $n \to \infty$, this expression tends to $\frac{1}{1-1/a} = \frac{a}{a-1}$, which is between 1 and 2. As $a \to \infty$, the expected number of fixed points goes to 1, which is the expected number of fixed points for a random permutation.

8. PUTTING IT ALL TOGETHER

Let us now combine our two major results of the last section to get a formula for $R^{(k)}$, the probability density for the riffle shuffle done k times. This is just k 2-shuffles, one after another. So, by the multiplication theorem, this is equivalent to a single $2 \cdot 2 \cdot 2 \cdots 2 = 2^k$-shuffle. Hence, in the $R^{(k)}$ density, there is a $\dbinom{2^k + n - r}{n} / 2^{nk}$ chance of a permutation with r rising sequences occurring, by our rising sequence formula. This now allows us to work on the variation distance $\|R^k - U\|$. For a permutation π with r rising sequences, we see that

$$|R^k(\pi) - U(\pi)| = \left| \binom{2^k + n - r}{n} \Big/ 2^{nk} - \frac{1}{n!} \right|.$$

We must now add up all the terms like this, one for each permutation. We can group terms in our sum according to the number of rising sequences. If we let $A_{n,r}$ stand for the number of permutations of n cards that have r rising sequences, each of which have the same probabilities, then the variation distance is given by

$$\|R^k - U\| = \frac{1}{2} \sum_{r=1}^{n} A_{n,r} \left| \binom{2^k + n - r}{n} \Big/ 2^{nk} - \frac{1}{n!} \right|.$$

The only thing unexplained is how to calculate the $A_{n,r}$. These are called the Eulerian numbers, and various formulas are given for them (e.g., see [8]). One recursive one is $A_{n,1} = 1$ and $A_{n,r} = r^n - \sum_{j=1}^{r-1} \binom{n+r-j}{n} A_{n,j}$. (It is interesting to note that the Eulerian numbers are symmetric in the sense that $A_{n,r} = A_{n,n-r+1}$. So there are just as many permutations with r rising sequences as there are with $n - r + 1$ rising sequences, which the reader is invited to prove directly.)

Now the expression for variation distance may seem formidable, and it is. But it is easy and quick for a computer program to calculate and graph $\|R^k - U\|$ versus k for any specific, moderately sized n. Even on the computer, however, this computation is tractable because we only have n terms, corresponding to each possible number of rising sequences. If we did not have the result on the invariance of the probability when the number of rising sequences is constant, we would have $|S_n| = n!$ terms in the sum. For $n = 52$, this is approximately 10^{68}, which is much larger than any computer could handle. Here is the graphical result of a short Mathematica program that does the calculations for $n = 52$. The horizontal axis is the number of riffle shuffles, and the vertical axis is the variation distance to uniform.

The answer is finally at hand. It is clear that the graph makes a sharp cutoff at $k = 5$, and gets reasonably close to 0 by $k = 11$. A good middle point for the cutoff seems to $k = 7$, and this is why seven shuffles are said to be enough for the usual deck of 52 cards. Additionally, asymptotic analysis in [2] shows that when n, the number of cards, is large approximately $k = \frac{3}{2} \log n$ shuffles suffice to get the variation distance through the cutoff and close to 0.

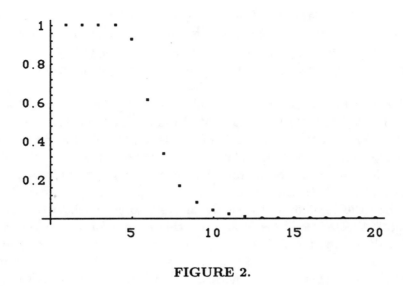

FIGURE 2.

We have now achieved our goal of constructing the variation distance graph, which explains why seven shuffles are "enough". In the remaining sections we present some other aspects to shuffling, as well as some other ways of approaching the question of how many shuffles should be done to deck.

9. THE INVERSE SHUFFLE

There is an unshuffling procedure which is in some sense the reverse of the riffle shuffle. It is actually simpler to describe, and some of the theorems are more evident in the reverse direction. Take a face-down deck, and deal cards from the bottom of the deck one at a time, placing the cards face-down into one of two piles. Make all the choices of which pile independently and uniformly, i.e., go 50/50 each way each time. Then simply put one pile on top of the other. This may be called the riffle unshuffle, and the induced density on S_n may be labeled $\hat{R}$. An equivalent process is generated by labeling the backs of all the cards with 0s and 1s independently and uniformly, and then pulling all the 0s to the front of the deck, maintaining their relative order, and pulling all the 1s the back of the deck, maintaining their relative order. This may quickly be generalized to an a-unshuffle, which is described by labeling the back of each card independently with a base a digit chosen uniformly. Now place all the cards labeled 0 at the front of the deck, maintaining their relative order, then all the 1s, and so on, up through the $(a-1)$s. This is the a-unshuffle, denoted by $\hat{R}_a$.

We really have a reverse or inverse operation in the sense that $\hat{R}_a(\pi) = R_a(\pi^{-1})$ holds. This is seen most easily by looking at n digit base a numbers. We have already seen in section 6 that each such n digit base a number may be treated as a code for a particular cut and interleaving in an a-shuffle; the above paragraph in effect gives a way of also treating each n digit base a numbers as code for a particular way of achieving an a-unshuffle. The two induced permutations we get when looking at a given n digit base a number in these two ways are inverse to one another, and this proves $\hat{R}_a(\pi) = R_a(\pi^{-1})$ since the uniform density on n digit base a numbers induces the right density on S_n.

We give a particular example that makes the general case clear. Take the 9 digit base 3 code 122020110 and apply it in the forward direction, i.e., treat it as directions for a particular 3-shuffle of the deck 123456789 in natural order. We get the cut structure 123|456|789 and hence the shuffled deck 478192563. Now apply the code to this deck order, but backward, i.e., treat it as directions for a 3-unshuffle of 478192563. We get the cards where the 0s are, 123, pulled forward; then the 1s, 456; and then the 2s, 789, to get back to the naturally ordered deck 123456789. It is clear from this example that, in general, the a-unshuffle directions for a given n digit base a number pull back the cards in a way exactly opposite to the way the a-shuffle directions from that code distributed them. This may be checked by applying the code both forward and backward to the unshuffled deck 123456789 and getting

$$\begin{pmatrix} 123456789 \\ 478192563 \end{pmatrix} \qquad \begin{pmatrix} 123456789 \\ 469178235 \end{pmatrix},$$

which inspection shows are indeed inverse to one another.

The advantage to using unshuffles is that they motivate the $A^B \& B$ formula in the proof of the multiplication theorem for an a-shuffle followed by a b-shuffle. Suppose you do a 2-unshuffle by labeling the cards with 0s and 1s in the upper right corner according to a uniform and independent random choice each time, and then sorting the 0s before the 1s. Then do a second 2-unshuffle by labeling the cards again with 0s and 1s, placed just to the left of the digit already on each card, and sorting these left-most 0s before the left-most 1s. Reflection shows that doing these two processes is equivalent to doing a single process: label each card with a 00, 01, 10, or 11 according to uniform and independent choices, sort all cards labeled 00 and 10 before all those labeled 01 and 11, and then sort all cards labeled 00 and 01 before all those labeled 10 and 11. In other words, sort according to the right-most digit, and then according to the left-most digit. But this is the same as sorting the 00s before the 01s, the 01s before the 10s, and the 10s before the 11s all at once. So this single process is equivalent to the following: label each card with a 0, 1, 2, or 3 according to uniform and independent choices, and sort the 0s before the 1s before the 2s before the 3s. But this is exactly a 4-unshuffle!

So two 2-unshuffles are equivalent to a $2 \cdot 2 = 4$-unshuffle and, generalizing in the obvious way, a b-unshuffle followed by an a-unshuffle is equivalent to an ab-unshuffle. (In the case of unshuffles we have orders reversed and write a b-unshuffle followed by an a-unshuffle, rather than vice-versa, for the same reason that one puts on socks and then shoes, but takes off shoes and then socks.) Since the density for unshuffles is the inverse of the density for shuffles [in the sense that $\hat{R}_a(\pi) = R_a(\pi^{-1})$], this means an a-shuffle followed by a b-shuffle is equivalent to an ab-shuffle. Furthermore, we are tempted to believe that combining the codes for unshuffles should be given by $A\&B$, where A and B are the sequences of 0s and 1s put on the cards, encapsulated as n digit base 2 numbers, and $\&$ is the already described symbol for digitary concatenation. This $A\&B$ is not quite right, however; for when two 2-unshuffles are done, the second group of 0s and 1s will not be put on the cards in their original order, but will be put on the cards in the order they are in after the first unshuffle. Thus we must compensate in the formula if we wish to treat the 00s, 01s, 10s, and 11s as being written down on the cards in their original order at the beginning, before any unshuffling. We can do this by by having the second sequence of 0s and 1s permuted, according to the inverse of the permutation described by the first sequence of 0s and 1s. So we must use A^B instead of A. Clearly this works for all a and b and not just $a = b = 2$. This is why the formula for combined unshuffles, and hence shuffles, is $A^B\&B$ and not just $A\&B$. (The fact that it is actually $A^B\&B$ and not $A\&B^A$ or some such variant is best checked by looking at particular examples, as in section 7.2.)

10. ANOTHER APPROACH TO SUFFICIENT SHUFFLING

10.1 Seven is not enough

A footnote must be added to the choosing of any specific number, such as seven, as making the variation distance small enough. There are examples where this does not randomize the deck enough. Peter Doyle has invented a game of solitaire that shows this quite nicely. A simplified, albeit less colorful version is given here. Take a deck of 52 cards, turned face-down, that is labeled in top to bottom order $123\cdots(25)(26)(52)(51)\cdots(28)(27)$. Riffle shuffle seven times. Then deal the cards one at a time from the top of the deck. If the 1 comes up, place it face up on the table. Call this pile A. If the 27 comes up, place it face up on the table in a separate pile, calling this B. If any other card comes up that it is not the immediate successor of the top card in either A or B, then place it face up in the pile C. If the immediate successor of the top card of A comes up, place it face up on top of A, and the same for B. Go through the whole deck this way. When done, pick up pile C, turn it face down, and repeat the procedure. Keep doing so. End the game when either pile A or pile B is full, i.e., has twenty-six cards in it. Let us say the game has been won if A is filled up first, and lost if B is.

It turns out that the game will end much more than half the time with pile A being full, i.e the deck is not randomized 'enough'. Computer simulations indicate that we win about 81% of the time. Heuristically, this is because the rising sequences in the permuted deck after a $2^7 = 128$-shuffle can be expected to come from both the first and second halves of the original deck in roughly the same numbers and length. However, the rising sequences from the first half will be 'forward' in order and the ones from the second half will be 'backward'. The forward ones require only one pass through the deck to be placed in pile A, but the backward ones require as many passes through the deck as their length, since only the last card can be picked off and put into pile B each time. Thus pile A should be filled more quickly; what really makes this go is that a 128-shuffle still has some rising sequences of length 2 or longer, and it is faster to get these longer rising sequences into A than it is into to get sequences of the same length into B.

In a sense, this game is almost a worst case scenario. This is because of the following definition of variation distance, which is equivalent to the one given in section 4. (The reader is invited to prove this.) Given two densities Q_1 and Q_2 on S_n,

$$\|Q_1 - Q_2\| = \max_{S \subset S_n} |Q_1(S) - Q_2(S)|,$$

where the maximum on the r.h.s. is taken over all subsets S of S_n, and the $Q_i(S)$ are defined to be $\sum_{\pi \in S} Q_i(\pi)$. What this really means is that the variation distance is an upper bound (in fact a least upper bound) for the difference of the probabilities of an event given by the two densities. This can be directly applied to our game. Let S be the set of all permutations for which pile A is filled up first, i.e., the event that we win. Then the variation distance $\|R^{(7)} - U\|$ is an upper bound for the difference between the probability of a permutation in S occurring after 7 riffles, and the probability of such a permutation occurring truly randomly. Now such winning permutations should occur truly randomly only half the time (by symmetry), but the simulations indicate that they occur 81% percent of the time after 7 riffle shuffles. So the probability difference is $|.81 - .50| = .31$. On the other hand, the variation distance $\|R^{(7)} - U\|$ as calculated in section 8 is .334, which is indeed greater than .31, but not by much. So Doyle's game of solitaire is nearly as far away from being a fair game as possible.

10.2 Is variation distance the right thing to use?

The variation distance has been chosen to be the measure of how far apart two densities are. It seems intuitively reasonable as a measure of distance, just taking the differences of the probabilities for each permutation, and adding them all up. But the game of the last section might indicate that it is too forgiving a measure, rating a shuffling method as nearly randomizing, even though in some ways it clearly is not. At the other end of the spectrum, however, some examples, as modified from [1] and [4], suggest that variation distance may be too harsh a measure of distance. Suppose that you are presented with a face-down deck, with n even, and told that it has been perfectly randomized, so that as far as you know any ordering is equally as likely as any other. So you simply have the uniform density $U(\pi) = 1/n!$ for all $\pi \in S_n$. But now suppose that the top card falls off, and you see what it is. You realize that to put the card back on the top of the deck would destroy the complete randomization by restricting the possible permutations, namely to those that have this paricular card at the first position. So you decide to place the card back at random in the deck. Doing this would have restored complete randomization and hence the uniform density. Suppose, however, that you realize this, but also figure superstitiously that you shouldn't move this original top card too far from the top. So instead of placing it back in the deck at random, you place it back at random subject to being in the *top half* of the deck.

How much does this fudging of randomization cost you in terms of variation distance? Well, the number of restricted possible orderings of the deck, each equally likely, is exactly half the possible total, since we want those orderings where a given card is in the first half, and not those where it is in the second half. So this density is given by $\bar{U}$, which is $2/n!$ for half of the permutations and 0 for the other half. So the variation distance is

$$\|U - \bar{U}\| = \frac{1}{2}\left(\frac{n!}{2}\left|\frac{2}{n!} - \frac{1}{n!}\right| - \frac{n!}{2}\left|0 - \frac{1}{n!}\right|\right) = \frac{1}{2}.$$

This seems a high value, given the range between 0 and 1. Should a good notion of distance place this density $\bar{U}$, which most everyone would agree is very nearly random, half as far away from complete randomness as possible?

10.3 The birthday bound

Because of some of the counterintuitive aspects of the variation distance presented in the last two subsections, we present another idea of how to measure how far away repeated riffle shuffling is from randomness. It turns out that this idea will give an upper bound on the variation distance, and it is tied up with the well-known birthday problem as well.

We begin by first looking at a simpler case, that of the top-in shuffle, where the top card is taken off and reinserted randomly anywhere into the deck, choosing among each of the n possible places between cards uniformly. Before any top-in shuffling is done, place a tag on the bottom card of the deck, so that it can be identified. Now start top-in shuffling repeatedly. What happens to the tagged card? Well, the first time a card, say a, is inserted below the tagged card, and hence on the bottom of the deck, the tagged card will move up to the penultimate position in the deck. The next time a card, say b, is inserted below the tagged card, the tagged card will move up to the antepenultimate position. Note that all possible orderings of a and b below the tagged card are equally likely, since it was equally likely that b went above or below a, given only that it went below the tagged card. The next time a card, say c, is put below the tagged card, its equal likeliness of being put anywhere among the order of a and b already there, which comes from a uniform choice among all orderings of a and b, means that all orders of a, b, and c are equally likely. Clearly as this process continues the tagged card either stays in its position in the deck or it moves up one position and, when this happens, all orderings of the cards below the tagged card are equally likely. Eventually the tagged card gets moved up to the top of the deck by having another card inserted underneath it. Say this happens on the $T' - - - 1$st top-in shuffle. All the cards below the tagged card, i.e., all the cards but the tagged card, are now randomized, in the sense that any order of them is equally likely. Now take the tag off the top card and top-in shuffle for the T'th time. The deck is now completely randomized, since the formerly tagged card has been reinserted uniformly into an ordering that is a uniform choice of all ones possible for the remaining $n - 1$ cards.

Now $\mathbf{T}'$ is really a random variable, i.e., there are probabilities that $\mathbf{T}' = 1, 2, \ldots$, and by convention we write it in boldface. It is a particular example of a *stopping time*, when all orderings of the deck are equally likely. We may consider its expected value $E(\mathbf{T}')$, which clearly serves well as an intuitive idea of how randomizing a shuffle is, for $E(\mathbf{T}')$ is just the average number of top-in shuffles needed to guarantee randomness by this method. The reader may wish to show that $E(\mathbf{T}')$ is asymptotic to $n \log n$. This is sketched in the following: Create random variables $\mathbf{T}_j$ for $2 \le j \le n$, which stand for the difference in time between when the tagged card first moves up to the jth position from the bottom and when it first moves up to the $j - 1$st position. (The tagged card is said to have moved up to position 1 at step 0.) Then $\mathbf{T}' = \mathbf{T}_2 + \mathbf{T}_3 + \cdots + \mathbf{T}_n + 1$. Now the $\mathbf{T}_j$ are all independent and have densities

$$P[\mathbf{T}_j = i] = \frac{j - 1}{n} \left(\frac{n - j + 1}{n} \right)^{i - 1}.$$

Calculating the expected values of these geometric densities gives $E(\mathbf{T}_j) = n/(j - 1)$. Summing over j and adding one shows $E(\mathbf{T}') = 1 + n \sum_{j=1}^{n-1} j^{-1}$, which, with a little calculus, gives the result.

$\mathbf{T}'$ is good for other things as well. It is a theorem of Aldous and Diaconis [1] that $P[\mathbf{T}' > k]$ is an upper bound for the variation distance between the density on S_n after k top-in shuffles and the uniform density corresponding to true randomness. This is because $\mathbf{T}'$ is what's known as a strong uniform time.

Now we would like to make a similar construction of a stopping time for the riffle shuffle. It turns out that this is actually easier to do for the 2-unshuffle; but the property of being a stopping time will hold for both processes since they are exactly inverse in the sense that $\hat{R}_a(\pi) = R_a(\pi^{-1})$. To begin, recall from section 9 that an equivalent way of doing a 2-unshuffle is to place a sequence of n 0s and 1s on the deck, one on each card. Subsequent 2-unshuffles are done by placing additional sequences of 0s and 1s on the deck, one on each card, each time placing a new 0 or 1 to left of the 0s and 1s already on the card. Here is an example of the directions for 5 particular 2-unshuffles, as written on the cards of a size $n = 7$ deck before any shuffling is done:

card#	unshuffle# 54321	base 32
1	01001	9
2	10101	21
3	11111	31
4	00110	6
5	10101	21
6	11000	24
7	00101	5

The numbers in the last column are obtain by using the digitary concatenation operator & on the five 0s and 1s on each card, i.e., they are obtained by treating the sequence of five 0s and 1s as a base $2^5 = 32$ number. Now we know that doing these 5 particular 2-unshuffles is equivalent to doing one particular 32-unshuffle by sorting the cards so that the base 32 labels are in the order 5, 6, 9, 21, 21, 24. Thus we get the deck ordering 741256.

Now we are ready to define a stopping time for 2-unshuffling. We will stop after $\mathbf{T}$ 2-unshuffles if $\mathbf{T}$ is the first time that the base $2^{\mathbf{T}}$ numbers, one on each card, are all distinct. Why in the world should this be a guarantee that the deck is randomized? Well, consider all orderings of the deck resulting from randomly and uniformly labeling the cards, each with a base $2^{\mathbf{T}}$ number, conditional on all the numbers being distinct. Any two cards in the deck before shuffling, say i and j, having received different base $2^{\mathbf{T}}$ numbers, are equally as likely to have gotten numbers such that is are greater than js as they are to have gotten numbers such that js are greater than is. This means after $2^{\mathbf{T}}$-unshuffling, i is equally as likely to come after j as to come before j. Since this holds for any pair of cards i and j, it means the deck is entirely randomized!

John Finn has contructed a counting argument that directly shows the same thing for 2-shuffling. Assume $2^{\mathbf{T}}$ is bigger than n, which is obviously necessary to get distinct numbers. There are $2^{\mathbf{T}}!/(2^{\mathbf{T}} - n)!$ ways to make a list of n distict $\mathbf{T}$ digit base 2 numbers, i.e., there are that many ways to 2-shuffle using distinct numbers, each equally likely. But every permutation can be achieved by $\binom{2^{\mathbf{T}}}{n}$ such ways, since we need only choose n different numbers from the $2^{\mathbf{T}}$ ones possible (so we have n nonempty packets of size 1) and arrange them in the necessary order to achieve the permutation. So the probability of any permutation under 2-shuffling with distinct numbers is

$$\binom{2^{\mathbf{T}}}{n} \bigg/ \left[\frac{2^{\mathbf{T}}!}{(2^{\mathbf{T}} - n)!} \right] = \frac{1}{n!},$$

which shows we have the uniform density, and hence that **T** actually is a stopping time.

Looking at the particular example above, we see that $T > 5$, since all the base 32 numbers are not distinct. The 2 and 5 cards both have the base 32 number 21 on them. This means that, no matter how the rest of the deck is labeled, the 2 card will always come before the 5, since all the 21s in the deck will get pulled somewhere, but maintaining their relative order. Suppose, however, that we do a 6th 2-unshuffle by putting the numbers 0100000 on the naturally ordered deck at the beginning before any shuffling. Then we have $T = 6$ since all the base 64 numbers are distinct:

card#	unshuffle# 654321	base 64
1	001001	9
2	110101	53
3	011111	31
4	000110	6
5	010101	21
6	011000	24
7	000101	5

Again, **T** is really a random variable, as was **T'**. Intuitively **T** really gives a necessary number of shuffles to get randomness; for, if we have not reached the time when all the base $2^{\mathbf{T}}$ numbers are distinct, then those cards having the same numbers will necessarily always be in their original relative order, and hence the deck could not be randomized. Also analogous to **T'** for the top-in shuffle is the fact that $P[\mathbf{T} > k]$ is an upper bound for the variation distance between the density after k 2-unshuffles and true randomness, and hence between k riffle shuffles and true randomness. So let us calculate $P[\mathbf{T} > k]$.

The probability that $\mathbf{T} > k$ is the probability that an n digit base 2^k number picked at random does not have distinct digits. Essentially this is just the birthday problem: given n people who live in a world that has a year of m days, what is the probability that two or more people have the same birthday? (Our case corresponds to $m = 2^k$ possible base 2^k digits/days.) It is easier to look at the complement of this event, namely that no two people have the same birthday. There are clearly m^n different and equally likely ways to choose birthdays for everybody. If we wish to choose distinct ones for everyone, the first person's may be chosen in m ways (any day), the second's in $m - 1$ ways (any but the day chosen for the first person), the third's in $m - 2$ ways (any but the days chosen for the first two people), and so on. Thus the probability of distinct birthdays being chosen is

$$\frac{\prod_{i=0}^{n-1}(m-i)}{m^n} = \frac{m!}{(m-n)!\,m^n} = \binom{m}{n}\frac{n!}{m^n},$$

and hence the probability of two people having the same birthday is one minus this number. (It is is interesting to note that for $m = 365$, the probability of matching birthdays is about 50% for $n = 23$ and about 70% for $n = 30$. So for a class of more than 23 students, it's a better than fair bet that two or more students have the same birthday.) Transferring to the setting of stopping times for 2-unshuffles, we have

$$P[\mathbf{T} > k] = 1 - \binom{2^k}{n} \frac{n!}{2^{kn}}$$

by taking $m = 2^k$. Here is a graph of $P[\mathbf{T} > k]$ (solid line), along with the variation distance $\|R^k - U\|$ (points) that it is an upper bound for.

It is interesting to calculate $E(\mathbf{T})$. This is given by

$$E(\mathbf{T}) = \sum_{k=0}^{\infty} P[\mathbf{T} > k] = \sum_{k=0}^{\infty} \left[1 - \binom{2^k}{n} \frac{n!}{2^{kn}} \right].$$

This is approximately 11.7 for $n = 52$, which means that, according to this viewpoint, we expect on average 11 or 12 shuffles to be necessary for randomizing a real deck of cards. Note that this is substantially larger than 7.

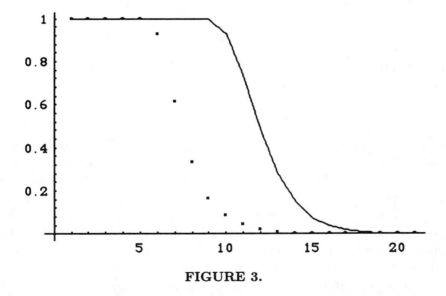

FIGURE 3.

11. STILL ANOTHER VIEWPOINT: MARKOV CHAINS

An equivalent way of looking at the whole business of shuffling is through Markov chains. A Markov chain is a stochastic process (meaning that the steps in the process are governed by some element of randomness) that consists of bouncing around among some finite set of states S, subject to certain restrictions. This is described exactly by a sequence of random variables $\{\mathbf{X}_t\}|_{t=0}^{\infty}$, each taking values in S, where $\mathbf{X}_t = i$ corresponds to the process being in state $i \in S$ at discrete time t. The density for $\mathbf{X}_0$ is arbitrary, meaning you can start the process off any way you wish. It is often called the *initial density*. In order to be a Markov chain, the subsequent densities are subject to a strong restriction: the probability of going to any particular state on the next step only depends on the current state, not on the time or the past history of states occupied. In particular, for each i and j in S there exists a fixed *transition probability* p_{ij} independent of t, such that $P[\mathbf{X}_t = j \mid \mathbf{X}_{t-1} = i] = p_{ij}$ for all $t \geq 1$. The only requirements on the p_{ij} are that they can actually be probabilities, i.e., they are nonnegative and $\sum_j p_{ij} = 1$ for all $i \in S$. We may write the p_{ij} as a *transition matrix* $p = (p_{ij})$ indexed by i and j, and the densities of the $\mathbf{X}_t$ as row vectors $(P[\mathbf{X}_t = j])$ indexed by j.

It turns out that, once the initial density is known, the densities at any subsequent time can be exactly calculated (in theory), using the transition probabilities. This is accomplished inductively by conditioning on the previous state. For $t \geq 1$,

$$P[\mathbf{X}_t = j] = \sum_{i \in S} P[\mathbf{X}_t = j \mid \mathbf{X}_{t-1} = i] \cdot P[\mathbf{X}_{t-1} = i].$$

There is a concise way to write this equation, if we treat $(P[\mathbf{X}_t = j])$ as a row vector. Then we get a matrix form for the above equation:

$$(P[\mathbf{X}_t = j]) = (P[\mathbf{X}_{t-1} = j]) \cdot p,$$

where the $\cdot$ on the r.h.s. stands for matrix multiplication of a row vector times a square matrix. We may of course iterate this equation to get

$$(P[\mathbf{X}_t = j]) = (P[\mathbf{X}_0 = j]) \cdot p^t,$$

where p^t is the tth power of the transition matrix. So the distribution at time t is essentially determined by the tth power of the transition matrix.

For a large class of Markov chains, called *regular*, there is a theorem that as $t \to \infty$, the powers p^t will approach a limit matrix, and this limit matrix has all rows the same. This row (i.e., any one of the rows) gives a density on S, and it is known as the *stationary density*. For these regular Markov chains, the stationary density is a unique limit for the densities of $\mathbf{X}_t$ as $t \to \infty$, regardless of the initial density. Furthermore, the stationary density is aptly named in the sense that, if the initial density $\mathbf{X}_0$ is taken to be the stationary one, then the subsequent densities for $\mathbf{X}_t$ for all t are all the same as the initial stationary density. In short, the stationary density is an equilibrium density for the process. We still need to define a regular chain. It is a Markov chain whose transition matrix raised to some power consists of all strictly positive probabilities. This is equivalent to the existence of some finite number t_0 for the Markov chain such that one can go from any state to any other state in exactly t_0 steps.

To apply all this to shuffling, let S be S_n, the set of permutations on n cards, and let Q be the type of shuffle we are doing (so Q is a density on S). Set X_0 to be the identity with probability one. In other words, we are choosing the intial density to reflect not having done anything to the deck yet. The transition probabilities are given by $p_{\pi\tau} = P[X_t = \tau \mid X_{t-1} = \pi] = Q(\pi^{-1} \circ \tau)$, since going from π to τ is accomplished by composing π with the permutation $\pi^{-1} \circ \tau$ to get τ. An immediate consequence of this is that the transition matrix for unshuffling is the transpose of the transition matrix for shuffling, since $\hat{p}_{\pi\tau} = \hat{R}(\pi^{-1} \circ \tau) = R((\pi^{-1} \circ \tau)^{-1}) = R(\tau^{-1} \circ \pi) = p_{\tau\pi}$.

Let us look at the example of the riffle shuffle with $n = 3$ from section 3 again, this time as a Markov chain. For $Q = R$ we had

π	[123]	[213]	[231]	[132]	[312]	[321]
$Q(\pi)$	1/2	1/8	1/8	1/8	1/8	0

So the transition matrix p, under this ordering of the permutations, is

[123] [213] [231] [132] [312] [321]

$$
\begin{array}{c}
[123] \\ [213] \\ [231] \\ [132] \\ [312] \\ [321]
\end{array}
\begin{pmatrix}
1/2 & 1/8 & 1/8 & 1/8 & 1/8 & 0 \\
1/8 & 1/2 & 1/8 & 1/8 & 0 & 1/8 \\
1/8 & 1/8 & 1/2 & 0 & 1/8 & 1/8 \\
1/8 & 1/8 & 0 & 1/2 & 1/8 & 1/8 \\
1/8 & 0 & 1/8 & 1/8 & 1/2 & 1/8 \\
0 & 1/8 & 1/8 & 1/8 & 1/8 & 1/2
\end{pmatrix}
$$

Let us do the computation for a typical element of this matrix, say $p_{\pi\tau}$ with $\pi = [213]$ and $\tau = [132]$. Then $\pi^{-1} = [213]$ and $\pi^{-1} \circ \tau = [231]$ and $R([231]) = 1/8$, giving us $p_{[213][132]} = 1/8$ in the transition matrix. Although in this case, the $n = 3$ riffle shuffle, the matrix is symmetric, this is not in general true; the transition matrix for the riffle shuffle with deck sizes greater than 3 is always nonsymmetric.

The reader may wish to verify the following transition matrix for the top-in shuffle:

[123] [213] [231] [132] [312] [321]

$$
\begin{array}{c}
[123] \\ [213] \\ [231] \\ [132] \\ [312] \\ [321]
\end{array}
\begin{pmatrix}
1/3 & 1/3 & 1/3 & 0 & 0 & 0 \\
1/3 & 1/3 & 0 & 1/3 & 0 & 0 \\
0 & 0 & 1/3 & 0 & 1/3 & 1/3 \\
0 & 0 & 0 & 1/3 & 1/3 & 1/3 \\
1/3 & 0 & 0 & 1/3 & 1/3 & 0 \\
0 & 1/3 & 1/3 & 0 & 0 & 1/3
\end{pmatrix}
$$

The advantage now is that riffle shuffling k times is equivalent to simply taking the kth power of the riffle transition matrix, which for a matrix of size

6-by-6 can be done almost immediately on a computer for reasonable k. By virtue of the formula

$$(P[\mathbf{X}_t = j]) = (P[\mathbf{X}_0 = j]) \cdot p^t$$

for Markov chains and that fact that in our example .
$(P[\mathbf{X}_0 = j]) = \begin{pmatrix} 1 & 0 & 0 & 0 & 0 & 0 \end{pmatrix}$, we may read off the density of the permutations after k shuffles simply as the first row of the kth power of the transition matrix. For instance, Mathematica gives p^7 approximately:

	[123]	[123]	[123]	[123]	[123]	[123]
[123]	.170593	.166656	.166656	.166656	.166656	.162781
[213]	.166656	.170593	.166656	.166656	.162781	.166656
[231]	.166656	.166656	.170593	.162781	.166656	.166656
[132]	.166656	.166656	.162781	.170593	.166656	.166656
[312]	.166656	.162781	.166656	.166656	.170593	.166656
[321]	.162781	.166656	.166656	.166656	.166656	.170593

and therefore the density after 7 shuffles is the first row:

π	[123]	[213]	[231]	[132]	[312]	[321]
$Q(\pi)$	.170593	.166656	.166656	.166656	.166656	.162781

It is clear that seven shuffles of the three card deck gets us very close to the uniform density (noting, as always, that the identity is still the most likely permutation), which turns out to be the stationary density. We first must note, not surprisingly, that the Markov chains for riffle shuffling are regular, i.e., there is some number of shuffles after which any permutation has a positive probability of being achieved. [In fact we know, from the formula $\begin{pmatrix} 2^k + n - r \\ n \end{pmatrix} / 2^{nk}$ for the probability of a permutation with r rising sequences being achieved after k riffle shuffles, that any number of shuffles greater than $\log_2 n$ will do.] Since the riffle shuffle Markov chains are regular, we know they have a unique stationary density, and this is clearly the uniform density on S_n.

From the Markov chain point of view, the rate of convergence of the $\mathbf{X}_t$ to the stationary density, measured by variation distance or some other metric, is often asymptotically determined by the eigenvalues of the transition matrix. We will not go into this in detail, but rather will be content to determine the eigenvalues for the transition matrix p for riffle shuffling. We know that the entries of p^k are the probabilities of certain permutations being achieved under k riffle shuffles. These are of the form $\begin{pmatrix} 2^k + n - r \\ n \end{pmatrix} / 2^{nk}$. Now we may explicitly write out

$$\begin{pmatrix} x + n - r \\ n \end{pmatrix} = \sum_{i=0}^{n} c_{n,r,i} x^i,$$

an nth degree polynomial in x, with coefficients a function of n and r. It doesn't really matter exactly what the coefficients are, only that we can write a polynomial in x. Substituting 2^k for x, we see the entries of p^k are of the form

$$[\sum_{i=0}^{n} c_{n,r,i}(2^k)^i]/2^{nk} = \sum_{i=0}^{n} c_{n,r,n-i}(\frac{1}{2^i})^k.$$

This means the entries of the kth power of p are given by fixed linear combinations of kth powers of 1, 1/2, 1/4, ..., and $1/2^n$. It follows from some linear algebra the set of all eigenvalues of p is exactly 1, 1/2, 1/4, ..., and $1/2^n$. Their multiplicities are given by the Stirling numbers of the first kind, up to sign: multiplicity$(1/2^i) = (-1)^{(}n-i)s1(n,i)$. This is a challenge to prove, however. The second highest eigenvalue is the most important in determining the rate of convergence of the Markov chain. For riffle shuffling, this eigenvalue is 1/2, and it is interesting to note in the variation distance graph of section 8 that once the distance gets to the cutoff, it decreases approximately by a factor of 1/2 each shuffle.

References

[1] **Aldous, David and Diaconis, Persi,** Strong Uniform Times and Finite Random Walks, *Advances in Applied Mathematics*, **8**, 69–97, 1987.

[2] **Bayer, Dave and Diaconis, Persi,** Trailing the Dovetail Shuffle to its Lair, *Annals of Applied Probability*, **2(2)**, 294–313, 1992.

[3] **Diaconis, Persi,** *Group Representations in Probability and Statistics,* Hayward, Calif: IMS, 1988.

[4] **Harris, Peter,** The Mathematics of Card Shuffling, senior thesis, Middlebury College, 1992.

[5] **Kolata, Gina,** In Shuffling Cards, Seven is Winning Number, *New York Times*, Jan. 9, 1990.

[6] **Reeds, Jim,** unpublished manuscript, 1981.

[7] **Snell, Laurie,** *Introduction to Probability,* New York: Random House Press, 1988.

[8] **Tanny, S.,** A Probabilistic Interpretation of the Eulerian Numbers, *Duke Mathematical Journal*, **40**, 717–722, 1973.

Chapter 10

STOCHASTIC GAMES AND OPERATORS

A. Maitra and W. Sudderth*
School of Statistics
University of Minnesota

ABSTRACT

Imagine two players engaged in a competition that consists of a sequence of games. Play begins at some state x and at every stage n the players select actions that determine the conditional distribution of the next state x_{n+1}. One player wants to reach a certain set of states or reach it repeatedly, while the other player wants exactly the opposite to occur. Techniques are presented for solving such games when the state space is countable and the action sets for the players are finite. A number of examples are given and some open questions are posed.

1. INTRODUCTION

Suppose Mary has $5 and Tom has $10. They decide to play a sequence of games involving both luck and skill until one of them has all $15. Such a competition corresponds to a "stochastic game" in which each player seeks to reach a certain goal. Either player might be more conservative and try to avoid ever going bankrupt. This would lead to a stochastic game in which the player tries to stay in the set of positive fortunes rather than trying for a certain goal. In principle a player could have much more complicated aspirations such as that of visiting a certain set of fortunes infinitely many times. We want to find methods for calculating how well a player can do and how a player should play in these problems of getting to a set, staying in a set, or visiting a set infinitely often. As it turns out, our methods correspond to iterating certain basic operators and solutions will be fixed points.

The next two sections are devoted to the necessary preliminaries on games and stochastic games. After them we will begin on our operator solutions.

*The research of W. Sudderth was supported by National Science Foundation Grant DMS - 9123358.

2. TWO-PERSON, ZERO-SUM GAMES

All of the games considered here will have two players I and II and will be zero-sum, which means that whatever I wins, II loses, and vice versa. Formally, such a game is a triple (A, B, φ) where A and B are nonempty sets of *actions* for I and II, respectively, and φ is a real-valued function on $A \times B$ that represents the *payoff* from II to I. That is, if I chooses action a and II chooses action b, then $\varphi(a, b)$ is the amount paid to I by II. If A and B are finite sets, we can represent φ by a matrix. Such games are often called *matrix games.*

Example 2.1 *Odd or even.* Mary and Tom simultaneously put up 1 or 2 fingers. Mary pays Tom the sum of the digits if the sum is even; Tom pays Mary the sum if it is odd. Formally, $A = B = \{1, 2\}$, $\varphi(1, 1) = -2$, $\varphi(1, 2) = \varphi(2, 1) = 3$, $\varphi(2, 2) = -4$. Here is the matrix representation:

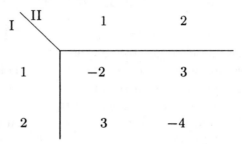

or, more simply,

$$\begin{bmatrix} -2 & 3 \\ 3 & -4 \end{bmatrix}.$$

Rows always correspond to actions of player I and columns to actions of player II.

Example 2.2 *A matching game* . Again Mary and Tom put up 1 or 2 fingers each. Mary wins \$1 if they put up the same number; she loses \$1 if not. Here's the matrix.

$$\begin{bmatrix} 1 & -1 \\ -1 & 1 \end{bmatrix}$$

Consider now a game (A, B, φ) from player I's point of view. If I chooses action a, then her payoff will be at least $\inf_b \varphi(a, b)$. By selecting a with circumspection, she can achieve a payoff of at least $\underline{v} - \varepsilon$ for any $\varepsilon > 0$ where $\underline{v}$ is the quantity

$$\underline{v} = \sup_a \inf_b \varphi(a, b).$$

In a similar fashion, player II can, through his choice of b, keep the payoff below $\bar{v} + \varepsilon$ where

$$\bar{v} = \inf_{b} \sup_{a} \varphi(a, b).$$

The quantities $\underline{v}$ and $\bar{v}$ are called the *lower* and *upper values of the game*, respectively. To see that

$$\underline{v} \leq \bar{v},$$

notice that $\sup_{a} \varphi(a, b) \geq \varphi(a, b)$ for every (a, b), and then take the infimum over b and supremum over a If $\bar{v} = \underline{v}$, then the game is said to have the *value* $v = \underline{v} = \bar{v}$.

If the game has value v, an action a^* is *optimal* for player I if

$$\varphi(a^*, b) \geq v \text{ for all } b.$$

Thus a^* guarantees player I a payoff no less than the value against any action of II. Likewise an action b^* is optimal for II if

$$\varphi(a, b^*) \leq v \text{ for all } a.$$

Neither of the games in the two examples above has a value. In example 2.2, $\underline{v} = -1$ and $\bar{v} = 1$. However, it follows from a theorem of J. von Neumann that these games do have a value if the players are allowed to choose actions at random and independently of each other. To state the theorem, assume that A and B are finite and let $P(A)$ and $P(B)$ be the sets of all probability distributions on A and B, respectively. Regard $P(A)$ and $P(B)$ as the sets of actions for a new game with payoff function defined, for each $\mu \in P(A)$ and $\nu \in P(B)$, as the expected value

$$E_{\mu,\nu}\varphi = \sum_{a} \sum_{b} \varphi(a, b)\mu\{a\}\nu\{b\}.$$

This new game does have a value.

Theorem 2.1 *von Neumann's minimax theorem.* If A and B are finite, then

$$\inf_{\nu} \sup_{\mu} E_{\mu,\nu}\varphi = \sup_{\mu} \inf_{\nu} E_{\mu,\nu}\varphi. \tag{2.1}$$

Furthermore, there exist optimal randomized actions $\mu^* \in P(A)$ and $\nu^* \in P(B)$ so that

$$E_{\mu^*,\nu}\varphi \geq \inf_{\nu} \sup_{\mu} E_{\mu,\nu}\varphi \geq E_{\mu,\nu^*}\varphi \tag{2.2}$$

for all $\mu \in P(A)$, $\nu \in P(B)$.

The quantity in (2.1) is, of course, the value of the game $(P(A), P(B),$ $E_{\mu,\nu}\varphi)$. However, from now on we will call it the value of the matrix game (A, B, φ) and sometimes, for brevity, the value of the matrix $(\varphi(a,b)), a \in A, b \in B$. There is little danger of confusion, because the value of the matrix game will be the same whenever it exists.

A proof of this fundamental result can be found in most books on game theory including the classic by von Neumann and Morgenstern [6]. Books on game theory also explain how to calculate the value and optimal actions using such techniques as linear programming (cf. [3] and [4]). We will explain a simple method for the special case when A has only two elements a_1 and a_2 and B is finite. Then each member μ of $P(A)$ corresponds to a vector $(\theta, 1-\theta)$ where $\theta = \mu\{a_1\}$ and $1 - \theta = \mu\{a_2\}$. So we can rewrite

$$E_{\mu,\nu}\varphi = \sum_b (\theta\varphi(a_1, b) + (1-\theta)\varphi(a_2, b))\nu\{b\}.$$

For fixed θ, this expression is minimized by taking ν to put all its probability at that b for which $\theta\varphi(a_1, b) + (1 - \theta)\varphi(a_2, b)$ is a minimum. So

$$\inf_\nu E_{\mu,\nu}\varphi = \min_b \{\theta\varphi(a_1, b) + (1-\theta)\varphi(a_2, b)\}$$

and

$$\sup_\mu \inf_\nu E_{\mu,\nu}\varphi = \max_{0 \leq \theta \leq 1} \min_b \{\theta\varphi(a_1, b) + (1-\theta)\varphi(a_2, b)\}.$$

Suppose B also has two elements b_1 and b_2. Then the value becomes

$$\max_{0 \leq \theta \leq 1} \min_{b_1, b_2} \{\theta\varphi(a_1, b_1) + (1-\theta)\varphi(a_2, b_1), \theta\varphi(a_1, b_2) + (1-\theta)\varphi(a_2, b_2)\}. \quad (2.3)$$

As example 2.1 illustrates, the supremum will be achieved at that value of θ (if there is one) where the two linear functions of θ in (2.3) are equal.

Example 2.1 (continued). The expression in (2.3) becomes

$$\max_{0 \leq \theta \leq 1} \min\{-2\theta + 3(1-\theta), 3\theta - 4(1-\theta)\}$$

$$= \max_{0 \leq \theta \leq 1} \min\{3 - 5\theta, 7\theta - 4\}.$$

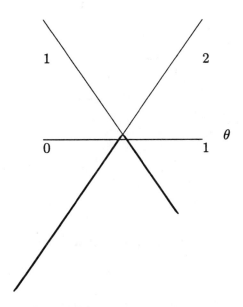

FIGURE 1.

As is illustrated in Figure 1, the line $3 - 5\theta$ is the expected payoff for I when I chooses the randomized action $(\theta, 1 - \theta)$ and II chooses action 1. Likewise the line $7\theta - 4$ is the payoff when I chooses $(\theta, 1 - \theta)$ and II chooses action 2. For each θ the dark line represents the least expected payoff for I and the maximum of these minima occurs at the intersection where

$$3 - 5\theta = 7\theta - 4$$

and $\theta = 7/12$. Thus the value is $3 - 5 \times 7/12 = 1/12$ and $(7/12, 5/12)$ is optimal for player I.

Similar calculations or a symmetry argument show that the matching game of example 2.2 has value zero and it is optimal for each player to put up 1 or 2 fingers with probability $1/2$ each.

Here is an asymmetric variation of the matching game in which the payoff is zero if Mary puts up two fingers and Tom puts up one.

Example 2.3

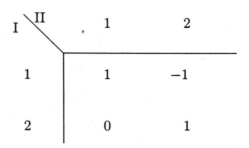

Argue as above to see that the optimal θ satisfies

$$\theta = -\theta + 1 - \theta$$

so that $\theta = 1/3$ and the value is also $1/3$. So the optimal randomized action for I is $\mu^* = (1/3,\ 2/3)$. A similar calculation shows II should use $\nu^* = (2/3,\ 1/3)$. If each player plays optimally (and independently) then the payoff $\varphi(a, b)$ will be a random variable which takes on the values -1, 0, and 1 with probabilities $(1/3)(1/3) = 1/9$, $(2/3)(2/3) = 4/9$, and $(1/3)(2/3) + (2/3)(1/3) = 4/9$, respectively. So

$$E_{\mu^*,\nu^*}\varphi = (1/9)(-1) + (4/9)(0) + (4/9)(1) = 1/3.$$

If A and B are infinite, then the game need not have a value even if randomized actions are allowed. Here is an example.

Example 2.4 Let $A = B = \{1, 2, ...\}$ and

$$\varphi(a, b) = \begin{cases} 1 & \text{if } a > b \\ 0 & \text{if } a \leq b. \end{cases}$$

Thus player I wins if her integer is larger than that of II and loses otherwise.

Now let μ be a randomized action for I; i.e., μ is a countably additive probability measure on A. Then, given any $\varepsilon > 0$, there is an integer a_1 so large that

$$\mu\{a_1, a_1 + 1, ...\} < \varepsilon.$$

So, if player II chooses action a_1, he wins with probability at least $1 - \varepsilon$. Thus

$$\sup_{\mu} \inf_{\nu} E_{\mu,\nu}\varphi \leq \varepsilon$$

and the lower value must be zero. A similar argument shows the upper value, $\inf\limits_{\nu}\sup\limits_{\mu} E_{\mu,\nu}\varphi$, to be one.

In most of the stochastic games considered below, each player will have an infinite set of possible choices and von Neumann's theorem will not apply directly. Nevertheless, we will see that many of the games do have a value that we can calculate.

3. STOCHASTIC GAMES

Stochastic games are played in stages. The two players start at some state x belonging to a state space X which we assume to be finite or countably infinite. Player I chooses an action a_1, possibly at random, from the finite set A of her possible actions and player II chooses b_1, possibly at random and independently of a_1, from his finite set B. The triple (x, a_1, b_1) determines the distribution $q(\cdot|x, a_1, b_1)$ of the next state x_1. The function q is called the *law of motion* for the game and assigns a probability distribution on X to each member of $X \times A \times B$. The actions a_1, b_1 and the new state x_1 are announced to the players. The players next choose actions a_2 and b_2 that, together with x_1, determine the distribution $q(\cdot|x_1, a_2, b_2)$ of x_2. The players continue choosing actions at each stage of play and thereby determine the distribution of a stochastic process $x_1, x_2, \ldots$ The payoff from II to I is the expected value of $g(x_1, x_2, \ldots)$ where g is a given real-valued function defined on sequences of states whose expectation is well-defined. Formally, a stochastic game can be viewed as the 5-tuple (X, A, B, q, g).

It is also possible to regard a stochastic game as being an ordinary game in the sense of the previous section. An action in the stochastic game will be called a strategy and will correspond to a completely specified plan for the choices to be made at every stage of play. Thus a *strategy* σ for player I is a sequence $\sigma_0, \sigma_1, \ldots$ where $\sigma_o \in P(A)$ is the first randomized action for I and, for $n \geq 1, \sigma_n$ is a function which assigns to each sequence $p_n = ((a_1, b_1, x_1), (a_2, b_2, x_2), \ldots, (a_n, b_n, x_n))$ of possible actions and states through stage n the randomized action $\sigma_n(p_n) \in P(A)$ for I at stage $n + 1$. A *strategy* $\tau = \tau_0, \tau_1, \ldots$ for player II is similarly defined with $P(B)$ replacing $P(A)$. For a given initial state x, strategies σ for I and τ for II (together with the law of motion q) determine the distribution $P = P_{x,\sigma,\tau}$ of $(a_1, b_1, x_1), (a_2, b_2, x_2), \cdots$. Namely, the marginal probability of the first triple is $\sigma_o\{a_1\}\tau_o\{b_1\}q(x_1|x, a_1, b_1)$ and the conditional probability of the $n + 1$st given p_n is $\sigma_n(p_n)\{a_{n+1}\}\tau_n(p_n)\{b_{n+1}\}q(x_{n+1}|x_n, a_{n+1}, b_{n+1})$.

Let $\sum$ and T be the sets of all strategies σ and τ for players I and II, respectively. For each initial state x, the stochastic game has action sets $\sum$ and T and payoff function $E_{x,\sigma,\tau}g$ where $E_{x,\sigma,\tau}$ denotes expectation with respect to $P_{x,\sigma,\tau}$. The upper and lower values of this game are, respectively,

$$\overline{V}(g)(x) = \inf_{\tau} \sup_{\sigma} E_{x,\sigma,\tau} g,$$

$$\underline{V}(g)(x) = \sup_{\sigma} \inf_{\tau} E_{x,\sigma,\tau} g.$$

In view of the simple example 2.4 of a game having no value, it seems too much to hope that stochastic games will always have a value in the sense that $\overline{V}(g) = \underline{V}(g)$ for all bounded measurable functions g. However, we do not know a counterexample.

Stochastic games were introduced by Shapley [21] who studied a payoff function of the form

$$g(x_1, x_2, \ldots) = \sum_{n=1}^{\infty} \beta^n r(x_n)$$

where $r : X \to R$ is a bounded function representing a daily reward and $\beta \in (0,1)$ is a discount factor. Shapley showed, for finite X, that these games with a discounted total reward have a value and his result has been extended to infinite (and even uncountable) X (cf. Nowak [18]). A number of authors worked on the problem with a long-run average reward of the form

$$g(x_1, x_2, \ldots) = \overline{\lim_{n}} \, n^{-1} \sum_{i=1}^{n} r(x_i).$$

It was shown by Mertens and Neyman [16] that such games have a value for X finite, and their result has been extended to infinite X by Maitra and Sudderth [14, 15].

The three problems mentioned in the introduction correspond to payoff functions which are the indicator functions of the sets

$$[\text{reach } G] = \{(x_1, x_2, \ldots) : x_n \in G \text{ for some } n\},$$
$$[\text{stay in } G] = \{(x_1, x_2, \ldots) : x_n \in G \text{ for all } n\}, \tag{3.1}$$
$$[G \text{ i.o.}] = \{(x_1, x_2, \ldots) : x_n \in G \text{ for infinitely many } n\},$$

where G is a given subset of X.

The reason why stochastic games are often more tractable than general games with infinite action sets is that they have a natural recursive structure. A strategy $\sigma = (\sigma_0, \sigma_1, \ldots\ldots)$ for player I can be thought of as a randomized action σ_0 in an initial game together with a *conditional strategy* $\sigma[p_1]$ that specifies all future choices for player I given the initial triple $p_1 = (a_1, b_1, x_1)$. Likewise a strategy $\tau = (\tau_0, \tau_1, \ldots)$ can be regarded as τ_0 together with a conditional strategy $\tau[p_1]$. Furthermore the probability distribution $P = P_{x,\sigma,\tau}$ for $(a_1, b_1, x_1), (a_2, b_2, x_2), \ldots$ has a marginal distribution $P_0 = P_{x,\sigma_0,\tau_0}$ for $p_1 = (a_1, b_1, x_1)$ and conditional distribution $P_{x_1, \sigma[p_1], \tau[p_1]}$ for $(a_2, b_2, x_2), (a_3, b_3, x_3), \ldots$ given p_1. Thus the familiar conditioning formula

$$Eg = E[E(g|p_1)]$$

can be written more explicitly as

$$E_{x,\sigma,\tau}g = \sum_{p_1} E_{x_1,\sigma[p_1],\tau[p_1]}(gx_1)P_0\{p_1\} \qquad (3.2)$$

where gx_1 is the x_1-section of g defined by

$$(gx_1)(x_2, x_3, \ldots) = g(x_1, x_2, x_3, \ldots).$$

The problem of playing well in a stochastic game is related, in view of (3.2), to that of playing well in an initial one-day game and then continuing to play well in another stochastic game starting from a new position. To exploit this idea we introduce a class of auxiliary one-day games.

Let Ψ be a bounded, real-valued function on X, and, for each x, let $Q(\Psi)(x)$ be the *one-day game* (A, B, φ) where

$$\varphi(a, b) = \sum_{x_1} \Psi(x_1)q(x_1|x, a, b)$$

is the expected value of Ψ under the law of motion when player I chooses action a and II chooses b. For fixed x and randomized actions $\mu \in P(A), \nu \in P(B)$, let

$$
\begin{aligned}
E_{x,\mu,\nu}\Psi &= E_{\mu,\nu}\varphi \qquad\qquad\qquad\qquad (3.3)\\
&= \sum_a \sum_b \sum_{x_1} \Psi(x_1)q(x_1|x, a, b)\mu\{a\}\nu\{b\}.
\end{aligned}
$$

By Theorem 2.1, the game $Q(\Psi)(x)$ has the value

$$(S\Psi)(x) = \inf_\nu \sup_\mu E_{x,\mu,\nu}\Psi. \qquad (3.4)$$

Notice that $S\Psi$ is, like Ψ, a bounded, real-valued function on X. The operator S is called the *one-day operator* for the stochastic game.

Here are two examples.

Example 3.1 *Additive games.* Suppose Mary and Tom agree to play and replay the same matrix game (A, B, φ). Let the current state x represent Mary's current fortune in dollars. We allow for the possibility that x is negative and take X to be the set of integers. If Mary chooses action a and Tom chooses b, then Mary's next fortune is $x_1 = x + \varphi(a, b)$. That is, the law of motion is deterministic with $q(\cdot|x, a, b)$ assigning probability 1 to $x+\varphi(a, b)$. In their interesting paper on "games of survival" Milnor and Shapley [17] studied additive games in which each player sought to drive the other into bankruptcy.

We will treat two such games in the next section, but, at present, only wish to calculate the one-day operator S.

To be specific, suppose the matrix game is the matching game of Example 2.2. Then, for any x and any bounded function Ψ on X, $(S\Psi)(x)$ is the value of the matrix game:

I \ II	1	2
1	$\Psi(x+1)$	$\Psi(x-1)$
2	$\Psi(x-1)$	$\Psi(x+1)$

Using the methods of section 2, you can calculate

$$(S\Psi)(x) = \max_{0\leq\theta\leq 1} \min\{\theta(\Psi(x+1) - \Psi(x-1)) + \Psi(x-1),$$
$$\theta(\Psi(x-1) - \Psi(x+1)) + \Psi(x+1)\}.$$

Two cases arise. If $\Psi(x+1) \neq \Psi(x-1)$, the maximum occurs at $\theta = 1/2$ as can be seen by equating $\theta[\Psi(x+1) - \Psi(x-1)] + \Psi(x-1)$ and $\theta[\Psi(x-1) - \Psi(x+1)] + \Psi(x+1)$. Thus

$$
\begin{aligned}
(S\Psi)(x) &= (\Psi(x+1) - \Psi(x-1))/2 + \Psi(x-1) \qquad (3.5)\\
&= (\Psi(x+1) + \Psi(x-1))/2.
\end{aligned}
$$

If $\Psi(x+1) = \Psi(x-1)$, the matrix has constant entries and (3.5) is trivially true. So, for this game, S is a linear operator. On the other hand, if the matrix game is the asymmetric matching game of example 2.3, you can use the same method to see that

$$(S\Psi)(x) = \frac{\Psi(x+1)^2 - \Psi(x)\Psi(x-1)}{2\Psi(x+1) - \Psi(x) - \Psi(x-1)} \qquad (3.6)$$

when the denominator is not zero.

Example 3.2 *Plus or minus one.* Let X be the set of integers and take $A = B = \{1,2\}$. Assume that, for every x, the law of motion q assigns all its probability to $\{x-1, x+1\}$ and that the probabilities do not depend on x. In other words, q satisfies

$$q(x+1|x,i,j) = p_{ij} = 1 - q(x-1|x,i,j) \qquad (3.7)$$

for all x, i, j and some $p_{ij} \in [0, 1]$. (Some additive games, like the one generated by the matching game, have this structure.) This time the one-day operator S is "piecewise linear." To see why, fix x and suppose $\Psi(x + 1) \geq \Psi(x - 1)$. Then player I wants to maximize and player II wants to minimize the chance p_{ij} that the next state is $x + 1$. Set p^* equal to the value of the matrix game

$$\begin{bmatrix} p_{11} & p_{12} \\ p_{21} & p_{22} \end{bmatrix}.$$

Now for randomized actions $\mu \in P(A)$ and $\nu \in P(B)$, the expectation $E_{\mu,\nu}\varphi$ in (3.3) is of the form

$$E_{\mu,\nu}\varphi = p\Psi(x + 1) + (1 - p)\Psi(x - 1) = p[\Psi(x + 1) - \Psi(x - 1)] + \Psi(x - 1)$$

where

$$p = p(\mu, \nu) = \sum_i \sum_j q(x + 1 | x, i, j)\mu\{i\}\nu\{j\}$$

is the probability the next state is $x + 1$. If μ^* is optimal for I in the game (p_{ij}), then, for all ν, $p(\mu^*, \nu) \geq p^*$ and, consequently,

$$E_{\mu^*,\nu}\varphi \geq p^*\Psi(x + 1) + (1 - p^*)\Psi(x - 1).$$

when $\Psi(x + 1) \geq \Psi(x - 1)$. Therefore, the value $(S\Psi)(x)$ of the one-day game $Q(\Psi)(x)$ satisfies

$$(S\Psi)(x) \geq p^*\Psi(x + 1) + (1 - p^*)\Psi(x - 1).$$

A symmetric argument proves the opposite inequality. So

$$(S\Psi)(x) = p^*\Psi(x + 1) + (1 - p^*)\Psi(x - 1) \tag{3.8}$$

when $\Psi(x + 1) \geq \Psi(x - 1)$. Similarly, if $\Psi(x + 1) \leq \Psi(x - 1)$, player I seeks to maximize and player II to minimize $1 - p_{ij}$, and

$$(S\Psi)(x) = (1 - q^*)\Psi(x + 1) + q^*\Psi(x - 1) \tag{3.9}$$

where q^* is the value of the matrix game

$$\begin{bmatrix} 1 - p_{11} & 1 - p_{12} \\ 1 - p_{21} & 1 - p_{22} \end{bmatrix}.$$

Here is a lemma with two useful properties of the one-day operator S.

Lemma 3.1. Let $\Psi_1 \leq \Psi_2 \leq \ldots$ be uniformly bounded, real-valued functions on X. Then (a) $S\Psi_1 \leq S\Psi_2$ and (b) $\lim_n S\Psi_n = S(\lim_n \Psi_n)$.

Proof. (a) is immediate from the definition of S in (3.4). For (b), set $\Psi = \lim_n \Psi_n$. Fix x and use Theorem 2.1 to choose $\mu \in P(A)$ such that, for all $b \in B$,

$$\sum_a \sum_{x_1} \Psi(x_1) q(x_1 | x, a, b) \mu\{a\} \geq (S\Psi)(x).$$

Let $\varepsilon > 0$. Because B is finite, we have for n sufficiently large and all b,

$$\sum_a \sum_{x_1} \Psi_n(x_1) q(x_1 | x, a, b) \mu\{a\} \geq (S\Psi)(x) - \varepsilon.$$

Hence, for n sufficiently large, $(S\Psi_n)(x) \geq (S\Psi)(x) - \varepsilon$. So $\lim_n (S\Psi_n)(x) \geq (S\Psi)(x)$. The opposite inequality is a consequence of (a). □

4.　REACHING A SET OR STAYING IN IT

Let $G \subseteq X$, and consider first the stochastic game (X, A, B, q, g) in which the payoff function g is the indicator function of the set [reach G] of (3.1). Assume also that the law of motion is trivial on G in the sense that

$$q(x | x, a, b) = 1 \text{ for all } x \in G, a \in A, b \in B. \tag{4.1}$$

This assumption of no motion on G is natural when the object is to reach G, but it will be important that we drop the assumption later.

Define a sequence of functions on X by

$$\begin{aligned} U_o &= 1_G \\ U_{n+1} &= SU_n \quad \text{for } n = 0, 1, \ldots \end{aligned} \tag{4.2}$$

where S is the one-day operator. All of the U_n have values in $[0,1]$ because U_o does and S is monotone (Lemma 3.1a). Also $U_n = 1$ on G for all n because of the no motion assumption (4.1). In particular, $U_o \leq U_1$. Now, assuming $U_n \leq U_{n+1}$, we can use the monotonicity of S to get $U_{n+1} = SU_n \leq SU_{n+1} = U_{n+2}$. So we conclude

$$0 \leq U_o \leq U_1 \leq \ldots \leq 1. \tag{4.3}$$

　　　Now set

$$U = \lim_n U_n$$

and we are ready for the main result of this section.

Theorem 4.1 Under the assumption of no motion on G, the game of reaching G from the initial state x has the value $U(x)$. Furthermore the value function U satisfies

$$SU = U,$$
$$U = 1 \text{ on } G. \tag{4.4}$$

To see that U is a fixed point of S, apply Lemma 3.1b to get

$$SU = \lim_n SU_n = \lim_n U_{n+1} = U.$$

The other assertion of (4.4), that $U = 1$ on G, is immediate because $U_n = 1$ on G for all n.

To see that $U(x)$ is the value of the game, it suffices to show

$$\overline{U}(x) \leq U(x) \leq \underline{U}(x) \tag{4.5}$$

where $\overline{U}(x)$ and $\underline{U}(x)$ are the upper and lower values:

$$\overline{U}(x) = \inf_\tau \sup_\sigma P_{x,\sigma,\tau}[\text{ reach } G],$$
$$\underline{U}(x) = \sup_\sigma \inf_\tau P_{x,\sigma,\tau}[\text{ reach } G].$$

Since the lower value is never larger than the upper value, it will follow from (4.5) that the inequalities there are, in fact, equalities and the game has a value.

We present the proof of (4.5) in two lemmas, which will also show how the two players can construct good strategies.

Lemma 4.1 *For every $x \in X$ and $n = 0, 1, \ldots$, there is a strategy σ for player I such that, for all strategies τ for player II,*

$$P_{x,\sigma,\tau}[\text{ reach } G] \geq U_n(x). \tag{4.6}$$

Hence, $\underline{U}(x) \geq \lim U_n(x) = U(x)$.

Proof: The final assertion is an immediate consequence of the first and the definitions of U and $\underline{U}$. In view of (4.1), the first assertion is trivial when $n = 0$. In fact, (4.6) holds for all x, σ, τ if $n = 0$. So, to do an induction, assume the result for n and, for each x, let $\tilde{\sigma}(x)$ be a strategy for I that makes (4.6) hold for all τ. Now fix an x and we must find σ so that (4.6) holds with n replaced by $n + 1$. Choose the initial action σ_0 for σ to be optimal in the one-day game $Q(U_n)(x)$ and, for each x_1, let the conditional strategy $\sigma[p_1] = \sigma[(a_1, b_1, x_1)]$ be $\tilde{\sigma}(x_1)$. Let τ be any strategy for player II and set $P = P_{x,\sigma,\tau}$. We will calculate $P[\text{reach } G]$ by conditioning on $p_1 = (a_1, b_1, x_1)$. If $x_1 \notin G$, then

$$P[\text{reach } G \mid p_1] = P_{x_1, \sigma[p_1], \tau[p_1]}[\text{reach } G]$$
$$\geq U_n(x_1)$$

because $\sigma[p_1] = \tilde{\sigma}(x_1)$. If $x_1 \in G$, then

$$P[\text{reach } G \mid p_1] = 1 \geq U_n(x_1).$$

So, as in (3.2),

$$
\begin{aligned}
P[\text{reach } G] &= E_{x,\sigma_o,\tau_o}(P[\text{reach } G \mid p_1]) \\
&\geq E_{x,\sigma_o,\tau_o} U_n.
\end{aligned}
$$

The final expression is at least $(SU_n)(x) = U_{n+1}(x)$ because σ_0 is optimal for I in $Q(U_n)(x)$. □

The proof of Lemma 4.1 shows how player I can construct a strategy that assures her of winning at least $U_n(x)$ and, therefore, $U(x) - \varepsilon$ for any $\varepsilon > 0$. It can happen that player I has no optimal strategy. However, the next lemma shows that player II always has an optimal strategy.

For each x, let $\nu_x \in P(B)$ be optimal for II in the one-day game $Q(U)(x)$ so that

$$E_{x,\mu,\nu_x} U \leq SU(x) = U(x)$$

for all $\mu \in P(A)$. Define τ^x to be that strategy for II in the stochastic game with initial state x which uses action ν_y whenever the current state is y.

Lemma 4.2 *For every $x \in X$ and every strategy σ for player I,*

$$P_{x,\sigma,\tau^x}[\text{reach } G] \leq U(x). \tag{4.7}$$

Hence, $\overline{U}(x) \leq U(x)$.

Proof: The final assertion is immediate from (4.7) and the definition of $\overline{U}(x)$. To prove (4.7), define, for $n = 1, 2, \ldots,$

$$
\begin{aligned}
G_n &= [\text{reach } G \text{ by time } n] \\
&= \{(x_1, x_2, \ldots) : x_i \in G \text{ for some } i \leq n\}.
\end{aligned}
$$

The sets G_n increase to $[\text{reach } G]$. So (4.7) will follow from

$$P_{x,\sigma,\tau^x}(G_n) \leq U(x), \tag{4.8}$$

which we will prove by induction on n.

Set $P = P_{x,\sigma,\tau^x}$. Now $U \geq U_o = 1_G$. So

$$P(G_1) = E_{x,\sigma_o,\nu^x} U_o \leq E_{x,\sigma_o,\nu_x} U \leq U(x).$$

Next make the inductive assumption that (4.8) holds for n and all x and σ. We will calculate $P(G_{n+1})$ by conditioning on $p_1 = (a_1, b_1, x_1)$. If $x_1 \notin G$, then

$$P[G_{n+1} \mid p_1] = P_{x_1,\sigma[p_1],\tau^{x_1}}(G_n) \le U(x_1)$$

by the inductive assumption. If $x_1 \in G$, then

$$P[G_{n+1} \mid p_1] = 1 = U(x_1).$$

So

$$
\begin{aligned}
P(G_{n+1}) &= E_{x,\sigma_o,\nu^x}(P[G_{n+1} \mid p_1]) \\
&\le E_{x,\sigma_o,\nu^x} U \\
&\le U(x).
\end{aligned}
$$

The proof of Theorem 4.1 is now complete and we turn to some examples.

Example 4.1 *Plus or minus one with a goal.* Suppose $X = \{0, 1, ..., N\}$. Think of the current state x as being Mary's fortune and think of $N - x$ as being Tom's fortune. Mary wants to reach N; so we take the set G to be the singleton $\{N\}$. As in example 3.2, take $A = B = \{1, 2\}$ and take the law of motion as in (3.7) for $x = 1, ..., N - 1$. The states 0 and N are assumed to be absorbing in the sense that

$$q(0 \mid 0, i, j) = 1, \ q(N \mid N, i, j) = 1 \qquad (4.9)$$

for all i, j. Obviously, Mary's chances can only improve as x increases. So the value function is nondecreasing and, as in (3.8),

$$(SU)(x) = p^*U(x+1) + (1 - p^*)U(x-1)$$

for $x = 1, ..., N-1$. But U is a fixed point of S by (4.4). So, for $x = 1, ..., N-1$,

$$U(x) = p^*U(x+1) + (1 - p^*)U(x-1) \qquad (4.10)$$

and, by (4.9),

$$U(0) = 0, \ U(N) = 1.$$

These equations for U are the same as those obtained when $U(x)$ is the "gambler's ruin" probability — namely, the probability that a simple random walk, which moves to the right with probability p^* and to the left with probability $1 - p^*$, will reach N before 0. The formula for this probability and, therefore, for the value function U is

$$
U(x) = \begin{cases}
\dfrac{1 - \left(\dfrac{1-p^*}{p^*}\right)^x}{1 - \left(\dfrac{1-p^*}{p^*}\right)^N}, & p^* \ne \dfrac{1}{2} \\[4ex]
\dfrac{x}{N}, & p^* = \dfrac{1}{2}.
\end{cases} \qquad (4.11)
$$

See Feller ([29], section XIV.2). In this game, both players have optimal strategies that correspond to always playing optimal actions in the one-day game $\mathcal{Q}(U)(x)$ when the current state is x.

Suppose next that X is the set of all integers and player I wants to reach $G = \{0\}$. Assume that the law of motion satisfies (3.7) for $x \neq 0$ and is absorbing at 0. This time U is nondecreasing to the left of zero and nonincreasing to the right of zero. So it follows from (3.8), (3.9), and (4.4) that

$$U(x) = \begin{cases} p^*U(x+1) + (1 - p^*)U(x-1) & , \quad x < 0, \\ 1 & , \quad x = 0, \\ (1 - q^*)U(x+1) + q^*U(x-1) & , \quad x > 0. \end{cases} \tag{4.12}$$

For $x \leq 0$, these equations are the same as when $U(x)$ is the probability that a simple random walk, which moves to the right with probability p^*, will reach 0 from x. As is shown in the same section of Feller's book [29], this probability is

$$U(x) = \begin{cases} \left(\dfrac{p^*}{1 - p^*}\right)^{-x} & \text{for } x \leq 0, p^* < \dfrac{1}{2}, \\ \\ 1 & \text{for } x \leq 0, p^* \geq \dfrac{1}{2}. \end{cases} \tag{4.13}$$

Likewise,

$$U(x) = \begin{cases} \left(\dfrac{q^*}{1 - q^*}\right)^{x} & \text{for } x \geq 0, q^* < \dfrac{1}{2}, \\ \\ 1 & \text{for } x \geq 0, q^* \geq \dfrac{1}{2}. \end{cases} \tag{4.14}$$

As before, optimal strategies for each player correspond to the use of optimal actions in the one-day games at every stage of play.

It would be interesting to generalize the plus or minus one game to higher dimensions. In two dimensions we could allow the motion to be a step of length one to the right or left or up or down. To be specific, take X to be the two-dimensional lattice of points with integer coordinates, $A = B = \{1, 2\}$, and, for $x \in X, i \in A, j \in B$,

$$q(x + e_k \mid x, i, j) = p_{ijk}, \text{ for } k = 1, ..., 4,$$

where $e_1 = (1, 0)$, $e_2 = (-1, 0)$, $e_3 = (0, 1)$, $e_4 = (0, -1)$, and $\sum_k p_{ijk} = 1$ for every pair (i, j). Which of these games has the property that player I can reach the origin with probability one from every x?

Blackwell [7] treats an interesting class of higher dimensional problems and his techniques could be useful for this one also.

Example 4.2 *Two games of survival.* Suppose, as in example 3.1, Mary and Tom agree to play the same matrix game (A, B, φ) repeatedly, but assume now that the game ends with the bankruptcy of either player. Milnor and Shapley [17] give an elegant and quite general treatment of this problem. We will consider two specific cases. In both we take $X = \{0, 1, ..., N\}$ and regard x and $N - x$ as the current fortunes of Mary and Tom, respectively. Mary's goal is to reach $G = \{N\}$. The law of motion is the additive one of example 3.1 for $x = 1, ..., N - 1$ but is taken to be absorbing at 0 and N as in (4.9).

Assume first that the matrix game is the matching game of example 2.2. Then by (3.5)

$$(SU)(x) = \frac{U(x + 1) + U(x - 1)}{2}$$

for $x = 1, ..., N - 1$. So, by (4.4) and (4.9),

$$U(x) = \frac{U(x + 1) + U(x - 1)}{2}, \quad x = 1, ..., N - 1$$
$$U(0) = 0, \quad U(N) = 1.$$

This is just (4.10) with $p^* = 1/2$ and thus the solution is

$$U(x) = \frac{x}{N}.$$

Optimal strategies for the two players correspond to optimal day-by-day play in the matrix game until one of the players is bankrupt.

Now assume that the matrix game is the asymmetric matching game of example 2.3. In this game we will see that Mary can win, from any positive x, with probability arbitrarily close to one. So

$$U(x) = 1, x = 1, ..., N.$$

It is no longer optimal for Mary to play optimally day-by-day in the matrix game, for, if she does, the conditional probability that $x_1 = x + 1$ given $x_1 \neq x$ is 4/5 for $x = 1, ..., N - 1$. (See the discussion of example 2.3.) However, Mary can make this conditional probability arbitrarily near 1 by choosing action 1 with probability ε and action 2 with probability $1 - \varepsilon$ where ε is small and positive. In fact, the conditional probability is 1 when Tom chooses action 1 and is $1 - \varepsilon$ when Tom chooses action 2. By continuing to choose her actions in this fashion, Mary will reach N from any $x \in \{1, .., N - 1\}$ with probability at least $(1 - \varepsilon)^{N-x}$ and this approaches 1 as ε goes to 0.

As Milnor and Shapley show, the crucial difference in the two cases is that one of the matrices has a 0 and the other does not.

Suppose now that we drop the assumption (4.1) of no motion on G but retain as payoff function the indicator of [reach G]. So if the initial state x is in G, the game is no longer trivial because the payoff is 1 only if some x_n with $n \geq 1$ belongs to G. However, it remains true that the game essentially ends once some $x_n \in G$ has been reached. So, after the first move, the game is as before. This suggests we can analyze the new game in terms of the old. To do so introduce a new law of motion q^* defined by

$$\begin{aligned} q^*(x_1 \mid x, a, b) &= q(x_1|x, a, b) \quad &\text{if } x \notin G, \\ q^*(x \mid x, a, b) &= 1 \quad &\text{if } x \in G. \end{aligned} \tag{4.15}$$

Thus q^* agrees with q off G but satisfies the no motion assumption on G. Let U^* be the value function for the game when q^* is the law of motion.

Theorem 4.2 *Without the assumption of no motion on G, the game of reaching G from the initial state x has the value*

$$U(x) = (SU^*)(x).$$

(Here S is the one-day operator for the law of motion q.)

Proof: To see that $\underline{U}(x) \geq (SU^*)(x) - \varepsilon$ for $\varepsilon > 0$, choose $\sigma_o \in P(A)$ to be optimal in the one-day game with payoff U^* and, for each $p_1 = (a_1, b_1, x_1)$, choose $\sigma[p_1]$ to be ε-optimal in the game of reaching G from x_1 when the law of motion is q^*. Then, for any strategy τ for player II, set $P = P_{x,\sigma,\tau}$ and calculate

$$\begin{aligned} P[\text{reach } G] &= E_{x,\sigma_o,\tau_o}[P[\text{reach } G \mid p_1]] \\ &\geq E_{x,\sigma_o,\tau_o}U^* - \varepsilon \\ &= (SU^*)(x) - \varepsilon. \end{aligned}$$

The proof that $\overline{U}(x) \leq (SU^*)(x) + \varepsilon$ is similar.   □

Here is an example adapted from Orkin [19].

Example 4.3 Let $X = \{g_1, g_2, b_1, b_2\}$ and $G = \{g_1, g_2\}$. Suppose $A = B = \{0, 1\}$ and that states g_1 and b_1 are absorbing. At states g_2 and b_2, the law of motion is deterministic and given by

$$\begin{aligned} q(g_1 \mid g_2, 1, 1) &= q(g_1 \mid b_2, 1, 1) = 1 \\ q(b_1 \mid g_2, 1, 0) &= q(b_1 \mid b_2, 1, 0) = 1 \\ q(g_2 \mid g_2, 0, 0) &= q(g_2 \mid b_2, 0, 0) = 1 \\ q(b_2 \mid g_2, 0, 1) &= q(b_2 \mid b_2, 0, 1) = 1. \end{aligned}$$

The law of motion q^* defined as in (4.15) will agree with q except that g_2 will become an absorbing state.

To simplify notation, we write functions on X in vector form. Thus the indicator function 1_G is written $(1,1,0,0)$. Now, because g_1, g_2, and b_1 are absorbing states under q^*, the function $U_1^* = S^* 1_G$ is of the form $(1, 1, 0, x_1)$ where x_1 is the value of the matrix game

$$\begin{bmatrix} 1 & 0 \\ 0 & 1 \end{bmatrix}.$$

So $x_1 = 1/2$. Similarly, given that $U_n^* = (1,1,0,x_n)$, $U_{n+1}^* = S^* U_n = (1,1,0,x_{n+1})$ where x_{n+1} is the value of

$$\begin{bmatrix} 1 & x_n \\ 0 & 1 \end{bmatrix}.$$

This value is $x_{n+1} = (2-x_n)^{-1}$. The x_n are increasing and $\lim x_n = 1$. Hence, $U^* = \lim U_n^* = (1,1,0,1)$.

Return now to the original law of motion q. Because g_1 and b_1 are absorbing under q, $U = SU^*$ (by Theorem 4.2) is of the form $(1, x, 0, y)$ where $x = y$ is the value of

$$\begin{bmatrix} 1 & 1 \\ 0 & 1 \end{bmatrix}.$$

This matrix game has value 1. So $U = U^* = (1,1,0,1)$.

To conclude this section, consider the game in which player I seeks to stay in a set G. Notice that

$$[\text{stay in } G]^c = [\text{reach } G^c]$$

where G^c is the complement of G. By reversing the roles of the players, we see that this game has a value and the value function is just $1 - U$ where U is the value function for $[\text{reach } G^c]$ when the players are reversed.

5. A GENERALIZATION AND AN OPERATOR OF BLACKWELL

Let Ψ be a function from X to the unit interval $[0,1]$ and consider a generalization of the game of reaching G in which the payoff to player I is $\Psi(x_n)$ if G is first reached on day n and the payoff is zero if G is never reached. So it is now desirable for player I not only to reach G, but to reach G at states where Ψ is large.

Let q^* be the modification of the given law of motion q as in (4.15) so that q^* satisfies the assumption of no motion on G. By analogy with (4.2) define

$$\begin{aligned} U_o &= \Psi 1_G, \\ U_{n+1} &= S^* U_n, \quad n = 0, 1, \ldots, \end{aligned} \tag{5.1}$$

where S^* is the one-day operator corresponding to q^*. The functions U_n satisfy (4.3) for the same reasons as before and we set

$$U^* = \lim_{n \to \infty} U_n.$$

A proof similar to that of Theorem 4.1 shows that our new game starting from x with law of motion q^* has value $U^*(x)$ and the function U^* satisfies

$$\begin{aligned} S^*U^* &= U^*, \\ U^* &= \Psi \text{ on } G. \end{aligned} \tag{5.2}$$

When the original law of motion q is used, the game has the value function

$$U = SU^*, \tag{5.3}$$

where S is the one-day operator corresponding to q. The proof of (5.3) is similar to that of Theorem 4.2.

Now define an operator T on the collection of functions Ψ from X to $[0,1]$ by letting $T\Psi$ be the value function for the game of this section. That is,

$$T\Psi = SU^*. \tag{5.4}$$

The operator T was introduced by Blackwell [9].

In order to write T more explicitly, first define $t_1 = t_1(x_1, x_2, \ldots)$ to be the least n (if any) such that $x_n \in G$ and to be ∞ if there is no such n. Then the payoff function g_Ψ for the game of this section is just

$$g_\Psi = \begin{cases} \Psi(x_{t_1}) & \text{if } t_1 < \infty, \\ 0 & \text{if } t_1 = \infty. \end{cases}$$

So the value of the game starting from x is

$$(T\Psi)(x) = \inf_\tau \sup_\sigma E_{x,\sigma,\tau} g_\Psi. \tag{5.5}$$

Here is a lemma with three properties of T.

Lemma 5.1. *Let $\Psi_1, \Psi_2, \ldots$ be a sequence of functions from X to the unit interval.*

(a) *If Ψ_1 agrees with Ψ_2 on G, then $T\Psi_1 = T\Psi_2$.*

(b) *If $\Psi_1 \geq \Psi_2$, then $T\Psi_1 \geq T\Psi_2$.*

(c) *If $\Psi_1 \geq \Psi_2 \geq \Psi_3 \geq \ldots$ and Ψ_n converges uniformly, then $T(\lim_n \Psi_n) = \lim_n T\Psi_n$.*

Proof: (a) If Ψ_1 agrees with Ψ_2 on G, then $g_{\Psi_1} = g_{\Psi_2}$. Now use (5.5).

(b) If $\Psi_1 \geq \Psi_2$, then $g_{\Psi_1} \geq g_{\Psi_2}$. Now use (5.5) again.

(c) Set $\Psi = \lim_n \Psi_n$. By (b), $\lim_n T\Psi_n \geq T\Psi$. To get the opposite inequality, let $\varepsilon > 0$ and use the uniform convergence to obtain n_o such that, for $n \geq n_o$, $\Psi_n \leq \Psi + \varepsilon$. Then, for $n \geq n_o$, $g_{\Psi_n} \leq g_{\Psi} + \varepsilon$ and

$$E_{x,\sigma,\tau}(g_{\Psi_n}) \leq E_{x,\sigma,\tau}(g_{\Psi}) + \varepsilon$$

for all x, σ, τ. By (5.5), $T\Psi_n \leq T\Psi + \varepsilon$ for $n \geq n_o$. $\qquad\square$

6.   VISITING A SET INFINITELY OFTEN

Let $G \subseteq X$ and consider now the stochastic game in which the payoff function g is the indicator function of the set $[G \text{ i.o.}]$ of (3.1). Our analysis of this game is taken from Blackwell [9] and features the operator T introduced in the previous section.

Define a sequence of functions Q_n on X by

$$\begin{aligned} Q_0 &= T1, \\ Q_{n+1} &= TQ_n \text{ for } n = 0, 1, \ldots. \end{aligned} \tag{6.1}$$

Notice that Q_o is the value function U for the game of reaching G as in Theorem 4.2 and each Q_{n+1} is the value function for a generalized game as in section 5. By Lemma 5.1 (b) and induction, all of the Q_n have values in [0,1] and satisfy

$$1 \geq Q_o \geq Q_1 \geq \ldots \geq 0. \tag{6.2}$$

Now define

$$Q_\infty = \lim_n Q_n.$$

It is tempting to guess that, by analogy with Theorem 4.1, the value function for the game is Q_∞. This is correct if the state space X is finite.

Theorem 6.1 *If X is finite, the $[G \text{ i.o.}]$ game starting from x has the value $Q_\infty(x)$ and the function Q_∞ is a fixed point of T.*

Because X is finite, Q_n converges to Q_∞ uniformly and, by Lemma 5.1 (c),

$$TQ_\infty = T(\lim Q_n) = \lim TQ_n = \lim Q_{n+1} = Q_\infty.$$

So it only remains to be shown that Q_∞ is the value function. That is, we need only show

$$\overline{Q}(x) \leq Q_\infty(x) \leq \underline{Q}(x) \tag{6.3}$$

where $\overline{Q}(x)$ and $\underline{Q}(x)$ are the upper and lower values:

$$\begin{aligned} \overline{Q}(x) &= \inf_\tau \sup_\sigma P_{x,\sigma,\tau}[G \text{ i.o.}], \\ \underline{Q}(x) &= \sup_\sigma \inf_\tau P_{x,\sigma,\tau}[G \text{ i.o.}]. \end{aligned} \tag{6.4}$$

The proof of (6.3) will be given in two lemmas.

Lemma 6.1 *If φ is a function from X to the unit interval and $T\varphi \geq \varphi$, then $Q \geq \varphi$. In particular, since $TQ_\infty = Q_\infty$, $Q \geq Q_\infty$.*

Proof: Fix $\varepsilon > 0$ and $x \in X$. It suffices to find a strategy σ^* for player I such that, for every strategy τ for player II,

$$P_{x,\sigma^*,\tau}[\text{G i.o.}] \geq \varphi(x) - \varepsilon. \tag{6.5}$$

Use (5.5) and our hypothesis that $T\varphi \geq \varphi$ to obtain, for each $y \in X$ and $n = 1, 2, ...,$ a strategy $\sigma^{(n)}(y)$ for I which is $\varepsilon/2^n$-optimal in the game with payoff g_φ starting from y. Then

$$E_{x,\sigma^{(n)}(y),\tau}\varphi(x_{t_1}) > (T\varphi)(y) - \frac{\varepsilon}{2^n} \geq \varphi(y) - \frac{\varepsilon}{2^n}. \tag{6.6}$$

Now let σ^* be the strategy for I which follows $\sigma^{(1)}(x)$ up to the time t_1 of the first visit to G and, if $t_1 < \infty$, σ^* follows $\sigma^{(2)}(x_{t_1})$ from time t_1 up to the time t_2 of the second visit to G and, in general, if t_n, the time of the nth visit to G, is finite, σ^* follows $\sigma^{(n+1)}(x_{t_n})$ from time t_n up to time t_{n+1}. To verify (6.5), define for $n = 1, 2, \ldots$

$$Y_n = \begin{cases} \varphi(x_{t_n}) & \text{if } t_n < \infty \\ 0 & \text{if } t_n = \infty, \end{cases}$$

and let $G^{(n)}$ be the event $[t_n < \infty]$ that G is visited at least n times. Fix a strategy τ for II and write E for $E_{x,\sigma^*,\tau}$. By the construction of σ^*,

$$EY_1 \geq \varphi(x) - \frac{\varepsilon}{2}. \tag{6.7}$$

For $n = 1, 2, \ldots$

$$E[Y_{n+1} \mid Y_n = y] \geq y - \frac{\varepsilon}{2^{n+1}}.$$

This is obvious if $y = 0$. If $y > 0$, then $t_n < \infty$ and the inequality is a restatement of (6.6). So

$$E[Y_{n+1} \mid Y_n] \geq Y_n - \frac{\varepsilon}{2^{n+1}}$$

and, taking expectations on both sides, we have

$$EY_{n+1} \geq EY_n - \frac{\varepsilon}{2^{n+1}}.$$

Combine this with (6.7) to get

$$EY_n \geq \varphi(x) - \varepsilon$$

for all $n = 1, 2, \ldots$ Now use the fact that $Y_n \leq 1_{G^{(n)}}$ to get

$$P_{x,\sigma^*,\tau}(G^{(n)}) \geq EY_n \geq \varphi(x) - \varepsilon.$$

Then (6.5) follows when n approaches infinity. $\qquad\square$

Lemma 6.2 *For $n = 0, 1, \ldots, \overline{Q} \leq Q_n$ and, consequently, $\overline{Q} \leq Q_\infty$.*

Proof: The set $[G \text{ i.o.}]$ is a subset of $G^{(1)} = [t_1 < \infty]$. So, for all x, σ, τ,

$$P_{x,\sigma,\tau}[G \text{ i.o.}] \leq P_{x,\sigma,\tau}[t_1 < \infty].$$

Take the "inf sup" as in (5.5) and (5.4) to get $\overline{Q}(x) \leq (T1)(x) = Q_o(x)$.

Now make the inductive assumption that $\overline{Q} \leq Q_n$. Fix x and $\varepsilon > 0$. To prove that $\overline{Q}(x) \leq Q_{n+1}(x)$, it suffices to find a strategy τ^* for II such that, for every strategy σ for I,

$$P_{x,\sigma,\tau^*}[G \text{ i.o.}] \leq Q_{n+1}(x) + \varepsilon. \tag{6.8}$$

To simplify notation, we set $Q_n(x_{t_1}) = 0$ if $t_1 = \infty$. Then $Q_n(x_{t_1}) = g_{Q_n}$. Now choose a strategy τ_1 for II which is $\varepsilon/2$-optimal in the game of section 5 with payoff g_{Q_n} and initial position x. Then, for every σ for I,

$$\begin{aligned} E_{x,\sigma,\tau_1} Q_n(x_{t_1}) &\leq (TQ_n)(x) + \frac{\varepsilon}{2} \\ &= Q_{n+1}(x) + \frac{\varepsilon}{2}. \end{aligned} \tag{6.9}$$

Next use the inductive assumption and the definition of $\overline{Q}$ to get, for every $y \in X$, a strategy $\tau_2(y)$ for II such that, for every σ for I,

$$\begin{aligned} P_{y,\sigma,\tau_2(y)}[G \text{ i.o.}] &\leq \overline{Q}(y) + \frac{\varepsilon}{2} \\ &\leq Q_n(y) + \frac{\varepsilon}{2}. \end{aligned} \tag{6.10}$$

Finally take τ^* to be that strategy which follows τ_1 up to time t_1 and follows $\tau_2(x_{t_1})$ from time t_1 onward. Fix a strategy σ for I and write E for E_{x,σ,τ^*} and P for P_{x,σ,τ^*}. We will calculate $P[G \text{ i.o.}]$ by conditioning on p_{t_1}, the history up to time t_1. If $t_1 < \infty$, then $p_{t_1} = ((a_1, b_1, x_1), \ldots, (a_{t_1}, b_{t_1}, x_{t_1}))$ and

$$\begin{aligned} P[G \text{ i.o.}|p_{t_1}] &= P x_{t_1}, \sigma[p_{t_1}], \tau_2(x_{t_1})[G \text{ i.o.}] \\ &\leq Q_n(x_{t_1}) + \frac{\varepsilon}{2} \end{aligned}$$

by (6.10). If $t_1 = \infty$, we define p_{t_1} to be the infinite history $((a_1, b_1, x_1), (a_2, b_2, x_2), \ldots)$ so that

$$P[G \text{ i.o.}|p_{t_1}] = 0 = Q_n(x_{t_1}).$$

Now take expectations to get

$$
\begin{aligned}
P[G \text{ i.o.}] &= E(P[G \text{ i.o.}|p_{t_1}]) \\
&\le EQ_n(x_{t_1}) + \frac{\varepsilon}{2} \\
&\le Q_{n+1}(x) + \varepsilon.
\end{aligned}
$$

This establishes (6.8) and the lemma. □

The proof of Theorem 6.1 is now complete. Before going on to the case when X is infinite, we will return to Orkin's example.

Example 6.1 (Same as example 4.3) As was already shown $U = (U(g_1), U(g_2), U(b_1), U(b_2)) = (1,1,0,1)$. But $U = Q_o$. So $Q_o = 1$ on G and, by Lemma 5.1 (a), $Q_1 = TQ_o = T1 = Q_o$. By induction, $Q_n = Q_o$ for all n and, hence $Q_\infty = Q_o$. In particular, $Q_\infty(g_2) = 1$. So player I can visit G infinitely often with probability arbitrarily near 1 starting from g_2. You may find this less than obvious if you attempt a direct proof. A proof due to Blackwell is in Orkin [19].

If X is infinite, Q_∞ need not be the value function for the $[G$ i.o.$]$ game. Here is a deterministic example in which player II is a dummy with only one action.

Example 6.2 Let X be the set of integers and let G be the collection of strictly positive integers. Assume $A = \{1,2\}$ and $B = \{1\}$. We take 0 to be an absorbing state and assume that, from any positive integer x, the only possible motion is to $x - 1$. So

$$
q(x - 1 \mid x, a, b) = 1 \text{ for } x \in G \text{ and all } a, b.
$$

If x is negative, the next state can be either $x - 1$ or $-x$; i.e.,

$$
\begin{aligned}
q(x - 1 \mid x, 1, 1) &= 1, \\
q(-x|x, 2, 1) &= 1 \text{ for } x < 0.
\end{aligned}
$$

Now from any negative integer, player I can use action 1 sufficiently many times to reach an $x \le -n$ and then use action 2 to go to $-x$. Thus I is guaranteed to have at least n visits to G before being absorbed at 0. This heuristic argument is reflected in the fact that $Q_n(x) = 1$ for all $x < 0$ and $n = 0, 1, \ldots$, as you can check. If $x > 0$, then there are sure to be exactly $x - 1$ visits to G before absorption at 0 and $Q_n(x) = 0$ for $n \ge x$. Thus

$$
Q_\infty(x) = \begin{cases} 1, & x < 0, \\ 0, & x \ge 0. \end{cases}
$$

It is obviously impossible to visit G infinitely often and the value of the game is 0 starting from any x. Notice that TQ_∞ is identically zero and equal to the value function.

For countably infinite X it is still possible to calculate the value of the $[G$ i.o.$]$ game by iterating the operator T. However, the iteration must be transfinite. So, for every countable ordinal number ξ, define

$$Q_\xi = T(\inf_{\eta < \xi} Q_\eta)$$

and set

$$Q = \inf_\xi Q_\xi$$

(See Halmos [30] for information about the ordinals.) A simple transfinite induction shows $Q_\xi \leq Q_\eta$ whenever $\xi > \eta$. So, because X is countable, there will be some countable ordinal ξ^* such that $Q = Q_{\xi^*}$. Hence $TQ = TQ_{\xi^*} = Q_{\xi^*+1} = Q$. (In example 6.2, ξ^* can be taken to be ω, the first infinite ordinal.) Since Q is a fixed point of T, it follows from Lemma 6.1 that $Q \leq \underline{Q}$ where $\underline{Q}$ is the lower value of the $[G$ i.o.$]$ game. A transfinite induction similar to the inductive proof of Lemma 6.2 shows $\overline{Q} \leq Q$ and establishes the result below.

Theorem 6.2 *If X is countable, the $[G$ i.o.$]$ game starting from x has the value $Q(x)$ and the function Q is a fixed point of T.*

We conclude this section with a final visit to one of our examples.

Example 6.3 *Plus or minus one.* Let $X, A, B,$ and q be the same as in example 3.2. Suppose first that $G = \{0, 1, 2, \ldots\}$. If $p^* \geq 1/2$, then, as in (4.13), player I can reach G with probability 1 and $Q_o(x) = 1$ for all x. So $TQ_o = T1 = Q_o$ and, by induction, $Q_\xi = Q_o$ for all x. Therefore the value function Q is identically equal to one. If $p^* < 1/2$, we will argue that Q is identically equal to zero. It is clear that Q is nondecreasing and that, for an initial state $x \geq 1$, the set G is always reached on the first day. It follows that, for $x \geq 1$,

$$Q(x) = (TQ)(x) = (SQ)(x) = p^*Q(x+1) + (1-p^*)Q(x-1).$$

The possible solutions are

$$Q(x) = \left(\frac{1-p^*}{p^*}\right)^x Q(0), \quad x \geq 1 \tag{6.11}$$

and

$$Q(x) = Q(0), \quad x \geq 1. \tag{6.12}$$

Suppose (6.11) holds. If $Q(0) > 0$, then $Q(x) > 1$ for large x. So we must have $Q(0) = 0$ and so $Q(x) = 0$ for all x. Next assume (6.12). Calculate

$Q_o(0) = p^* \cdot 1 + (1 - p^*)(p^*/1 - p^*) = 2p^* < 1$. So $Q(0) \le Q_o(0) < 1$. Because Q is constant on G, $Q(0) = (TQ)(0) = Q(0)(T1)(0) = Q(0)^2$. Hence, $Q(0) = 0$ and Q must be identically zero.

Next consider the case when $G = \{0\}$. In this case, you can show that Q is identically equal to one if $p^* \ge 1/2$ and $q^* \ge 1/2$, and is identically zero otherwise.

It would be of interest to know analogous results for higher dimensions, and the techniques of Blackwell in [7] could be helpful.

7.   REMARKS AND QUESTIONS

a) Suppose that player II is a dummy with only one action and, consequently, only one strategy. The problem faced by player I of choosing a strategy σ so as to maximize the expected payoff $E_{x,\sigma,\tau}g$ becomes a *stochastic control* problem. The theory of stochastic control [23] is a large and interesting subject on its own and is known also as dynamic programming ([24], [25]), Markov decision theory [28], and gambling theory [26]. The problem of reaching a set G can be viewed as a *positive* dynamic programming problem [25] in which the payoff is one when the set G is first reached or as a *leavable* gambling problem [26] in which the player can stop play when the goal set is reached. The problem of visiting G infinitely often corresponds to a *nonleavable* gambling problem where the player must continue to reach G over and over.

The equation $SU = U$ of Theorem 4.1 is just the optimality equation of dynamic programming or gambling when player II is a dummy. The equation $TQ = Q$ of Theorems 6.1 and 6.2 is a fairly new optimality equation in nonleavable gambling theory [27].

b) Return now to the setting of stochastic games. Give the countable state space X the discrete topology and endow the infinite product $X^N = X \times X \times \ldots$ with the product topology. Then the set

$$[\text{reach } G] = \bigcup_{n=1}^{\infty} \{(x_1, x_2, \ldots) : x_n \in G\}$$

is the union of open sets and, consequently, is itself open. On the other hand, it can be shown that every open subset O of X^N can be written in the form

$$O = \bigcup_{n=1}^{\infty} \{(x_1, x_2, \ldots) : (x_1, \ldots, x_n) \in G_n\}$$

where $G_n \subseteq X^n$ for every n. One can use this representation to adapt the techniques of section 4 to show that the stochastic game with payoff the indicator function of O has a value.

Next write

$$[G \text{ i.o.}] = \bigcap_{n=1}^{\infty} \bigcup_{k=n}^{\infty} \{(x_1, x_2, \ldots) : x_k \in G\}$$

as the countable intersection of open sets, i.e., as a G_δ set. It can also be shown that every G_δ subset E of X^N can be written in the form

$$E = \bigcap_{n=1}^{\infty} \bigcup_{k=n}^{\infty} \{(x_1, x_2, \ldots) : (x_1, \ldots, x_k) \in E_k\}$$

where $E^k \subseteq X_k$ for every k, and one can adapt the techniques of section 6 to show that the stochastic game with payoff 1_E has a value.

The next sets in the Baire hierarchy are the $G_{\delta\sigma}$s that are countable intersections of G_δs. We do not know whether stochastic games whose payoff functions are indicators of $G_{\delta\sigma}$s always have a value. If X is the set of integers, then the set

$$\{(x_1, x_2, \ldots) : \frac{x_1 + \ldots + x_n}{n} \nrightarrow 0\}$$

is a particular $G_{\delta\sigma}$ for which the question is open. (Blackwell [7] showed this game has a value in an interesting special case.)

c) Let u be a bounded, real-valued function defined on X and set

$$u^*(x_1, x_2, \ldots) = \limsup_n u(x_n).$$

Operator methods similar to those used above show that the stochastic game with this payoff has a value (Maitra and Sudderth [14]). Notice that, if $u = 1_G$, then $u^* = [G \text{ i.o.}]$. For a general u and any real number a,

$$[u^* \geq a] = \bigcap_{k=1}^{\infty} \{(x_1, x_2, \ldots) : u(x_n) \geq a - \frac{1}{k} \text{ i.o.}\}$$

is a G_δ. As noted above, we do not know whether a stochastic game with payoff g has a value when $[g \geq a]$ is a $G_{\delta\sigma}$ for all a.

d) Under appropriate topological and measurability assumptions, all of our results on stochastic games extend to uncountable state spaces and action sets (cf. Maitra and Sudderth [15]). The major new difficulties are measure theoretic.

References

General game theory

[1] **Blackwell, David and Girshick, M. A.**, *Theory of Games and Statistical Decisions*, Dover, New York, 1979.

[2] **Gale, David**, *Theory of Linear Economic Models*, McGraw-Hill, New York, 1960.

[3] **Luce, R. Duncan and Raiffa, Howard,** *Games and Decisions*, Dover, New York, 1989.

[4] **McKinsey, J. C. C.,** *Introduction to the Theory of Games*, McGraw-Hill, New York, 1952.

[5] **Owen, Guillermo,** *Game Theory*, Academic Press, New York, 1982.

[6] **von Neumann, J. and Morgenstern, O.,** *Theory of Games and Economic Behavior*, Princeton University Press, Princeton, New Jersey, 1947.

Stochastic games

[7] **Blackwell, David,** An analog of the minimax theorem for vector payoffs, *Pacific Journal of Mathematics*, **6**, 1–8, 1956.

[8] —— Infinite G_δ games with imperfect information, *Zastosowania Matematyki, Hugo Steinhaus Jubilee* Volume X, 99–101, 1969.

[9] —— Operator solution of G_δ games of imperfect information, *Probability, Statistics, and Mathematics*, Papers in Honor of S. Karlin, ed. by T. W. Anderson, K. B. Athreya, and D. L. Iglehart, Academic Press, New York, 83–87, 1989.

[10] **Blackwell, D. and Ferguson, T. S.,** The big match, *Annals of Math. Statist.* **39**, 159–163, 1968.

[11] **Everett, H.,** Recursive games, *Ann. Math. Studies*, No. 39, Princeton University Press, Princeton, New Jersey, 47–58, 1957.

[12] **Gillette, D.,** Stochastic games with zero stop probabilities, *Ann. Math. Studies* No. 39, Princeton University Press, Princeton, New Jersey, 179–187, 1957.

[13] **Hoffman, A. D. and Karp, R. M.,** On non-terminating stochastic games, *Management Science*, **12**, 359–370, 1966.

[14] **Maitra, A. and Sudderth, W.,** An operator solution of stochastic games, *Israel J. Math.*, **78**, 33–49, 1992.

[15] —— Borel stochastic games with limsup payoff, *Annals of Probability*, **21**, 861–885, 1993.

[16] **Mertens, J.-F. and Neyman, A.,** Stochastic games, *International J. Game Theory*, **10**, 53–66, 1981.

[17] **Milnor, J. and Shapley, L. S.,** On games of survival, *Ann. Math. Studies*, No. 39, Princeton University Press, Princeton, New Jersey, 15–45, 1957.

[18] **Nowak, A. S.,** Universally measurable strategies in zero-sum stochastic games, *Annals of Probability*, **13**, 269–287, 1985.

[19] **Orkin, M.,** Infinite games with imperfect information, *Transactions Amer. Math. Soc.*, **171**, 501–507, 1972.

[20] **Parthasarathy, T. and Stern, M.,** Markov games — a survey, *Lecture Notes in Pure and Applied Math.*, No. 30, Marcel Dekker, New York, 1–46, 1977.

[21] **Shapley, L.,** Stochastic games, *Proc. Nat. Acad. Sci. U.S.A.*, **39**, 1095–1100, 1953.

[22] **Zachrisson, L. E.,** Markov games, *Ann. Math. Studies*, No.52, Princeton University Press, Princeton, New Jersey, 211–253, 1964.

Stochastic control

[23] **Bertsekas, D.P. and Shreve, S.,** *Stochastic Optimal Control: The Discrete Time Case*, Academic Press, New York, 1978.

[24] **Blackwell, David,** Discounted dynamic programming, *Annals Math. Statist.*, **36**, 226–235, 1965.

[25] —— Positive dynamic programming, *Proc. of 5th Berkeley Symp. on Math. Statist. and Prob.*, University of California Press, Berkeley, 415–418, 1966.

[26] **Dubins, Lester E. and Savage, Leonard J.,** *Inequalities for Stochastic Processes: How to Gamble If You Must*, Dover, New York, 1976.

[27] **Dubins, L., Maitra, A., Purves, R. and Sudderth, W.,** Measurable, nonleavable gambling problems, *Israel J. Math.*, **67**, 257–271, 1989.

[28] **Mine, H. and Ósaki, S.,** *Markovian Decision Processes*, Elsevier, New York, 1970.

Additional references

[29] **Feller, William,** *An Introduction to Probability Theory and Its Applications*, Vol. I, 3rd edition, John Wiley & Sons, New York, 1968.

[30] **Halmos, Paul,** *Naive Set Theory*, Van Nostrand, Princeton, 1960.

Chapter 11

"DECISIONS, DECISIONS": THE BANDIT MODEL FOR DECISION PROCESSES, OPTIMAL STRATEGY, AND COMPUTER IMPLEMENTATION

Christopher P. Thron*

Department of Mathematics and Physics

King College

I returned, and saw under the sun, that the race is not to the swift, nor the battle to the strong, neither yet bread to the wise, nor yet riches to men of understanding, nor yet favour to men of skill; but time and chance happeneth to them all. (Ecclesiastes 9:11)

ABSTRACT

We introduce the "bandit model" as a simple example of a decision process. The model is introduced heuristically, then presented in a precise mathematical fashion, in terms of random variables. We indicate how the model may be programmed on a computer. Next, we formulize the concept of a decision strategy and its value: an optimum strategy has the highest possible value. We show how strategies can be derived from indices, which are real-values functions of the state of the bandit; and show that an optimal strategy is determined by a particular index, the so-called "Gittins index". Finally, we show how the Gittins index function may be computed to arbitrary accuracy.

*With grateful acknowledgement of support from J. L. Snell and M. K. Hudson.

1. BANDIT PROCESSES – THEIR PRACTICAL ROOTS

One of life's greatest privileges — and aggravations — is the opportunity to make decisions. "Which is the best course of action to take, given what I already know about the situation?" This question, in various forms, is weighed daily by billions the world over in all stations of life: from the Chinese peasant considering what to plant in the spring; to the Christmas shopper searching for just the right gift; to the president of General Motors determining next year's model.

Think of how much indecision, how much mental agony, and how much waste of time and resources could be avoided if a systematic strategy for making the very best possible choice under a given set of circumstances could be found.

This is a very complex and general problem, which this short article could not begin to address in its full generality. Instead, it is more fruitful to analyze in depth a simple type of decision-making situation, and hope that the results thus obtained might extend also to more general and realistic situations. This paper will thus have a somewhat narrower focus, and will treat specifically processes that require repeated decision-making among a fixed set of options. Consider for instance the folowing examples:

Example 1.1   A high school guidance counselor is confronted by a student who wants to decide on a career. In the past, the counselor has faced students with similar backgrounds, who have since then gone on to different careers with varying degrees of success. Based on the previous experiences of these former graduates, what recommendation should the guidance counselor make?

Example 1.2   A large company is opening up operations in a new city. In the past, in similar cities, various marketing techniques have been employed (such as newspaper ads, radio spots, TV, door-to-door, telephone surveys, etc.) with varying degrees of success. Based on the company's track record, which sales techniques should be used?

Example 1.3   At a more mundane level, consider buying a car. In the past, you (and other friends you consult with) have bought different makes (e.g., Ford, Toyota, GM, Volkswagen, ...) with varying degrees of success. Based on your combined experience, which car should you buy?

In each of the above three examples, the same choice situation is encountered repeatedly, each time featuring the same limited set of options. Because of the "time and chance" factors involved, any given option may lead sometimes to success and sometimes to failure; but the options are static in the sense that the probabilities of success and the rewards gained from a given option remain fixed. Obviously, then, the best option to choose is the one with the highest rate of success. Unfortunately, most often the actual success probabilities are unknown. Usually, the only information available is the decision-maker's own record of options chosen, together with the resulting successes and failures. The decision-maker must weigh the relative advantages of "sticking with a proven winner" or "shopping around" to see if an even better option with higher return can be found.

It is commonly the case that immediate reward is more valuable than a delayed reward, particularly if the reward is financial. Money acquired immediately can be put in the bank, where it gains interest. The same money, if obtained later on, will yield less total interest. To account for this, the rewards of success should be multiplied by a compounded interest factor or, what amounts to the same thing, future rewards should be multiplied by a compounded discounting factor because they are worth less than immediate rewards.

In the statistics literature, the general situation sketched above has been compared to a compulsive gambler using a slot machine. Each choosable option can be visualized as one arm of a cosmic "multi-armed bandit" gambling machine. The gambler repeatedly chooses an option, pulls the corresponding arm, and lets the grinding gears of chance decide the outcome.

In the next section, a mathematical model of the type of decision-making situation described above will be formulated. This simple model, which will be phrased in the "bandit" terminology introduced in the preceding paragraph, is such that a strategy can be found that yields the maximum possible expected reward. It is hoped that this simplified solution can serve as a take-off point in the investigation of more complicated situations.

Readers who have some background in stochastic processes will recognize many concepts that are used implicitly in the following discussion (e.g., stopping time, increasing σ-family, measurability). This text is written with the intent that readers who lack any such background should nonetheless be able to follow. Only elementary analysis and a few basic concepts of probability theory are assumed: the notions of random variable [which will be distinguished by boldface type, e.g. $\boldsymbol{\Theta}$, $\boldsymbol{w}_i(t)$, etc.], independence, expected value, and basic formulas for conditional probability and expectation.

2. PRESENTATION OF THE n-ARMED BANDIT MODEL

What follows is a semi-formal description of the n-armed bandit model. The intention is to make the concepts precise enough to use in a rigorous way, without getting entangled in technical details.

A convenient graphical representation of the proposed bandit model is shown in Figure 2.1 for the case $n = 2$. The reader may want to refer to the figure while reading the description of the model, to motivate the whys and wherefores of the various definitions.

Consider a bandit machine with n arms. Arm i has success probability $\boldsymbol{\Theta}_i$, where $\{\boldsymbol{\Theta}_i\}_{i=1,...,n}$ are independent, identically distributed (i.i.d.) uniform

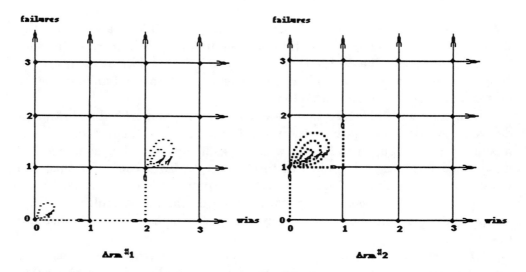

FIGURE 2.1. A picture of the 2-armed bandit model. Each grid represents one bandit arm. Nodes represent states of the bandit arms; edges represent transitions between states. The paths traced by dotted arrows show a particular possible history of play for $t = 6$.

random variables (r.v.s) on $[0,1]$. The values of $\{\Theta_i\}$ are fixed prior to the beginning of play, and are unknown to the player.

Define r.v.s $\{\alpha_{iN}\}_{i=1\ldots n; N=1\ldots\infty}$ as follows: given that r.v.s $\Theta_1\ldots\Theta_n$ take the specific values $\theta_1,\ldots,\theta_n$, the conditional r.v.s $\{\alpha_{iN}|\Theta_1 = \theta_1,\ldots\Theta_n = \theta_n\}$ are independent 0–1 random variables with

$$\Pr[\alpha_{iN} = 1|\Theta_1 = \theta_1,\ldots\Theta_n = \theta_n] = \theta_i,$$
$$\Pr[\alpha_{iN} = 0|\Theta_1 = \theta_1,\ldots\Theta_n = \theta_n] = 1 - \theta_i.$$

α_{iN} expresses the outcome of the Nth play of arm i: if $\alpha_{iN} = 1$ (respectively 0), the Nth play of arm i is a win (respectively a failure).

The variables $\{\alpha_{iN}\}$ are fundamental in the sense that they can be used to generate the other random variables used in the model. The notation for these additional random variables will now be introduced, together with their significance. Later it will be shown how to obtain them from the $\{\alpha_{iN}\}$s.

The r.v.s $w_i(t)$, $f_i(t)$ denote respectively the number of wins and failures for arm i prior to (but not including) the tth play (at time t). $N_i(t) \doteq w_i(t) + f_i(t)$ is the total number of plays of arm i prior to time t: thus $\sum_{i=1}^{n} N_i(t) = t$.

The *state* of the n-armed bandit at time t is denoted by $\sigma(t)$, where

$$\sigma(t) \doteq (w_1(t), f_1(t); w_2(t), f_2(t); \ldots; w_n(t), f_n(t)). \tag{2.1}$$

In Figure 2.1, states of the respective arms are represented as nodes on the corresponding grids.

The *history* of the bandit at time t is denoted by $\boldsymbol{\mu}(t)$, where

$$\boldsymbol{\mu}(t) \doteq [\boldsymbol{\sigma}(0), \boldsymbol{\sigma}(1), \ldots, \boldsymbol{\sigma}(t)]. \tag{2.2}$$

Let Ω_t be the set of all possible histories at time t: $\Omega_t \doteq \{\mu(t)\}$ where $\{\mu(t)\}$ (not boldface) refers to the possible values of the r.v. $\boldsymbol{\mu}(t)$ (boldface).

In Figure 2.1, a particular history for $t = 6$ is delineated by the two paths marked by dotted arrows. Circular loops indicate that the bandit arm remained in the same state for two consecutive tries. Thus, in the case shown, arm 1 stayed in state $(0,0)$ from $t = 0$ to $t = 2$; arm 2 remained in state $(1,1)$ from $t = 2$ to $t = 3$; and so on.

The random variables just introduced are intended to describe the chance factors intrinsic to the decision-making situation. The bandit model also contains a deterministic element that reflects the player's response to the circumstances that arise. This response is expressed in the form of a strategy function, defined as follows. Let $\Omega \doteq \bigcup_{t \geq 0} \Omega_t$ be the set of all possible histories for all times $t \geq 0$. A *strategy* is a function $\psi : \Omega \to \{1, \ldots N\}$ which assigns to each $\mu(t) \in \Omega_t$ a choice of arm $i = \psi(\mu(t))$ to be played at time t in the event $\boldsymbol{\mu}(t) = \mu(t)$. Figure 2.1 shows a particular strategy in action. At time $t = 0$, both arms were in state $(0,0)$, and arm #2 was played: thus the strategy ψ employed satisfies $\psi([(0,0); (0,0)]) = 2$. At time $t = 1$, the history was $\mu(1) = [((0,0); (0,0)), ((0,0); (1,0))]$, and arm #2 was played again; thus $\psi([((0,0); (0,0)); ((0,0); (1,0))]) = 2$.

Specifying a strategy may be a laborious process, because there are so many histories: each set Ω_t has 2^t elements, and strategies are defined on the union of all Ω_ts. One way to get around this difficulty is to specify a strategy as a general rule, dependent simply on certain general characteristics of the past history. For example, a very simple strategy would be the "alternating rule", under which the arms $1, 2, \ldots, n, 1, 2, \ldots$ are played cyclically one after the other regardless of the outcome of each play. A more complicated strategy is the "best record rule", under which the arm played is the one with the highest success ratio [wins÷(wins + failures)]. (Here, an arm with no wins or failures would be assigned a success ratio of $1/2$.) Yet another example would be the "play the winner" rule. Under this strategy, play begins with arm 1, which is played until the first failure occurs. Then arm 2 is chosen, and played until failure occurs, at which point play passes on to arm 3. When arm n fails during the course of play, play passes on to arm 1. The reader should convince himself that these rules do in fact specify strategies according to the above definition.

Note that the random variables $\boldsymbol{\sigma}(t)$ and $\boldsymbol{\mu}(t)$ will depend, not only on the $\{\alpha_{iN}\}$ defined above, but also on the particular strategy ψ employed, because it is the value of $\psi(\boldsymbol{\mu}(t))$ that determines which arm i is played at each time t.

It will now be demonstrated how the r.v.s $\boldsymbol{w}_i(t), \boldsymbol{f}_i(t), \boldsymbol{\sigma}(t)$ and $\boldsymbol{\mu}(t)$ may be generated inductively from the r.v.s $\{\alpha_{iN}\}$, under a given strategy ψ.

First, when $t = 0$, $\boldsymbol{w}_0(t) = \boldsymbol{f}_0(t) = 0$ for all $i, 1 \leq i \leq n$; and $\boldsymbol{\mu}(0) = \boldsymbol{\sigma}(0) = [(0,0);(0,0);\ldots(0,0)]$.

Suppose now that $\boldsymbol{w}_i(t), \boldsymbol{f}_i(t)$ $(1 \leq i \leq n)$, $\boldsymbol{\sigma}(s)$, and $\boldsymbol{\mu}(s)$ are all known for $0 \leq s \leq t$, and the same variables for time $t+1$ are to be expressed. In the remainder of this section, for convenience's sake we will violate our boldface convention and write

$$
\begin{aligned}
k &\doteq \psi(\boldsymbol{\mu}(t)), \\
N_k &\doteq \boldsymbol{w}_k(t) + \boldsymbol{f}_k(t).
\end{aligned}
\tag{2.3}
$$

Thus arm k is played at time t, and has been played a previously total of N_k times. It follows that the outcome of the next play of arm k is determined by α_{k,N_k+1}: if $\alpha_{k,N_k+1} = 1$, then $\boldsymbol{w}_k(t+1) = \boldsymbol{w}_k(t) + 1$ and all other $\boldsymbol{w}_i$'s and $\boldsymbol{f}_i$'s are unchanged from t to $t+1$, $1 \leq i \leq n$. These rules can be expressed in a mathematically concise fashion as:

$$
\begin{aligned}
\boldsymbol{w}_i(t+1) &= \begin{cases} \boldsymbol{w}_k(t) + \alpha_{k,N_k+1} & \text{if } i = k \\ \boldsymbol{w}_i(t) & \text{if } i \neq k \end{cases} \\
\boldsymbol{f}_i(t+1) &= \begin{cases} \boldsymbol{f}_k(t) + (1 - \alpha_{k,N_k+1}) & \text{if } i = k \\ \boldsymbol{f}_i(t) & \text{if } i \neq k, \end{cases}
\end{aligned}
\tag{2.4}
$$

(k and N_k are as in Equation 2.3.) The r.v.s $\boldsymbol{\sigma}(t+1)$ and $\boldsymbol{\mu}(t+1)$ may be obtained via Equations 2.1 and 2.2. The induction step is thus complete; and these r.v.s may be found for all t, using only the given ψ and the r.v.s $\{\alpha_{iN}\}$.

The progress of play has been described; it remains to express the player's reward. At time t, the player's success or failure depends on the value of α_{k,N_k}. It is assumed that a win brings reward 1, a failure brings reward 0. Thus the *reward* $\boldsymbol{R}(t)$ is defined by

$$
\boldsymbol{R}(t) \doteq \alpha_{k,N_k}.
$$

An equivalent definition is,

$$
\boldsymbol{R}(t) \doteq \sum_{i=1}^{n} [\boldsymbol{w}_i(t+1) - \boldsymbol{w}_i(t)].
\tag{2.5}
$$

(The equivalence is left as an exercise for the reader.) The player's *total reward* $\boldsymbol{R}$ is given by

$$
\boldsymbol{R} \doteq \sum_{t=0}^{\infty} \beta^t \boldsymbol{R}(t),
\tag{2.6}
$$

where a discounting factor $\beta < 1$ has been introduced in accordance with the preliminary discussion in Section 1; later rewards should contribute less to the total value, according to the usual laws of compounding interest.

Note that $\boldsymbol{R}(t)$ and $\boldsymbol{R}$ are r.v.s that also depend on the particular strategy ψ that is employed.

The model is now well-defined, enough so that computer simulations can be carried out. First, a particular strategy ψ is specified. Then, the values

$\{\Theta_i\}$ are initialized by choosing them randomly (and independently) according to a uniform distribution. Next, values of α_{iN} can also be generated as independent 0–1 r.v.s where $\Pr[\alpha_{iN} = 1] = \Theta_i$. Next, the progress of play is given by $w_i(t)$, $f_i(t)$, $\sigma(t)$, and $\mu(t)$, which are all found inductively via Equations 2.1–2.4; and finally, rewards and total reward can be computed according to Equations 2.5 and 2.6.

An enormous number of papers have been written on the bandit model; see [2] for a more advanced introduction and bibliography of the existing literature.

3. THE VALUE OF A STRATEGY

The whole point of developing the bandit model is to help the decision maker find the "best" strategy for decision making. The notion of "best" must be clarified, because a given strategy does not always bring the same reward, but only a probability distribution of rewards. We define a "best" or optimal strategy as one that maximizes the *expected* total reward, which will be called the *value* of the strategy.

For a given strategy ψ, its value $E[R]$ can be computed as follows.

$$E[R] = \sum_{t=0}^{\infty} \beta^t E[R(t)] \tag{3.1}$$

$$= \sum_{t=0}^{\infty} \beta^t \sum_{\mu(t) \in \Omega_t} E[R(t)|\mu(t) = \mu(t)] \Pr[\mu(t) = \mu(t)]. \tag{3.2}$$

Note that the sum in 3.1 will converge, since both β and $E[R(t)]$ are less than 1.

The first factor in 3.2 will be computed first. Given a particular history $\mu(t)$ (which is our choice, not a random variable, and thus is not boldfaced), let $k = \psi(\mu(t))$. Then the probability distribution for $R(t)$ given $\mu(t) = \mu(t)$ depends on Θ_k: with probability Θ_k, a reward of 1 is gained, and with probability $1 - \Theta_k$, no reward is gained. Thus

$$E[R(t)|\mu(t) = \mu(t)] = E[\Theta_k|\mu(t) = \mu(t)]. \tag{3.3}$$

The exact value of Θ_k is unknown to the player. However, the player does possess information relevant to Θ_k, namely, by time t he has played arm k $N_k(t)$ times, with $w_k(t)$ wins and $f_k(t)$ failures. The additional information contained in $\mu(t)$ pertains to arms other than k, and does not affect the player's estimate of Θ_k (by assumption, Θ_i and Θ_k are independent, $i \neq k$). From Equation 3.3 it follows

$$E[R(t)|\mu(t) = \mu(t)] = E[\Theta_k|\text{arm } k \text{ has } w_k(t) \text{ wins in } N_k(t) \text{ plays}]. \tag{3.4}$$

Recall that Θ_k is chosen initially according to a uniform distribution on $[0, 1]$. The conditional probability density for Θ_k may be obtained according to the laws of conditional probability: writing $w_k(t)$ and $N_k(t)$ as w and N respectively, it follows

$$\Pr[\Theta_k \in (\theta, \theta + d\theta))|w \text{ wins in } N \text{ trials}]$$
$$= \frac{\Pr[w \text{ wins in } N \text{ trials}, \Theta_k \in (\theta, \theta + d\theta)]}{\Pr[w \text{ wins in } N \text{ trials}]}$$
$$= \frac{\binom{N}{w}\theta^w(1-\theta)^{N-w} \cdot d\theta}{\int_0^1 \binom{N}{w}\theta^w(1-\theta)^{N-w}d\theta}$$
$$= \frac{\theta^w(1-\theta)^{N-w} \cdot d\theta}{\int_0^1 \theta^w(1-\theta)^{N-w}d\theta}.$$

Using this conditional density for Θ_i, the expected value on the right-hand side of Equation 3.4 may be computed:

$$
\begin{aligned}
E[\Theta_k|w \text{ wins in } N \text{ trials}] &= \int_0^1 \theta \Pr[\Theta_k \in (\theta, \theta + d\theta)]|w \text{ wins in } N \text{ trials}) \\
&= \int_0^1 \theta^{w+1}(1-\theta)^{N-w}d\theta \bigg/ \int_0^1 \theta^w(1-\theta)^{N-w}d\theta \\
&= \frac{(w+1)!(N-w)!}{(N+2)!} \bigg/ \frac{(w)!(N-w)!}{(N+1)!} \\
&= \frac{w+1}{N+2}.
\end{aligned}
$$

$$(3.5)$$

The integrals in Equations 3.5 were performed by integrating by parts repeatedly. In view of Equations 3.4 and 3.5, it follows that the first factor of the summand in Equation 3.2 is

$$E[\mathbf{R}(t)|\boldsymbol{\mu}(t) = \mu(t)] = \frac{w_k + 1}{N_k + 2} \qquad (\text{where } k = \psi(\mu(t))). \qquad (3.6)$$

The second factor in Equation 3.2 may be computed as follows: conditioning on values of $\{\Theta_i\}_{i=1...n}$,

$$
\begin{aligned}
\Pr[\boldsymbol{\mu}(t) = \mu(t)] &= \int_0^1 \cdots \int_0^1 \Pr[\boldsymbol{\mu}(t) = \mu(t)|\Theta_1 = \theta_1, \ldots, \Theta_n = \theta_n] \\
&\quad \cdot \Pr[\Theta_1 \in (\theta_1, \theta_1 + d\theta_1), \ldots \Theta_n \in (\theta_n, \theta_n + d\theta_n)].
\end{aligned}
$$

Since the $\{\Theta_i\}$ are uniformly distributed, it follows $\Pr[\Theta_i \in (\theta_i, \theta_i + d\theta_i)] = d\theta_i$ and

$$\Pr[\boldsymbol{\mu}(t) = \mu(t)] =$$
$$\int_0^1 d\theta_1 \ldots \int_0^1 d\theta_n \Pr[\boldsymbol{\mu}(t) = \mu(t)|\Theta_1 = \theta_1, \ldots, \Theta_n = \theta_n]. \qquad (3.7)$$

It is thus sufficient to compute $\Pr[\boldsymbol{\mu}(t) = \mu(t)]$ for each $\mu(t)$ given that the success probabilities $\{\Theta_i\}$ are known values $\{\theta_i\}$, integrate over $\{\theta_i\}$, multiply by $\frac{w_k+1}{N_k+2}$ and then sum over the possible histories $\{\mu(t)\}$ at time t to obtain the desired result.

The probability $\Pr[\boldsymbol{\mu}(t) = \mu(t)]$ for a given fixed $\mu(t)$ and given values $\{\theta_i\}$ may be computed as follows. Let $k_s = \psi(\mu(s)), s = 0, 1, \ldots, t$. If $w_{k_s}(s + 1) - w_{k_s}(s) = 1$, then let $p_s = \theta_{k_s}$; otherwise, if $f_{k_s}(s+1) - f_{k_s}(s) = 1$, then let $p_s = 1 - \theta_{k_s}$. The probability $\Pr[\boldsymbol{\mu}(t) = \mu(t)|\Theta_1 = \theta_1, \ldots, \Theta_n = \theta_n]$ will be the product $p_0 \ldots p_t$, and the integration in Equation 3.7 yields the second factor of the summand in Equation 3.2.

It has been shown that, in principle, the expected total reward $E[\boldsymbol{R}(\psi)]$ may be computed (at least approximately), though evidently the computation is quite complicated due to the large number of histories (recall Ω_t contains 2^t histories). The following sections are devoted to finding the best strategy, i.e., one that maximizes the expected total reward $E[\boldsymbol{R}]$.

4. INDEX-DETERMINED STRATEGIES

One class of bandit strategies has the following form. At each time t, each of the n bandit arms is assigned a "rating" or "index" ν based on the win/loss record of that arm: that is, $\boldsymbol{\nu}_i(t) = \nu(\boldsymbol{w}_i(t), \boldsymbol{f}_i(t)), i = 1, \ldots, n$. This index is intended to be a measure of the desirability of playing an arm; obviously the more wins and the fewer failures arm i has experienced, the larger should be the value of $\boldsymbol{\nu}_i(t)$. The strategy then consists of playing at each time t the arm i which has the largest value of $\boldsymbol{\nu}_i(t)$:

$$\psi(\boldsymbol{\mu}(t)) = \{k|\boldsymbol{\nu}_k(t) = \max_{i=1\ldots N} \boldsymbol{\nu}_i(t)\}.$$

(If more than one arm achieves the maximum value, any of the maximizing arms may be chosen, for instance the one with smallest i.)

An index-determined strategy has many advantages. In principle, the strategy for a 1000-armed bandit is no more complex than that for a 2-armed bandit — only the index must be calculated for more arms. The value of $\boldsymbol{\nu}_i(t)$ depends only on $\boldsymbol{\sigma}_i(t) = (\boldsymbol{w}_i(t), \boldsymbol{f}_i(t))$ and not explicitly on i, t, or $\boldsymbol{\sigma}_i(s)$ for $s < t$. Thus values of $\nu(w, f)$ can be tabulated, stored, and reused every time another "bandit" situation is encountered.

The million-dollar question is: Can an index be found that will determine an optimal strategy? Fortunately, the answer is yes, as will be shown presently.

5. DEFINITION OF THE GITTINS INDEX

A particular index function ν called the "Gittins index" (after its originator J. C. Gittins [1]) will now be introduced. The definition of the Gittins index will be made operationally, in terms of a slightly modified one-armed bandit model. This modified model differs in that of necessity there is only a single arm; the initial state is not necessarily $(0,0)$; and the gambler must "pay" each time he plays.

Eventually, it will be shown that $\nu(w_o, f_o)$ provides a good measure for rating the desirability of playing an arm in the n-armed bandit in state (w_o, f_o): in fact, the strategy determined (in the sense of Section 4) by the Gittins index is optimal (in the sense of maximizing expected total reward), as will be shown in Section 7. This property of the Gittins index was first proved by Gittins in [1] (see also [3]).

5.1. The modified one-armed bandit model, and graphical representation

The modified bandit model will also be expressed in terms of states, histories, and strategies, which will be denoted by $\boldsymbol{\sigma}(t) = (\boldsymbol{w}(t), \boldsymbol{f}(t))$, $\boldsymbol{\mu}(t) = [\boldsymbol{\sigma}(0), \boldsymbol{\sigma}(1), \dots, \boldsymbol{\sigma}(t)]$, and ψ respectively, just as in the regular bandit model. In the ensuing sections, the type of each bandit discussed (modified or unmodified) will always be clearly specified, in order to avoid confusion.

Modifications to be introduced are as follows. The initial state $\boldsymbol{\sigma}(0)$ is no longer $(0, 0)$, but rather (w_o, f_o). The net gain for a win is no longer 1, but $1 - P$, and the net gain for failure is $-P$ (P is to be thought of as a "fee" that the player must pay every time he plays the arm). The player's strategy no longer involves deciding which arm to play, since there is only a single arm; rather, the player is given the choice at each time t of continuing play or ceasing play.

Figure 5.1 provides a graphical representation of the modified one-armed bandit, in analogy to Figure 2.1. Grid nodes again represent states of the bandit; and edges joining the nodes represent possible transitions between states (again, from left to right or vertically upwards). The path indicated by dotted arrows gives a particular history $\mu(t)$ for $t = 3$. In contrast to Figure 2.1, the state $\boldsymbol{\sigma}(t)$ must change at each step: thus no circular loops are possible. Each state can only be realized at a well-defined time t indicated by the dashed lines on the figure.

At each time t, the gambler decides whether to play the arm or not, basing his decision only on the history $\boldsymbol{\mu}(t)$. Thus the gambler's strategy can be characterized by a function ψ defined on the set of all possible histories $\mu(t)$: $\psi(\boldsymbol{\mu}(t)) = 1$ or ∞ respectively depending on whether the gambler chooses to continue or quit, given $\boldsymbol{\mu}(t) = \mu(t)$. The gambler is required to play at $t = 0$, so $\psi(\boldsymbol{\sigma}(0)) = \psi(w_o, f_o) = 1$.

Let $\Omega_{w_o, f_o, t}$ denote the set of possible histories at time t for the modified bandit with initial state (w_o, f_o), and let $\Omega_{w_o, f_o} \doteq \bigcup_{t=0}^{\infty} \Omega_{w_o, f_o, t}$. Thus ψ can be characterized as a function $\psi : \Omega_{w_o, f_o} \rightarrow \{1, \infty\}$ with $\psi(w_o, f_o) = 1$.

Let the random variable τ_ψ (τ for short) denote the quitting time according to strategy ψ: hence, the gambler's last play occurs at time $\tau - 1$. Mathematically, τ can be expressed as

$$\tau_\psi \doteq \min\{t | \psi(\boldsymbol{\mu}(t)) = \infty\}.$$

For $t \geq \tau$, $\boldsymbol{w}(t)$, $\boldsymbol{f}(t)$, and $\boldsymbol{\sigma}(t)$ are all constant functions.

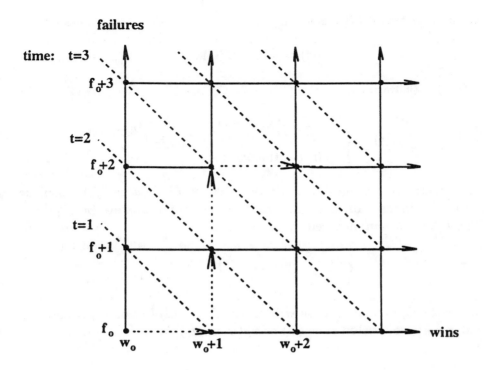

FIGURE 5.1. Graphical representation of the modified 1-armed bandit.

As before, $R(t)$ denotes the gambler's reward at time t. If $t \geq \tau$, then $R(t) = 0$; if $t < \tau$, then $R(t) = 1$ if the player wins at time t, and $R(t) = 0$ if he fails. In concise terms,

$$R(t) = \begin{cases} 1 & \text{if } w(t+1) - w(t) = 1, \\ 0 & \text{if } f(t+1) - f(t) = 1. \end{cases}$$

The total reward R is given as in (2.6) by $R = \sum_{t=0}^{\infty} \beta^t R(t)$.

New r.v.s must be introduced to describe the payment scheme. Let $P(t)$ denote the fee the gambler pays at time t. If $t \geq \tau$, then $P(t) = 0$, since the gambler has quit play; otherwise, $P(t) = P$, the fixed fee for playing. In concise terms,

$$P(t) = \begin{cases} P & \text{if } t < \tau, \\ 0 & \text{if } t \geq \tau, \end{cases} \tag{5.1}$$

or alternatively,

$$P(t) = \begin{cases} P & \text{if } w(t+1) - w(t) = 1 \text{ or } f(t+1) - f(t) = 1. \\ 0 & \text{otherwise} \end{cases}$$

The *total fee paid* P is given by

$$P \doteq \sum_{t=0}^{\infty} P(t) \cdot \beta^t.$$

The *net gain at time t* $G(t)$ is given by

$$G(t) \doteq R(t) - P(t).$$

Hence $G(t)$ satisfies

$$G(t) = \begin{cases} 1 - P & \text{if } w(t+1) - w(t) = 1, \\ -P & \text{if } f(t+1) - f(t) = 1, \\ 0 & \text{otherwise.} \end{cases}$$

$G(t)$ is sometimes written $G_P(t)$ or $G_P(t, \psi)$ or $G_P(t, \psi, w_o, f_o)$, to make explicit its dependence on the fee P, strategy ψ and initial state (w_o, f_o).
     The *total net gain*, denoted by G, is

$$G \doteq \sum_{t=0}^{\infty} G(t) \cdot \beta^t = R - P.$$

Just as for $G(t)$, G may also be written as G_P or $G_P(\psi)$ or $G_P(\psi, w_o, f_o)$.
The *value* v_P [also written as $v_P(\psi)$ or $v_P(\psi, w_o, f_o)$] is

$$v_P \doteq E[G_P].$$

5.2. Gittins index defined in terms of the modified bandit

Having elucidated the modified 1-armed bandit model, it is now possible to proceed with the definition of the Gittins index.
     Let Ψ be the set of allowed strategies for the modified bandit, that is $\Psi \doteq \{\psi | \psi(w_o, f_o) = 1\}$. Define $g(P, w_o, f_o)$ (or $g(P)$ for short) as

$$g(P, w_o, f_o) \doteq \sup_{\psi \in \Psi} v_P(\psi, w_o, f_o). \qquad (5.2)$$

$g(P)$ is the greatest possible value, maximized over all strategies, given that the playing fee is P. It can be shown for any ψ that $\frac{-P}{1-\beta} \leq E[G_P(\psi)] \leq \frac{1-P}{1-\beta}$ (this inequality is left as an exercise to the reader); and it follows that $g(P)$ exists and is finite for all P.
     Intuitively one would expect that, the greater the fee, the smaller the greatest possible expected total net gain. This intuition is verified in the following lemma.

Lemma 5.1 *If* $P_2 > P_1$, *then*

$$P_2 - P_1 \leq g(P_1, w_o, f_o) - g(P_2, w_o, f_o) \leq \frac{P_2 - P_1}{1 - \beta}.$$

Proof: Consider two players simultaneously playing the same modified 1-armed bandit, using the same strategy $\psi \in \Psi$, one paying fee P_1 and the other paying P_2. Thus each time the arm is played, win or lose, the former player receives an amount $P_2 - P_1$ greater than the latter. Recalling the definition of τ,

$$\boldsymbol{G}_{P_1}(\psi) - \boldsymbol{G}_{P_2}(\psi) = (P_2 - P_1) + \beta(P_2 - P_1) + \cdots + \beta^{\tau}(P_2 - P_1).$$

Since $1 \leq \tau \leq \infty$,

$$(P_1 - P_2) \leq \boldsymbol{G}_{P_1}(\psi) - \boldsymbol{G}_{P_2}(\psi) \leq \frac{P_2 - P_1}{1 - \beta}.$$

and taking expectations

$$(P_1 - P_2) \leq E[\boldsymbol{G}_{P_1}(\psi)] - E[\boldsymbol{G}_{P_2}(\psi)] \leq \frac{P_2 - P_1}{1 - \beta}.$$

Rearranging and taking the supremum over ψ, recalling the definition of $g(P)$, we have

$$(P_2 - P_1) + g(P_2) \leq g(P_1) \leq \frac{P_2 - P_1}{1 - \beta} + g(P_2),$$

and subtracting $g(P_2)$ from each expression in the inequality yields the lemma.

Lemma 5.1 implies that $g(P)$ is continuous and strictly decreasing in P, and thus has an inverse g^{-1}. It can be shown that $g(0) > 0$, $g(1) \leq 0$ (an exercise for the reader); so by the intermediate value theorem there exists a P with $0 < P \leq 1$ such that $g(P) = 0$, that is, $P = g^{-1}(0)$. The index $\nu(w_o, f_o)$ given by

$$\nu(w_o, f_o) \doteq g^{-1}(0). \tag{5.3}$$

is called the *dynamic allocation index* (so-called by Gittins) or the *Gittins index* (so-called by everyone else!). Intuitively the Gittins index can be conceived of as a kind of "break-even" fee. If the bandit arm begins in state (w_o, f_o), the gambler who employs the best possible strategy can just expect to break even if his fee for playing is $\nu(w_o, f_o)$. If the fee is larger (respectively smaller), the gambler stands to lose (respectively gain) by playing the arm.

6. OPTIMALITY OF THE GITTINS INDEX STRATEGY FOR THE MODIFIED 1-ARMED BANDIT

The modified bandit introduced in Section 5 is somewhat easier to work with than the full n-armed bandit. In fact, the n-armed bandit can be obtained by "pasting together" modified bandits, in a sense to be made precise in the next section. In this section, an optimal strategy for the modified 1-armed bandit (based on the Gittins index) will be established. Optimality will be proved by means of a pictorial representation of the modified bandit model using trees, which is presented in the first subsection below.

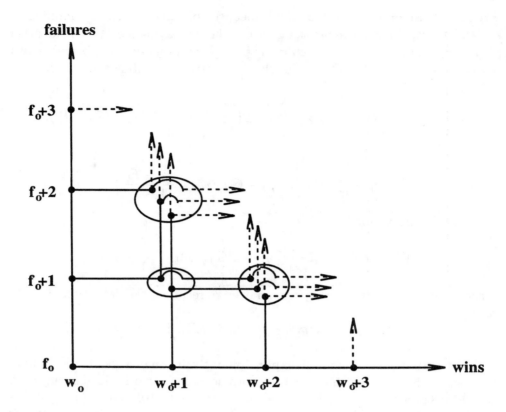

FIGURE 6.1. Histories in the modified 1-armed bandit. Circled groups of nodes correspond to single bandit states $\sigma(t)$.

6.1. "Pruned" tree representation of the modified bandit; tree values

The modified bandit has already been presented graphically in Section 5. An alternative picture which more clearly presents the role of histories in the process is given in Figure 6.1.

Here the graph is a tree (i.e., has no closed circuits), and thus there is a unique path from the root (w_o, f_o) to any given node. Each node now represents a history. Several histories can correspond to the same modified bandit state; in Figure 6.1, circled groups of nodes correspond to single states as labelled.

Strategies are conveniently represented in this picture as an assignment of 1s and ∞s to the nodes, as in Figure 6.2.

In fact, since the progress of the game stops once a node labelled "∞" is encountered, a strategy can be represented as a "pruned" tree whose terminal nodes are all labelled "∞" and whose non-terminal nodes are all labelled "1". The pruned tree corresponding to Figure 6.2 is shown in Figure 6.3.

The pruned tree corresponding to strategy ψ with initial node (w, f) is

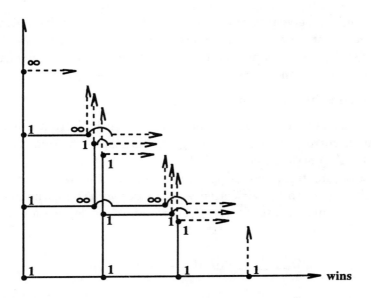

FIGURE 6.2. A strategy for the modified 1-armed bandit.

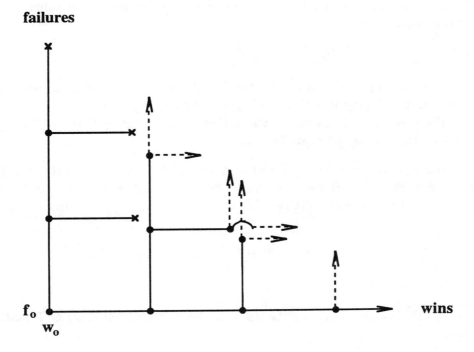

FIGURE 6.3. A "pruned" tree showing the strategy in Figure 6.2.

denoted (ψ, w, f), or simply (ψ) if the initial node is (w_o, f_o). The *value* of the pruned tree (ψ, w, f) is given by the expected total net gain $E[\mathbf{G}_P(\psi, w, f)]$ of the corresponding strategy, and is denoted by $v_P(\psi, w, f)$ [or $v_P(\psi)$ if the initial node is (w_o, f_o)]. Tree values can be computed in much the same fashion as strategy values were found in Section 3 for the n-armed bandit.

6.2. Tree branches and grafting

A tree *branch* consists of that part of the tree that springs from a given n-ode. The branch corresponding to strategy ψ that springs from the node corresponding to $\mu'(t)$ is denoted $(\psi, \mu'(t))$.

A branch springing from node $\mu'(t)$ is in fact itself a tree with root $\sigma'(t) = (w'(t), f'(t))$. This tree can be associated with a strategy $\psi_{\mu'}$ for the modified 1-armed bandit with initial state $(w'(t), f'(t))$ where $\psi_{\mu'}$ is defined for $\mu''(t') \in \Omega_{w'(t), f'(t), t'}$ as follows:

$$\psi_{\mu'}(\mu''(t')) \doteq \psi([\mu'(t-1), \mu''(t')]),$$

where $[\mu'(t-1), \mu''(t')]$ denotes the history formed by concatenating the two histories $\mu'(t)$ and $\mu''(t')$, namely

$$[\mu'(t-1), \mu''(t')] \doteq [\sigma'(0), \ldots, \sigma'(t-1), \sigma''(0), \sigma''(1), \ldots \sigma''(t')].$$

The *value* of the branch $(\psi, \mu'(t))$ is defined to be equal to the value of the strategy $\psi_{\mu'}$, namely

$$v_P(\psi, \mu'(t)) \doteq E[\mathbf{G}_P(\psi_{\mu'}, w'(t), f'(t))].$$

Trees can be modified by "cutting" and "grafting", that is, excising one branch and replacing it with another, as shown in Figure 6.4.

The following lemma elucidates further the relationship between the value of a tree and the values of its branches.

Lemma 6.1 *Given a tree (ψ) and a history $\mu'(t') \in \Omega_{w_o, f_o, t'}$. Let $(\psi, \mu'(t'))$ be the branch of (ψ) that springs from node $\mu'(t')$. Let (ψ') be the tree formed by cutting the branch $(\psi, \mu'(t'))$ off from (ψ) at node $\mu(t')$ (as shown in Figure 6.5). Then*

$$v_P(\psi) = v_P(\psi') + \beta^{t'} \Pr[\boldsymbol{\mu}(t') = \mu'(t')] v_P(\psi, \mu'(t')). \tag{6.1}$$

Proof: The value $v_P(\psi) = E[\mathbf{G}_P(\psi)]$ can be written in analogy to Equation 3.2 as

$$v_P(\psi) = \sum_{t=0}^{\infty} \beta^t \sum_{\mu(t) \in \Omega_{w_o, f_o, t}} E[\mathbf{G}_P(t, \psi) | \boldsymbol{\mu}(t) = \mu(t)] \Pr[\boldsymbol{\mu}(t) = \mu(t)].$$

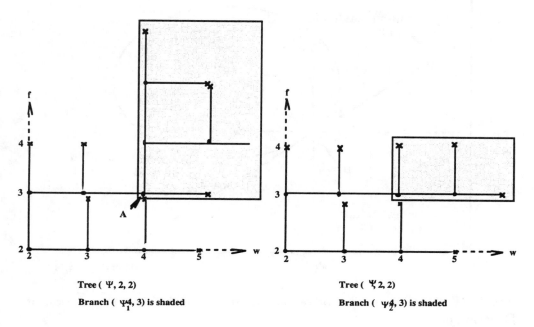

FIGURE 6.4. Cutting and grafting. Tree $(\psi', 2, 2)$ is obtained from tree $(\psi, 2, 2)$ by cutting branch $(\psi_1, 4, 3)$ at node A and grafting in branch $(\psi_2, 4, 3)$.

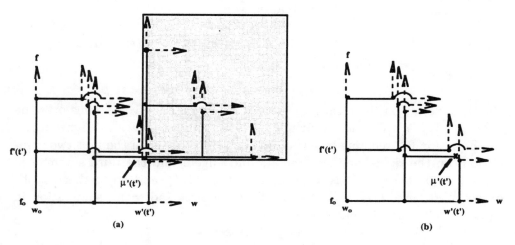

FIGURE 6.5. Relationship between (ψ) and (ψ') as defined in Lemma 6.1. (a) Tree shown is (ψ)-branch $(\psi, \mu'(t'))$ is shaded; (b) Tree shown is (ψ').

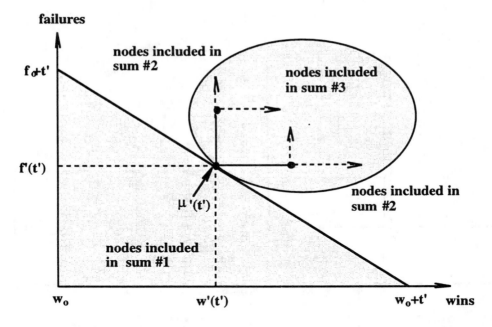

FIGURE 6.6. Tree nodes corresponding to each of the three summations in Equation 6.2.

Conditioning on the event $\boldsymbol{\mu}(t') = \mu'(t')$, this can be rewritten

$$
\begin{aligned}
v_P\ (\psi) = {} & \sum_{t=0}^{t'-1} \beta^t \sum_{\mu(t)\in\Omega_{w_o,f_o,t}} E[\boldsymbol{G}_P(t,\psi)|\boldsymbol{\mu}(t)=\mu(t)]\Pr[\boldsymbol{\mu}(t)=\mu(t)] \\
+ {} & \sum_{t=t'}^{\infty} \beta^t \sum_{\mu(t)\in\Omega_{w_o,f_o,t},\mu(t')\neq\mu'(t')} E[\boldsymbol{G}_P(t,\psi)|\boldsymbol{\mu}(t)=\mu(t)]\Pr[\boldsymbol{\mu}(t)=\mu(t)] \\
+ {} & \sum_{t=t'}^{\infty} \beta^t \sum_{\mu(t)\in\Omega_{w_o,f_o,t},\mu(t')=\mu'(t')} E[\boldsymbol{G}_P(t,\psi)|\boldsymbol{\mu}(t)=\mu(t)]\Pr[\boldsymbol{\mu}(t)=\mu(t)].
\end{aligned}
$$
(6.2)

The three terms on the right-hand side of Equation 6.2 involve sums over the nodes $\mu \in \Omega_{w_o,f_o}$. The nodes included in each sum are shown in Figure 6.6.

Equation 6.2 remains true if ψ is replaced by ψ'. Then the first two summations on the right-hand side remain unchanged: and the third summation is equal to 0, since $\mu'(t')$ is a terminal node for (ψ'), and no further payoffs are received once $\mu'(t')$ is attained. Thus the first and scond summation in Equation 6.2 added together give $v_P(\psi')$.

The third summation can be re-expressed as

$$
\sum_{t=t'}^{\infty} \beta^t \sum_{\mu''(t-t')\in\Omega_{w'(t'),f'(t'),t-t'}} E[\boldsymbol{G}_P(t,\psi)|\boldsymbol{\mu}''(t-t')=\mu''(t-t')]\cdot
$$

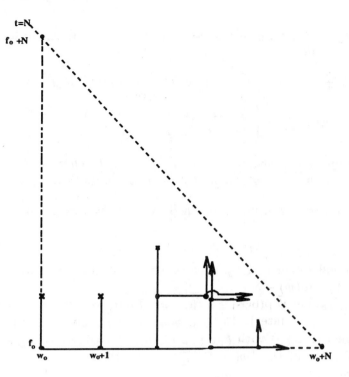

FIGURE 6.7. Pruned tree $(\bar{\psi})$ for the strategy $\bar{\psi}$.

$$\Pr[\boldsymbol{\mu}(t') = \mu(t')]\Pr[\boldsymbol{\mu}''(t-t') = \mu''(t-t')],$$

where $\boldsymbol{\mu}''(t-t')$ is the random variable

$$\boldsymbol{\mu}''(t-t') \doteq [\boldsymbol{\sigma}(t'), \boldsymbol{\sigma}(t'+1), \ldots, \boldsymbol{\sigma}(t)].$$

Letting $t'' \doteq t - t'$, the sum becomes

$$\beta^{t'} \cdot \Pr[\boldsymbol{\mu}(t') = \mu(t')] \cdot \sum_{t''=0}^{\infty} \beta^{t''} \cdot \sum_{\mu''(t'') \in \Omega_{w'(t'), f'(t'), t''}}$$

$$E[\boldsymbol{G}_P(t, \psi)|\boldsymbol{\mu}''(t'') = \mu''(t'')]\Pr[\boldsymbol{\mu}''(t'') = \mu''(t'')],$$

which upon inspection is found to be equal to

$$\beta^{t'} \Pr[\boldsymbol{\mu}(t') = \mu'(t')]v_P(\psi, \mu'(t')).$$

The lemma is thus proved.

The following is an immediate corollary to Lemma 6.1:

Corollary *Let ψ_1, ψ_2 be two modified bandit strategies which are identical apart from the branches $(\psi_1, \mu(t))$ and $(\psi_2, \mu(t))$. If $v_P(\psi_1, \mu(t)) \geq v_P(\psi_2, \mu(t))$, then $v_P(\psi_1) \geq v_P(\psi_2)$.*

6.3. Optimality of Gittins index-related strategy for modified 1-armed bandit

Theorem 6.1 *Let $P = \nu(w_o, f_o)$ and let $\bar{\psi}$ be the following strategy:*

$$\bar{\psi}(w, f) = \begin{cases} 1 & \text{if } \nu(w, f) > \nu(w_o, f_o), \\ \infty & \text{if } \nu(w, f) \leq \nu(w_o, f_o). \end{cases}$$

(The pruned tree for strategy $\bar{\psi}$ is shown in Figure 6.7). Then $\bar{\psi}$ is an optimal strategy for the modified 1-armed bandit with initial state (w_o, f_o) and fee P.

Proof: Optimality of $\bar{\psi}$ will be established by showing that for any strategy ψ and for any $\delta > 0$,

$$v_P(\bar{\psi}) > v_P(\psi) - \delta.$$

This inequality will be proven by constructing a strategy $\tilde{\psi}$ with $v_P(\tilde{\psi}) \geq v_P(\psi)$ and $v_P(\bar{\psi}) > v_P(\tilde{\psi}) - \delta$.

The construction of $\tilde{\psi}$ proceeds as follows. First, fix a large integer N (N will be chosen explicitly later). Next, choose (if one exists) a terminal node $\mu(t)$ of the pruned tree (ψ) with $t < N$ and $\bar{\psi}(\mu(t)) = 1$, as shown in Figure 6.8. By the definition of $\bar{\psi}$, $\nu(w(t), f(t)) > P$ and the left-hand inequality of Lemma 7.1 gives

$$\nu(w(t), f(t)) - P \leq g(P, w(t), f(t)) - g(\nu(w(t), f(t)), w(t), f(t)).$$

The definition of ν implies $g(\nu(w(t), f(t)), w(t), f(t)) = 0$; thus $g(P, w(t), f(t)) > 0$. Again by the definition of $g(P)$ there must exist a strategy ψ' such that $v_P(\psi', w(t), f(t)) > 0$. If $(\psi', w(t), f(t))$ is grafted onto (ψ) at node $\mu(t)$, the result is a tree (ψ_1) that by the corollary to Lemma 6.1 satisfies $v_P(\psi_1) > v_P(\psi)$.

This process can be iterated to produce a sequence of strategies $\psi_1, \psi_2, \ldots$ with $v_P(\psi) < v_P(\psi_1) < v_P(\psi_2) < \ldots$. The procedure must terminate after a finite number of steps, because there are only $2^{N+1} - 1$ histories $\mu(t)$ with $t < N$. The end result will be a strategy $\hat{\psi}$ where $v_P(\hat{\psi}) > v_P(\psi)$ and no terminal nodes $\mu(t)$ of $(\hat{\psi})$ with $t < N$ satisfy $\bar{\psi}(\mu(t)) = \infty$. It follows that $(\bar{\psi})$ is contained in $(\hat{\psi})$ as a "subtree" for $t < N$; that is, all nodes $\mu(t)$ with $t < N$ in the pruned tree $(\bar{\psi})$ are also contained in the pruned tree $(\hat{\psi})$, which is pictured in Figure 6.9.

Next, choose (if one exists) a non-terminal node $\mu(t)$ of $(\hat{\psi})$ with $t < N$ and $\bar{\psi}(\mu(t)) = \infty$, such as those shown in Figure 6.9. It follows from the definition of $\bar{\psi}$ that $\nu(w(t), f(t)) \leq P$; and again by the left-hand inequality of Lemma 7.1 it follows that

$$0 \leq P - \nu(w(t), f(t)) \leq 0 - g(P, w(t), f(t)),$$

so $g(P, w(t), f(t)) \leq 0$. From the definition of g it follows that any strategy with initial state $(w(t), f(t))$ will satisfy $E[\boldsymbol{G}_P(\psi)] \leq 0$; hence the definition of

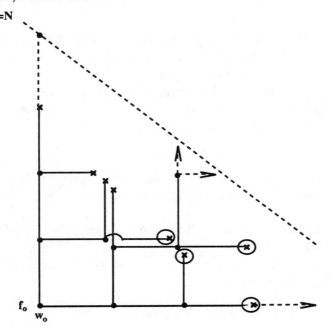

FIGURE 6.8. Tree (ψ), where circled nodes are terminal nodes of (ψ) with $t < N$ and $\bar{\psi}(\mu(t)) = 1$.

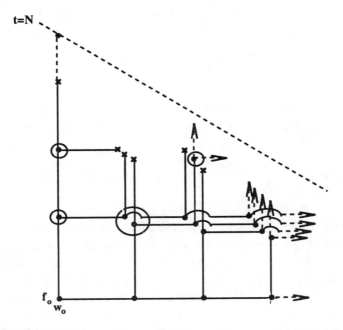

FIGURE 6.9. The pruned tree $(\hat{\psi})$, which contains $(\bar{\psi})$ as a subtree. Circled nodes $\mu(t)$ are non-terminal nodes of $(\hat{\psi})$ with $\bar{\psi}(\mu(t)) = \infty$.

v_P implies $v_P(\hat{\psi}, \mu(t)) \leq 0$. Let $(\hat{\psi}_1)$ be the tree obtained from $(\hat{\psi})$ by cutting the branch $(\hat{\psi}, \mu(t))$ and grafting in the null branch, which has value 0. Then by the corollary to Lemma 6.1, $v_P(\hat{\psi}_1) \geq v_P(\hat{\psi})$.

This process can be continued iteratively to produce a sequence $\hat{\psi}_1, \hat{\psi}_2, \ldots$ with $v_P(\hat{\psi}) \leq v_P(\hat{\psi}_1) \leq \ldots$. The process must terminate in a finite number of steps; the result is a strategy $\tilde{\psi}$ that satisfies

(i) $v_P(\tilde{\psi}) \geq v_P(\hat{\psi}) \geq v_P(\psi)$;
(ii) all non-terminal nodes $\mu(t)$ of $(\tilde{\psi})$ satisfy $\tilde{\psi}(\mu(t)) = 1$; and
(iii) all terminal nodes $\mu(t)$ of $(\tilde{\psi})$ satisfy $\tilde{\psi}(\mu(t)) = \infty$.

Thus $\tilde{\psi}$ and $\bar{\psi}$ are identical strategies for $t < N$, i.e., $\tilde{\psi}(\mu(t)) = \bar{\psi}(\mu(t))$ for all possible histories $\mu(t)$ with $t < N$.

It remains to show that $v_P(\bar{\psi}) > v_P(\tilde{\psi}) - \delta$, or equivalently, $v_P(\tilde{\psi}) - v_P(\bar{\psi}) < \delta$. Recalling the definition of v_P,

$$
\begin{aligned}
v_P(\tilde{\psi}) - v_P(\bar{\psi}) &= E[\boldsymbol{G}_P(\tilde{\psi}) - \boldsymbol{G}_P(\bar{\psi})] \\
&= E[\sum_{t=0}^{\infty} \beta^t \{\boldsymbol{G}_P(t, \tilde{\psi}) - \boldsymbol{G}_P(t, \bar{\psi})\}] \\
&= E[\sum_{t=0}^{N-1} \beta^t \{\boldsymbol{G}_P(t, \tilde{\psi}) - \boldsymbol{G}_P(t, \bar{\psi})\}] \\
&\quad + E[\sum_{t=N}^{\infty} \beta^t \{\boldsymbol{G}_P(t, \tilde{\psi}) - \boldsymbol{G}_P(t, \bar{\psi})\}] \\
&\leq 0 + \frac{\beta^N}{1-\beta}.
\end{aligned}
$$

(The last inequality is left as an exercise for the reader.) Since N was an arbitrary positive integer, the proof is completed by choosing N sufficiently large so that $\frac{\beta^N}{1-\beta} < \delta$.

7. OPTIMALITY OF THE GITTINS INDEX STRATEGY FOR THE n-ARMED BANDIT

The histories for the n-armed bandit represented in Figure 2.1 and those for the modified bandit in Figure 5.1 have a few evident differences. Circular loops may occur in the histories represented by the dotted arrow paths in Figure 2.1, while no such loops occur in the modified bandit representation. Such loops may be introduced into the modified bandit model by means of allowing time delay as a strategic option. This will be done in the first subsection below.

The "fee" introduced in the modified bandit model is intended to assign a worth to each play of an arm. In the n-armed bandit, the player's assessment of the worth of playing an arm will evidently change as play progresses. In view

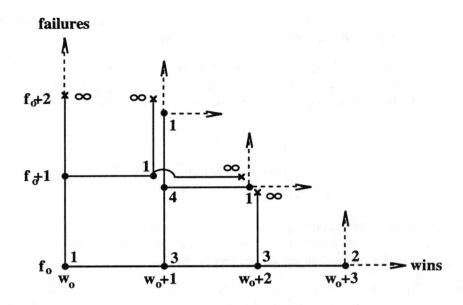

FIGURE 7.1. A delayed strategy for the modified 1-armed bandit. Each node $\mu(t)$ is labelled with the corresponding time delay $\psi^+(\mu(t))$.

of this situation, the second subsection following will introduce and discuss the possibility of adjusting the playing fee as play progresses. Finally, the third subsection below will show that the n-armed bandit is in some sense comparable to n modified 1-armed bandits where delayed strategies and fee adjustment are allowed, and the optimality of the Gittins index will follow directly.

7.1. Delayed strategies

In the foregoing discussion of the modified bandit, the player's strategy at time t consisted of choosing between quitting or continuing play. Now he will be given an additional option — he can wait for some time (of his choosing) before playing again. Mathematically, his strategy can be expressed by a function $\psi^+ : \Omega \to \mathbf{Z}^+ \bigcup \{\infty\}$, where $\psi^+(\mu(t)) = d$ means that there is a waiting period d between the tth and $t + 1$st play, given that the history at time t is $\mu(t)$. Such a function ψ^+ will be called a *delayed strategy*.

A delayed strategy ψ^+ may be conveniently represented as a pruned tree (ψ^+) where each node $\mu(t)$ is labelled by the corresponding delay $\psi^+(\mu(t))$, as shown in Figure 7.1.

The delays have no effect on the net gains $G_P(t, \psi^+)$, but do change the total net gain $G_P(\psi^+)$ due to their influence on the discounting factors. For $t > 0$, the tth play will occur at time $\psi^+(\mu(0)) + \ldots + \psi^+(\mu(t-1))$, and thus

$$G_P(\psi^+) = G_P(0, \psi^+) + \sum_{t=1}^{\infty} \beta^{\psi^+(\mu(0))+\ldots+\psi^+(\mu(t-1))} G_P(t, \psi^+). \qquad (7.1)$$

The *value* $v_P(\psi^+)$ is defined to be $E[\boldsymbol{G}_P(\psi^+)]$. An obvious analogy to Lemma 6.1 holds (proof is left as an exercise for the reader):

Lemma 7.1 *Given a delayed strategy ψ^+ with tree (ψ^+), and a history $\mu'(t') \in \Omega_{w_o, f_o, t'}$. Let $(\psi^+, \mu'(t'))$ be the branch of (ψ^+) which springs from node $\mu'(t')$. Let (ψ'^+) be the tree formed by cutting the branch $(\psi^+, \mu'(t'))$ off from (ψ^+) at node $\mu(t')$. Then*

$$v_P(\psi^+) = v_P(\psi'^+) + \beta^{\psi^+(\mu(0)) + \ldots + \psi^+(\mu(t-1))}.$$
$$\Pr[\boldsymbol{\mu}(t') = \mu(t')] v_P(\psi^+, \mu'(t')). \tag{7.2}$$

It is not immediately clear whether the Gittins Index strategy $\tilde{\psi}$ (defined in Theorem 6.1) is still optimal among all delayed strategies ψ^+ when $P = \nu(w_o, f_o)$. The following lemma settles the question.

Lemma 7.2 *If ψ^+ is any delayed strategy for the modified bandit with initial state (w_o, f_o) and fee $\nu(w_o, f_o)$, and $\tilde{\psi}$ is the Gittins index strategy defined in Theorem 6.1, then $v_P(\tilde{\psi}) \geq v_P(\psi^+)$.*

Proof: The procedure of proof is as follows. Given any delayed strategy ψ^+ and any $\delta > 0$, an undelayed strategy $\tilde{\psi}$ will be constructed with $v_P(\tilde{\psi}) \geq v_P(\psi^+) - \delta$. Theorem 6.1 implies that $v_P(\bar{\psi}) \geq v_P(\tilde{\psi})$; hence $v_P(\bar{\psi}) \geq v_P(\psi^+) - \delta$ which is the desired result.

It will be evident that the construction is similar to that in Lemma 6.1. Take N to be a fixed, large positive integer (to be chosen explicitly later). Choose (if one exists) a node $\mu(t)$ of the tree (ψ^+) with $\psi^+(\mu(t)) = d > 1$ and $t < N$. A new strategy ψ_1^+ will be constructed as follows (refer to Figure 7.2). If $v_P(\psi^+, \mu(t)) \leq 0$, then cut the branch $(\psi^+, \mu(t))$ from the tree (ψ^+) and call the result ψ_1^+. On the other hand, if $v_P(\psi^+, \mu(t)) > 0$, then the tree for ψ_1^+ is identical to that for ψ^+ except that the 'd' at node $\mu(t)$ is replaced by a '1'. It follows from Lemma 6.1 that

$$v_P(\psi^+) - v_P(\psi^+, \mu(t)) = v_P(\psi_1^+) - v_P(\psi_1^+, \mu(t)).$$

By construction, it follows that $v_P(\psi^+, \mu(t)) \leq v_P(\psi_1^+, \mu(t))$; hence $v_P(\psi_1^+) \geq v_P(\psi^+)$.

The procedure just used to obtain ψ_1^+ from ψ^+ can be applied iteratively to obtain ψ_2^+ from ψ_1^+, ψ_3^+ from ψ_2^+, ... such that $v_P(\psi^+) \leq v_P(\psi_1^+) \leq \ldots$. This process must terminate after a finite number of steps, since there are only a finite number of histories $\mu(t)$ with $t < N$. The result is a strategy $\hat{\psi}^+$ with $\hat{\psi}^+(\mu(t)) = 0$ or 1 for all $t < N$. Define $\tilde{\psi}$ by truncating $\hat{\psi}^+$ after the Nth play, that is

$$\tilde{\psi}(\mu(t)) = \begin{cases} \hat{\psi}^+(\mu(t)) & t < N, \\ \infty & t \geq N. \end{cases}$$

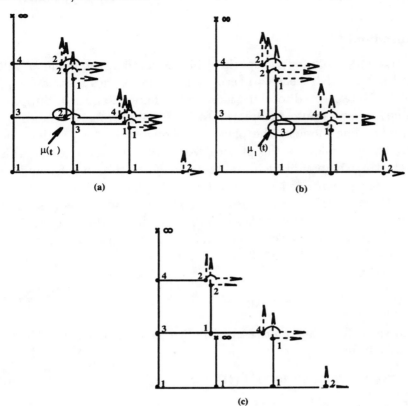

FIGURE 7.2. Modification of strategy ψ^+ in the proof of Lemma 7.1: (a) Tree representation for ψ^+. Modification will take place at circled node $\mu(t)$; (b) Tree for ψ_1^+, in the case $v_P(\psi^+, \mu(t)) \le 0$. Further modification will take place at the circled node $\mu_1(t)$; (c) Tree for ψ_1^+, in the case $v_P(\psi^+, \mu_1(t)) > 0$.

Then $\tilde{\psi}$ agrees with $\hat{\psi}^+$ up to the Nth play, and

$$|v_P(\tilde{\psi}) - v_P(\hat{\psi}^+)| \le \frac{\beta^N}{1 - \beta}.$$

(Verification is left as an exercise to the reader.) Choosing N sufficiently large so that $\frac{\beta^N}{1-\beta} < \delta$, it follows that

$$v_P(\tilde{\psi}) > v_P(\hat{\psi}^+) - \delta \ge v_P(\psi^+) - \delta.$$

But $\tilde{\psi}$ is an undelayed strategy; so by Theorem 6.1

$$v_P(\bar{\psi}) \ge v_P(\tilde{\psi}),$$

and the lemma follows.

7.2. Fee adjustment

Under the Gittins index strategy $\bar{\psi}$ specified in Lemma 6.1, the gambler will play a modified bandit with fee $P = \nu(w_o, f_o)$ until he reaches a state (w, f) with $\nu(w, f) \leq P$. His expected total net gain under this strategy is 0. Suppose now the "casino owner" wants the player to keep on playing the modified bandit indefinitely. The moment the player reaches a state (w, f) where he wants to quit, the owner lowers the playing fee to $\nu(w, f)$. Essentially, this amounts to restarting the bandit at initial state (w, f), with a new fee and an additional discounting factor reflecting the fact that play has been going on for some time. When the player again stops play, the fee will be readjusted in the same manner: and so on. Under these conditions, the fee that the gambler pays at time t is equal to the record low value of the Gittins index attained prior to time t, as shown in Figure 7.3.

In precise terms,

$$P(t) = \min_{s \leq t} \nu(\boldsymbol{w}(s), \boldsymbol{f}(s)). \tag{7.3}$$

Let the random variable $\boldsymbol{\tau}_i$ denote the time the ith fee change occurs. The $\boldsymbol{\tau}_i$ are given inductively by

$$\boldsymbol{\tau}_1 = \min_{t>0}\{t | \nu(\boldsymbol{w}(t), \boldsymbol{f}(t)) < \nu(w_o, f_o)\}$$

$$\boldsymbol{\tau}_{i+1} = \min_{t>\boldsymbol{\tau}_i}\{t | \nu(\boldsymbol{w}(t), \boldsymbol{f}(t)) < \nu(\boldsymbol{w}(\boldsymbol{\tau}_i), \boldsymbol{f}(\boldsymbol{\tau}_i))\}$$

Note that $P(t)$ given by equation 7.3 and $\boldsymbol{\tau}_i$ are related by

$$P(t) = \nu(\boldsymbol{w}(\boldsymbol{\tau}_i), \boldsymbol{f}(\boldsymbol{\tau}_i)) \text{ if } \boldsymbol{\tau}_i \leq t < \boldsymbol{\tau}_{i+1}.$$

Suppose first that the player pulls at consecutive times, without any delays. Then his expected total net gain (discounted sum of rewards minus fees paid) in the time interval $[0, \boldsymbol{\tau}_1 - 1]$ is 0, because $\boldsymbol{\tau}_1$ is exactly the "stopping time" which gives expected value 0 for the arm played with the initial Gittins index value as fee. Similarly, the expected total net gain in the time intervals $[\boldsymbol{\tau}_i, \boldsymbol{\tau}_{i+1} - 1]$ are 0 for each i. The player's expected total net gain is the sum of the expected net gains from 0 to $\boldsymbol{\tau}_1 - 1$, $\boldsymbol{\tau}_1$ to $\boldsymbol{\tau}_2 - 1$, etc. (This is true even if $\boldsymbol{\tau}_i = \infty$ for some i.) So the expected total net gain is 0.

If, instead of playing at consecutive times, the player is allowed delays between plays, (infinite delays are OK, too!) then Lemma 7.2 shows that the gambler's expected net gain in each time interval $[\boldsymbol{\tau}_i, \boldsymbol{\tau}_{i+1} - 1]$ will be nonpositive; thus his total expected net gain will be nonpositive. Hence follows

Lemma 7.3 *Consider the modified 1-armed bandit as in Lemma 6.1, except with a fee which is not fixed but rather is always set to the previous record low value of the Gittins index achieved by the arm (as given by Equation 7.3). Then for any delayed strategy ψ^+,*

$$E[\boldsymbol{G}(\psi^+)] \leq 0.$$

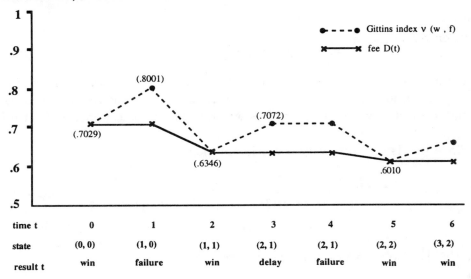

FIGURE 7.3. Fee adjustment scheme applied to the modified 1-armed bandit with $\beta = .9$, for a particular history of play. Gittins index values are obtained from reference [2].

Equality occurs when the strategy employed only has delays at times when fee changes occur.

7.3. Optimality for the n-armed bandit

The optimal strategy for the (unmodified) n-armed bandit may now be established in a relatively straightforward way, using the machinery developed in the previous subsections.

Theorem 7.1 *The expected total reward for an n-armed bandit is maximized by the following strategy: always pull the arm with maximum Gittins index.*

Proof: Fix a particular arm k. In order to focus in on the contibution of arm k to the expected total reward, the histories of all other arms will be fixed, and only arm k will be left to have random outcomes. To this end, fix values $\{\alpha_{iN}\}_{i \neq k}$ and define the condition $\mathcal{A}_{i \neq k}$ to be

$$\mathcal{A}_{i \neq k} \doteq \text{``}\alpha_{iN} = \alpha_{iN}, i \neq k, N = 0, 1, 2, \dots \text{''}. \tag{7.4}$$

That is, for all arms $i \neq k$, the outcome of every play is known in advance — only the outcome of the kth arm is unknown.

Probabilistically speaking, an n-armed bandit with the condition $\mathcal{A}_{i \neq k}$ under a fixed strategy ψ is identical to a 1-armed bandit with a delayed strategy. For this 1-armed bandit, using the notation of Section 5 (with the addition of the subscript "k" to denote that it is arm k being considered), it follows from Lemma 7.3 that

$$E[G_k(\psi)|\mathcal{A}_{i \neq k}] \leq 0.$$

Equivalently,

$$E[\boldsymbol{P}_k(\psi)|\mathcal{A}_{i\neq k}] \geq E[\boldsymbol{R}_k(\psi)|\mathcal{A}_{i\neq k}],$$

where here, as in the following discussion, $\boldsymbol{P}_k(\psi)$ refers to the adjusted fee given by Equation 7.3. The expected reward due to a single bandit arm is dominated by the expected sum of fees which would be paid under a "Gittins index" payment scheme.

Since this is true for every condition $\mathcal{A}_{i\neq k}$, it follows

$$
\begin{aligned}
E[\boldsymbol{R}_k(\psi)] &= \sum E[\boldsymbol{R}_k(\psi)|\mathcal{A}_{i\neq k}]\Pr[\mathcal{A}_{i\neq k}] \\
&\leq E[\boldsymbol{P}_k(\psi)|\mathcal{A}_{i\neq k}]\Pr[\mathcal{A}_{i\neq k}] \\
&= E[\boldsymbol{P}_k(\psi)].
\end{aligned}
$$

Since this is true for every arm k, it follows

$$
\begin{aligned}
E[\boldsymbol{R}(\psi)] = E[\sum_k \boldsymbol{R}_k] &= \sum E[\boldsymbol{R}_k] \\
&\leq \sum E[\boldsymbol{P}_k] = E[\sum \boldsymbol{P}_k] = E[\boldsymbol{P}(\psi)],
\end{aligned} \tag{7.5}
$$

where $\boldsymbol{P}(\psi) \doteq \sum \boldsymbol{P}_k$. When $\psi = \bar{\psi}$, it is clear that the only delays in the course of play of arm k will take place when the fee for arm k changes. Thus Lemma 7.3 implies

$$E[\boldsymbol{R}_k(\bar{\psi})|\mathcal{A}_{i\neq k}] = E[\boldsymbol{P}_k(\bar{\psi})|\mathcal{A}_{i\neq k}],$$

and summing over all possible values of $\mathcal{A}_{i\neq k}$,

$$E[\boldsymbol{R}_k(\bar{\psi})] = E[\boldsymbol{P}_k(\bar{\psi})].$$

Summing over k from 1 to n,

$$E[\boldsymbol{R}(\bar{\psi})] = E[\boldsymbol{P}(\bar{\psi})]. \tag{7.6}$$

The random variable $\boldsymbol{P}_k$ in fact depends on the strategy employed. To evaluate the effect of the strategy on $\boldsymbol{P}_k$, fix values $\{\alpha_{iN}\}$ and define the condition $\mathcal{A} \doteq$ "$\alpha_{iN} = \alpha_{iN}$ for all i". That is, the entire history of play is known and specified beforehand. In this case, Equation 7.3 shows what the fee on the Nth play of the ith arm will be, namely the Gittins index calculated on the basis of the known win/loss record of arm i prior to the Nth play. It may be seen that fees paid for each arm will be a decreasing sequence. These decreasing sequences are known and fixed. A strategy effectively chooses a sequence from among these decreasing sequences, and the total fee paid under the strategy is the discounted sum of the chosen sequence. In order to maximize the total fee paid, the terms of the sequence should be chosen as large as possible, larger terms occurring earlier in the sequence so that they are discounted less. The Gittins index-based strategy $\bar{\psi}$ by its very definition will choose first the arm

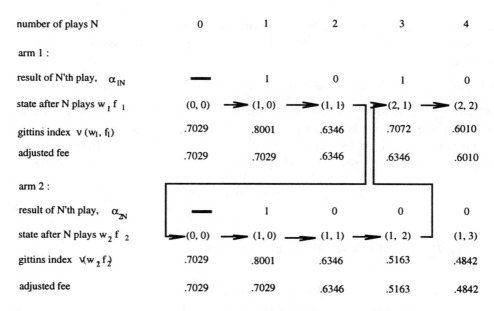

FIGURE 7.4. Sequence of plays as determined by the Gittins index strategy for a particular event $\mathcal{A}$. Arrows show the sequence of play.

with largest fee: then successively thereafter will choose the largest remaining fee, as shown in Figure 7.3 . Thus, applying the Gittins index strategy leads to the sequence of plays that maximizes the total fees paid, and for any strategy ψ we can write

$$E[\boldsymbol{P}(\bar{\psi})|\mathcal{A}] \geq E[\boldsymbol{P}(\psi)|\mathcal{A}].$$

Since this is true for any given $\mathcal{A}$, it follows

$$E[\boldsymbol{P}(\bar{\psi})] \geq E[\boldsymbol{P}(\psi)],$$

and from Equation 7.5,

$$E[\boldsymbol{P}(\bar{\psi})] \geq E[\boldsymbol{R}(\psi)].$$

Thus from Equation 7.6 above,

$$E[\boldsymbol{R}(\bar{\psi})] \geq E[\boldsymbol{R}(\psi)],$$

and the optimality of the Gittins strategy is established.

8. CALCULATION OF THE GITTINS INDEX

This section presents the theoretical foundation for a practical method of calculating the Gittins index $\nu(w, f)$ for different values of w and f. The first subsection introduces an alternative characterization of the Gittins index, following Whittle [3]. The second subsection presents the notion of "finite horizon", which reduces the number of possible states from infinite to finite, thus making the problem accessible to calculation. The final subsection establishes the convergence of the finite horizon approximation as the horizon moves out to infinity.

8.1. Whittle's alternative characterization of the Gittins index

Returning to the modified one-armed bandit, consider now a slightly different situation where there is no fee for playing the arm, but a stream of "pension payments" is received by the player after he "retires" from playing. To this end, define the random variable $\tilde{P}(t) \doteq P - P(t)$. From Equation 5.1 it follows

$$\tilde{P}(t) = \begin{cases} 0 & \text{if } t < \tau_\psi; \\ P & \text{if } t \geq \tau_\psi. \end{cases} \tag{8.1}$$

Let $\tilde{P} \doteq \sum_{t=0}^\infty \beta^t \tilde{P}(t)$. Equation 8.1 implies

$$\tilde{P} = \sum_{t=\tau}^\infty \beta^t P = P \frac{\beta^\tau}{1-\beta},$$

which is equivalent to a single lump-sum "retirement payment" of $\frac{P}{1-\beta}$ made at time τ.

Define $\tilde{G} \doteq R + \tilde{P}$. It is clear that $\tilde{G} = G + \sum_{t \geq 0} \beta^t P = G + \frac{P}{1-\beta}$ (exercise for the reader). Recall the definition of $\nu(w_o, f_o)$ given in Equations 5.2 and 5.3:

$$\nu(w_o, f_o) = \{P | \sup_{\psi \in \Psi} E[G_P(\psi, w_o, f_o 0)] = 0\}.$$

It follows that $\nu(w_o, f_o)$ can also be characterized as

$$\nu(w_o, f_o) = \{P | \sup_{\psi \in \Psi} E[\tilde{G}_P(\psi, w_o, f_o)] = \frac{P}{1-\beta}\}. \tag{8.2}$$

Keeping in mind the "retirement payment" interpretation of $\tilde{P}$, Equation 8.2 can be understood as follows: the player can with equal advantage retire immediately and receive a retirement payment $\lambda \doteq \frac{\nu(w_o, f_o)}{1-\beta}$, or he can play at least once and retire optimally thereafter, receiving his retirement payment λ when he ceases play.

The above discussion is summarized in the following lemma.

Lemma 8.1 *Given a modified 1-armed bandit with initial state (w_o, f_o) and $P = 0$, but with the addition of a single retirement payment made at the quitting time τ. Let λ be the retirement payment that makes the following two choices equally lucrative: (1) retire immediately; (2) play the arm at least once and retire optimally thereafter. Then $\lambda(1 - \beta)$ is equal to the Gittins index $\nu(w_o, f_o)$.*

8.2. Finite horizon, and calculation of Gittins index for horizon N problem

If the bandit arm starts in state $(0,0)$, then a potentially infinite number of states can be reached, namely any (w, f) with $w \geq 0, f \geq 0$. It is therefore impossible to solve the bandit stopping problem via an iterative method that loops through all the states. A finite problem can be formed by imposing a horizon N, that is, no more than N plays are allowed. As $N \to \infty$, one might expect convergence (in some sense) to the original problem without horizon. This convergence will be examined in the next subsection.

Consider now the horizon N problem, where the player receives a lump-sum payment λ upon retirement. If the initial state is $(w, N - w - 1)$, then at most one more play can be made. The player ponders: should he play, or simply collect his payment λ without playing? If he does play, Equation 3.5 shows that he wins with expected probability p where

$$p \doteq \frac{w + 1}{N + 2}, \tag{8.3}$$

and his expected total discounted reward is $p + \beta\lambda$. So for initial state $(w, N - w - 1)$ under the best strategy, the expected value of the total reward is $v_{N,\lambda}(w, N - w - 1)$, where

$$v_{N,\lambda}(w, N - w - 1) = \max(\lambda, p + \beta\lambda). \tag{8.4}$$

This argument can be generalized to arbitrary (w, f) with $w + f < N$. Let $v_{N,\lambda}(w, f)$ be the expected total reward under the optimal strategy for the horizon N problem with initial state (w, f). Then

$$v_{N,\lambda}(w, f) = \max\{\lambda, p[1 + \beta v_{N,\lambda}(w + 1, f)] + (1 - p)\beta v_{N,\lambda}(w, f + 1)\} \tag{8.5}$$

Since Equation 8.5 expresses $v_{N,\lambda}(w, f)$ in terms of $v_{N,\lambda}(w+1, f)$ and $v_{N,\lambda}(w, f+1)$, it is possible to find $v_{N,\lambda}(w, f)$ for all (w, f) where $w + f \leq N - 1$ using "backward induction". A Gittins index $\nu_N(w, f)$ for the horizon N problem can be defined following Lemma 8.1 as

$$\nu_N(w, f) \doteq (1 - \beta)\lambda_N, \tag{8.6}$$

where

$$\lambda_N \doteq \min\{\lambda | v_{N,\lambda}(w, f) = \lambda\}. \tag{8.7}$$

λ_N can be found to a given accuracy by successively trying larger values of λ, using an increment equal to the desired accuracy. Since there are on the order of m^2 entries in a $m \times m$ table of values for $v_{N,\lambda}(w, f)$, approximately on the order of m^2 values of λ should be tried, so that each different state (w, f) will be assigned a different value of $\nu_N(w, f)$ [Note that the exact value of $\nu_N(w, f)$ is unimportant in multi-armed bandit applications. Rather, it is the relative sizes of the two indices $\nu_N(w_1, f_1)$, $\nu_N(w_2, f_2)$ that indicates which of the states (w_1, f_1) or (w_2, f_2) is advantageous to play].

Since the computation of a $N \times N$ table of values $v_{N,\lambda}(w, f)$ for a given λ takes on the order of N^2 computations (actually, the entire table need not be computed), altogether the computation of a $m \times m$ table of $\nu(w, f)$ values for the horizon N problem takes on the order of $m^2 N^2$ computations. As far as storage requirements, they are not stringent; two vectors of size N are also required for the computation.

8.3. The effect of horizon on Gittins index estimates

It remains to show that $\nu_N(w, f)$ is actually a good approximation to $\nu(w, f)$ as N becomes large.

Lemma 8.2 *Let $\nu_N(w, f)$ be the Gittins index for an arm with horizon N. Then $\nu_N(w, f)$ is increasing with respect to N.*

Proof: In the notation of the previous subsection, it is clear that $v_{N,\lambda}(w, N - w) = \lambda$, since the state $(w, N - w)$ is at the horizon already and the player must retire immediately. Also, $v_{N+M,\lambda}(w, N - w) \geq \lambda$, since the player with horizon $N + M$ has at least the option of immediate retirement. Thus

$$v_{N,\lambda}(w, N - w) \leq v_{N+M,\lambda}(w, N - w). \tag{8.8}$$

Using Equations 8.4, 8.5, 8.8, and backward induction, it follows that for all states (w, f)

$$0 \leq v_{N,\lambda}(w, f) \leq v_{N+M,\lambda}(w, f).$$

Equation 8.7 leads to

$$\lambda_N(w, f) \leq \lambda_{N+M}(w, f).$$

(See Figure 8.1.) In view of the definition of ν_N in Equation 8.6, the lemma is thus proved.

It remains to estimate the magnitude of the increase in ν_N as N increases. In light of the definition of λ_N in terms of $v_{N,\lambda}$, it is reasonable to investigate the behavior of $v_{N,\lambda}(w, f)$ as N increases, which turns out to be relatively easy to quantify. For states of the form $(w, N - w)$,

$$v_{N,\lambda}(w, N - w) \;=\; \lambda$$

$$v_{N+M,\lambda}(w, N - w) \;\leq\; \max\left\{\lambda, 1 + \beta\lambda, \ldots, \left(\sum_{j=0}^{m-1} \beta^j\right) + \beta^m \lambda\right\}$$

$$\leq\; \lambda + \frac{1}{1 - \beta}.$$

Thus

$$v_{N+M,\lambda}(w, N - w) - v_{N,\lambda}(w, N - w) \leq \frac{1}{1 - \beta}.$$

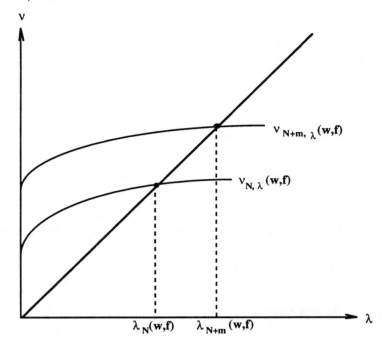

FIGURE 8.1. Illustration of the proof of Lemma 8.2.

Using this and Equation 8.5 with $f = N - w - 1$, it follows that

$$v_{N+M,\lambda}(w, N - w - 1) - v_{N,\lambda}(w, N - w - 1) \leq \frac{\beta}{1 - \beta},$$

and, continuing in this fashion, an expression for the general state (w, f) can be derived:

$$v_{N+M,\lambda}(w, f) - v_{N,\lambda}(w, f) \leq \frac{\beta^{N-w-f}}{1 - \beta}. \tag{8.9}$$

On the other hand, an increase in λ affects $v_{N+M,\lambda}(w, f)$ as follows:

Lemma 8.3 *If $\lambda_{N+M}(w, f) \geq \lambda' > \lambda$, then*

$$v_{N+M,\lambda'}(w, f) - v_{N+M,\lambda}(w, f) \leq \beta(\lambda' - \lambda). \tag{8.10}$$

Proof: Recall the interpretation of $v_{N+M,\lambda'}(w, f)$ as the expected reward from a one-armed bandit under an optimal strategy with retirement payment λ'. If $\lambda_{N+M} \geq \lambda'$, then an optimal strategy exists without immediate retirement. If the same strategy is used with retirement payment λ rather than λ', then the total discounted reward is reduced by at most $\beta(\lambda' - \lambda)$ since the retirement payment is claimed at some time $t \geq 1$. Hence the lemma is proved.

What consequence does this hold for the Gittins index estimates $\lambda_N(w, f)$ and $\lambda_{N+M}(w, f)$? Equations 8.9 and 8.10 can be combined to yield

$$v_{N+M,\lambda'}(w, f) - v_{N,\lambda}(w, f) - \frac{\beta^{N-w-f}}{1 - \beta} \leq \beta(\lambda' - \lambda),$$

when $\lambda_{N+M}(w, f) \geq \lambda' \geq \lambda$. Using $\lambda' \doteq \lambda_{N+M}(w, f)$, $\lambda \doteq \lambda_N(w, f)$ and recalling Equation 8.7 it follows

$$\lambda_{N+M}(w, f) - \lambda_N(w, f) \leq \frac{\beta^{N-w-f}}{(1-\beta)^2}.$$

Letting $M \to \infty$ and recalling $\nu_N = \lambda_N(1 - \beta)$, it follows

$$\nu_N(w, f) \leq \nu(w, f) \leq \nu_N(w, f) + \frac{\beta^{N-w-f}}{1-\beta}.$$

Thus ν_N does converge to ν as $N \to \infty$.

As a not-so-easy exercise, the reader can try to prove the sharper estimate

$$\nu_N(w, f) \leq \nu_{N+M}(w, f) \leq \nu_N(w, f) + \frac{\beta^{N-w-f}}{(1-\beta)(1+\beta p)},$$

where p is given by Equation 8.3.

References

[1] **Gittins, J. C.,** Bandit processes and dynamic allocation indices, *J. R. Statist. Soc.*, Ser. B **41(2)**, 148–177, 1979.

[2] **Gittins, J. C.,** Multi-armed Bandit Allocation Indices, John Wiley & Sons, New York, 1989.

[3] **Whittle, P.,** Multi-armed bandits and the Gittins index, *J. R. Statist. Soc.*, Ser. B **42**, 143–149, 1980.

Chapter 12

THREE BEWITCHING PARADOXES

J. Laurie Snell
Department of Mathematics and Computer Science
Dartmouth College

Robert Vanderbei
Department of Civil Engineering and Operations Research
Princeton University

ABSTRACT

For hundreds of years paradoxes dealing with questions of probability have ranked among some of the most perplexing of all mathematical paradoxes. In this chapter, we will discuss in detail three probabilistic paradoxes, each of which involves a decision to "switch" or "not to switch".

1. THE "LET'S MAKE A DEAL" PARADOX

This paradox has been around for a long time, but recently it has generated a great deal of interest following its appearance in syndicated puzzler Marilyn vos Savant's weekly brain teaser article that appears in many newspapers throughout the United States (see Marilyn vos Savant [8]).

Here's how the paradox goes. In the game show *Let's Make a Deal*, the following scenario arises frequently. On the stage there are three large doors. The host, Monty Hall, tells the contestant that behind one of the doors is a nice prize, but the other two doors have nothing of value behind them. The contestant is offered the chance to select one of the doors. Let's call this door A. Before showing the contestant what is behind door A, Monty Hall, who knows which door actually conceals the nice prize, shows the contestant what is behind one of the two doors that the contestant didn't choose. Let's call this door B and the door that he doesn't show you we will call door C. Monty Hall always picks for door B a door that does not have the prize. At this point, Monty asks the contestant whether he or she would like to switch to door C. Should the contestant switch? In other words, what is the probability that the prize is behind door A given that there is nothing behind door B?

When first presented with this question, most students (and professors) of probability say that it does not matter; either way, the probability of success

0-8493-8073-1/95/$0.00+$.50

is 1/2. After all, there are two choices that remain and knowing that door B is empty doesn't say anything about doors A and C. Right?

Actually, it's wrong. But the conviction that 1/2 is the correct answer can be very strong. In fact, on a recent visit to an Ivy League school one of us got into a rather intense discussion about this problem with three distinguished mathematicians (one of whom is even a probabilist). All three of them were convinced that 1/2 was the success probability whether they stick with their original door or switch.

All sorts of reasoning was applied to convince them of their error, but it was to no avail. Experimental evidence was offered first. They were told that a computer program had been written that simulates this game. The program was written so that the contestant always switches and was set to simulate 10 million plays of the game. Of the 10 million simulated contestants, only 3,332,420 of them lost. This gives an empirical success rate of 0.6667580. It was suggested that this number looks suspiciously close to 2/3. Of course, this piece of evidence did not alter their convictions. They preferred to believe that the program was flawed rather than their logic.

After experimental evidence failed, raw logic was employed. It was argued that the probability of success for door A was 1/3 before door B was revealed and, since Monty Hall is always able to find some door to be door B, how can the opening of door B say anything about door A. Hence, its probability must still be 1/3. Since door C is the only remaining door, its success probability must then be 2/3. Amazingly, this still did not convince our three distinguished friends.

Finally, they were offered the best explanation. They were shown an actual definition of a probability space that models this game and were then shown the computations that lead to 1/3 and 2/3. Here is the model. First we need a random variable P that describes which door the prize is behind (i.e., $P = 1, 2$, or 3, each with probability 1/3). Without loss of generality, we may assume that the contestant always picks door 1 (so that door 1 is door A). Now, if the prize is behind door 3, Monty Hall will open door 2 (so that door 2 becomes door B and door 3 becomes door C). Similarly, if the prize is behind door 2, Monty Hall will open door 3. For these two cases Monty Hall had no choice.

However, what if the prize is actually behind door 1. Now what is Monty going to do? Maybe he always shows door 2 in this case. Or maybe he always shows door 3. Or maybe he uses some complicated secret algorithm for deciding which door to open. In any case the contestant has no knowledge as to how Monty will behave in this situation and regards Monty's two possibilities as equally likely. Hence, we may as well assume that Monty tosses a fair coin: if it comes up heads he opens door 2 while if it comes up tails he opens door 3. This random coin toss is independent of P. Therefore, our sample space consists of six points as shown in the table below. If the contestant switches, then success will correspond to the four sample points (H,2), (H,3), (T,2) and (T,3). Hence the probability of success is 4/6 = 2/3. On the other hand, if the

	1	2	3
H	2	3	2
T	3	3	2

The columns represent the values of P. The two rows represent the coin that Monty flips to decide what to say in the case that $P = 1$. Each of the six sample points has probability $1/6$. The numbers given in the table indicate which door Monty will show.

contestant holds onto door A, then success will correspond to sample points (H,1) and (T,1) and the success probability will be only $1/3$.

Though they were unable to find any flaw in this line of argument, they were left quite puzzled since their intuition had failed them so miserably.

2. THE "OTHER PERSON'S ENVELOPE IS GREENER" PARADOX

Here is another paradox having to do with switching from one choice to another.

Two envelopes each contain an IOU for a specified amount of gold. One envelope is given to Ali and the other to Baba and they are told that the IOU in one envelope is worth twice as much as the other. However, neither knows who has the larger prize. Before anyone has opened their envelope, Ali is asked if she would like to trade her envelope with Baba. She reasons as follows. With 50 percent probability Baba's envelope contains half as much as mine and with 50 percent probability it contains twice as much. Hence, its expected value is

$$1/2(1/2) + 1/2(2) = 1.25,$$

which is 25 percent greater than what I already have and so yes, it would be good to switch. Of course, Baba is presented with the same opportunity and reasons in the same way to conclude that he too would like to switch. So they switch and each thinks that his/her net worth just went up by 25 percent. Of course, since neither has yet opened any envelope, this process can be repeated and so again they switch. Now they are back with their original envelopes and yet they think that their fortune has increased 25 percent twice. They could continue this process ad infinitum and watch their expected worth zoom off to infinity.

Clearly, something is wrong with the above reasoning, but where is the mistake? This paradox is quite puzzling until one carefully writes down a probabilistic model that describes the situation. Here is one possible model. Let X_0 denote the smaller amount of money between the two envelopes. This is a random variable taking values in the positive reals, but we (and more importantly, Ali and Baba) know nothing about its distribution. Let X_1 denote the larger amount of money so that $X_1 = 2X_0$. To select one of the two envelopes at random and give it to Ali means that we toss a fair coin and deliver to Ali either the envelope containing X_0 or the one containing X_1 depending on whether heads or tails appears. Mathematically, this means that we have another random variable N independent of X_0 (and hence of X_1) and taking values 0 and 1, each with probability 1/2. The envelope that Ali receives contains $Y = X_N$ and the envelope that Baba receives contains $Z = X_{1-N}$.

Ali's reasoning about Baba's envelope starts by conditioning on the two possible values of N and bearing in mind that on the event $\{N = 0\}$, $Z = 2Y$ and on the event $\{N = 1\}$, $Z = Y/2$. Hence,

$$E[Z] = \frac{1}{2}E[2Y|N = 0] + \frac{1}{2}E[\frac{1}{2}Y|N = 1]. \tag{1}$$

At this point she mistakenly assumes that Y and N are independent and continues her argument as follows:

$$E[Z] = \frac{1}{2}E[2Y] + \frac{1}{2}E[\frac{1}{2}Y] = \frac{5}{4}E[Y].$$

Of course, the correct way to complete the analysis is to first note that

$$E[2Y|N = 0] = E[2X_0|N = 0] = 2E[X_0]$$

and

$$E[\frac{1}{2}Y|N = 1] = E[\frac{1}{2}X_1|N = 1] = \frac{1}{2}E[X_1] = E[X_0].$$

Then, substituting these into (1) we see that

$$E[Z] = \frac{1}{2}2E[X_0] + \frac{1}{2}E[X_0] = \frac{3}{2}E[X_0] = E[Y].$$

3. "CHOOSING THE BIGGER NUMBER" PARADOX

Here is another paradox closely related to the previous one. Ali and Baba are again given two envelopes with an IOU for a specified amount of gold in each envelope. This time they know nothing about the amounts other than that they are non-negative numbers. After opening her envelope, Ali is offered the chance to switch her envelope with that of Baba. Can Ali find a strategy for deciding whether to switch which will make her chance of getting the envelope with the larger of the two numbers greater than one half? At first blush, this would appear to be impossible.

But consider the following strategy: Ali does an auxiliary experiment of choosing a number U by some chance device that makes all non-negative numbers possible. For example, choose U according to a exponential distribution with mean 10. If the IOU given Ali is greater than U she keeps this envelope, if it is less than U she switches to the other envelope.

Let's see why this auxiliary experiment helps. As before, let X_0 be the smaller and X_1 the larger of the two numbers in the envelopes. Assume first that U is less than both numbers, that is $U < X_0 < X_1$. Then Ali will not switch and, since she chose an envelope at random, she has a fifty percent chance of getting the larger number X_1. Assume next that her auxiliary number is between the two numbers in the envelopes, that is, $X_0 < U < X_1$. In this case, she is certain to get the larger number, since if her envelope has X_0 she will switch, and if it has X_1 she will keep it giving her in both cases the bigger number. Thus, in either case she ends up with the larger number. Finally, consider the case $X_0 < X_1 < U$. Then she will switch envelopes and, since her original choice of envelopes was random she again has a fifty percent chance of having the bigger number. Thus, in two of the cases Ali has a fifty percent chance and in the third case, which can happen, she is certain to get the largest number. Thus, her overall probability is greater than $1/2$. Note that, to have this probability well defined, we would have to assume that there is some probability distribution that describes the probability that any two particular numbers are put in the envelopes.

This problem arose in work of David Blackwell on estimating translation parameters. His example 1 on page 397 of [4] was the following: W is an unknown integer and X is a random variable with values 1 or -1 with probability $1/2$ each. You observe $Y = W + X$ and want to estimate W. Using method we just described, Blackwell showed that you can guess the value of W with a probability greater than $1/2$ of being correct.

4. COMMENTS ON THE MONTY HALL PROBLEM

The Monty Hall problem is a conditional probability problem that is very similar to other such problems that have puzzled students of probability throughout its history. Before commenting on the Monty Hall problem itself, we consider some other variants of this conditional probability problem. A more complete discussion of these many variants can be found in Barbeau [2] and Bar-Hillel and Falk [1].

One of the first conditional probability paradoxes is the *Box Paradox* formulated by Bertrand [3].

1	G G
2	S S
3	S G

A cabinet has three drawers. In the first drawer there are two gold balls, in the second drawer there are two silver balls, and in the third drawer one silver and one gold ball. A drawer is picked at random and a ball chosen at random from the two balls in the drawer. Given that a gold ball was drawn, what is the probability that the drawer with the two gold balls was chosen?

The intuitive answer might be 1/2 but, of course, that is wrong. The possible outcomes for this experiment are displayed in the tree diagram of Figure 1. We assign the appropriate branch probabilities and path probabilities as products of these branch probabilities along the path.

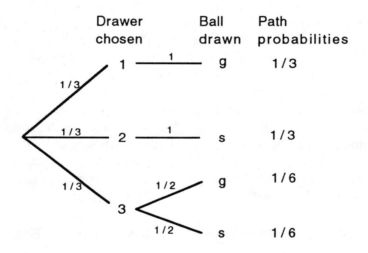

FIGURE 1. Unconditional probabilities for the box paradox.

This tree diagram provides the probabilities for the unconditional problem. To find the conditional probabilities, given that a gold ball was drawn, we need only delete the paths that do not result in a gold ball being drawn and renormalize the probabilities for the remaining paths of the tree to add to one. We can then compute the desired conditional probability by adding the conditional probabilities for the paths that give the desired outcome. We do this in Figure 2 and we see that the conditional probability that the drawer with two gold balls was drawn is 2/3 and not 1/2.

This is a particularly simple version of the conditional probability paradox because there is not a lot of argument about how to set up the model. It is pretty clear what "picking a box at random" and then "picking a ball at random" means. The next version of the problem called *the sibling problem* gets more complicated. In its simplest form this problem may be stated:

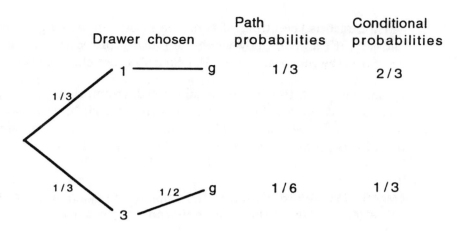

FIGURE 2. Conditional probabilities for the box paradox.

Consider a family of two children. Given that there is a boy in the family what is the probability that there are two boys in the family?

Again we are supposed to say 1/2 and find that we are wrong. The "text book" solution would be to draw the tree diagram and then form the conditional tree by deleting paths to leave only those paths that are consistent with the given information. The result is shown in Figure 3. We see that the probability of two boys given a boy in the family is not 1/2 but rather 1/3.

One often says that the more intuitive answer 1/2 is the correct answer if the given information is that the youngest child is a boy.

This problem and others like it are discussed in [1]. These authors stress that the answer to conditional probabilities of this kind can change depending upon how the information given was actually obtained. For example, they show that 1/2 is the correct answer for the following scenario presented in [5].

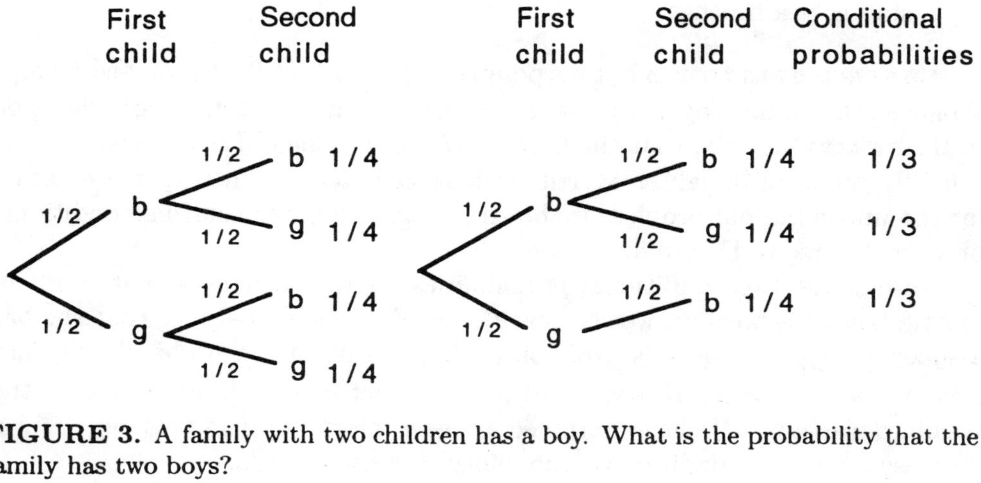

First child	Second child		First child	Second child	Conditional probabilities
1/2 b	1/2 b 1/4		1/2 b	1/2 b 1/4	1/3
	1/2 g 1/4			1/2 g 1/4	1/3
1/2 g	1/2 b 1/4		1/2 g	1/2 b 1/4	1/3
	1/2 g 1/4				

FIGURE 3. A family with two children has a boy. What is the probability that the family has two boys?

Mr. Smith is the father of two. We meet him walking along the street with a young boy whom he proudly introduces as his son. What is the probability that Mr. Smith's other child is also a boy?

As usual we have to make some additional assumptions. For example, we will assume that, if Mr. Smith has a boy and a girl, he is equally likely to choose either one to accompany him on his walk. In Figure 4 we show the tree analysis of this problem and we see that 1/2 is, indeed, the correct answer.

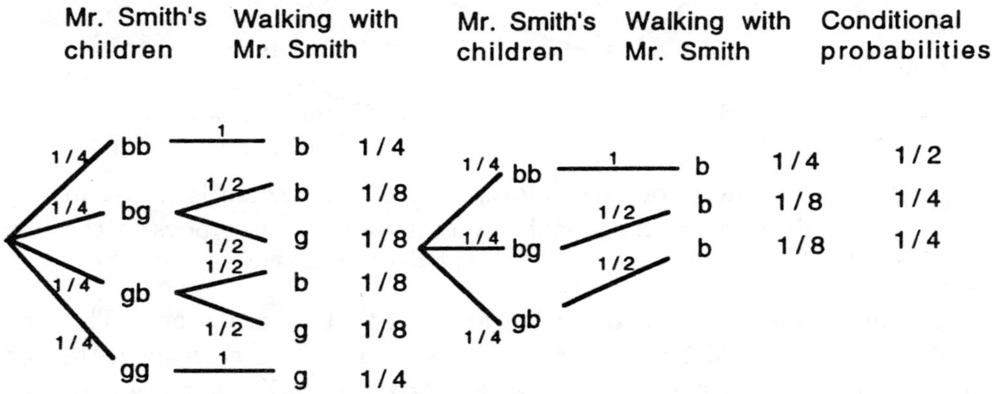

Mr. Smith's children	Walking with Mr. Smith		Mr. Smith's children	Walking with Mr. Smith	Conditional probabilities

FIGURE 4. You meet Mr. Smith out walking with a son, one of his two children. What is the probability that his other child is a boy?

In his popular book *Innumeracy* John (see John Paulos [10]) decided to jazz the problem up a bit by asking:

Consider some randomly selected family of four (i.e., two children) which is known to have at least one daughter. Say Myrtle is her name. Given this, what is the conditional probability that Myrtle's sibling is a brother?

He gave the answer 2/3 but, as pointed out to us by Bill Vinton and George Wolford, this is no longer a reasonable answer. There is a new consideration in the process, namely, that the family has a girl named Myrtle. Assume that a family names a daughter Myrtle with probability p. Then a tree diagram for the unconditional problem is shown in Figure 5 with resulting conditional probability tree in Figure 6.

Adding the two conditional probabilities for the sequences that result in Myrtle having a brother, we see that the conditional probability that she has a sister is $\frac{(2-p)}{(4-p)}$. Thus this probability depends upon the probability p that a family will name a girl Myrtle. If p is 1 we get a probability 1/3 as in the standard version. We have obtained no new information in this case. As p decreases to 0 this conditional probability increases to 1/2.

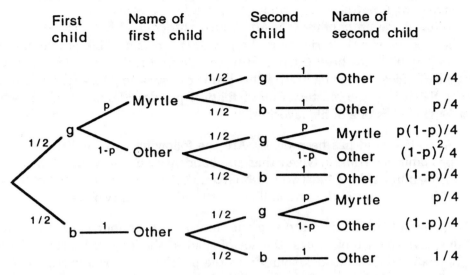

FIGURE 5. A family with two children has a daughter named Myrtle. What is the probability that Myrtle has a brother?

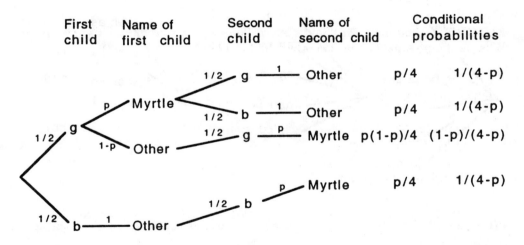

FIGURE 6. Conditional probabilities given a child named Myrtle.

It is not so easy to think of reasonable scenarios that would lead to the classical 1/3 answer. An attempt was made by Stephen Geller in proposing this problem to Marilyn vos Savant [9]. He writes:

A shopkeeper says she has two new baby beagles to show you, but she doesn't know whether they're male, female or a pair. You tell her that you want only a male, and she telephones the fellow who's giving them a bath. "Is at least one a male?" she asks. "Yes," she informs you with a smile. What is the probability that the *other* one is male?

The next version historically is the *two aces problem.* This problem, dating back to 1936, has been attributed to the English mathematician J. H. C. Whitehead. (see Gridgeman [7]). This problem was also submitted to Marilyn vos Savant by the master of mathematical puzzles Martin Gardner, who remarks that it is one of his favorites.

> A bridge hand has been dealt. Are the following two conditional probabilities equal? Given that the hand has an ace what is the probability that it has two aces? Given that the hand has the ace of hearts, what is the probability that the hand has two aces?

It is customary to choose two cards from a smaller four card deck that contains say: the ace of hearts, the ace of spades, the king of hearts and the king of spades. The textbook solution to the problem is shown in Figure 7.

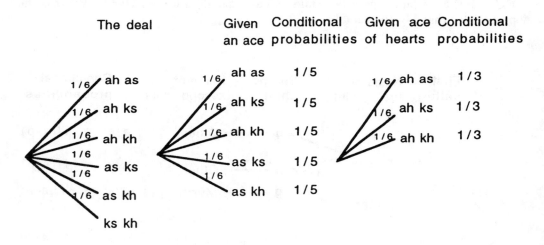

FIGURE 7. Probability of two aces given an ace, as compared to given the ace of hearts.

We see the somewhat surprising result that the conditional probability, given an ace, is 1/5 and, given the ace of hearts, is 1/3. It is natural to ask "how do we get the information that you have an ace?" Gridgeman [7] considers several different ways that we might get this information. For example, assume that the person holding the hand is asked to "name an ace in your hand" and answers "the ace of hearts." Then what is the probability that he has two aces? The tree analysis is shown in Figure 8.

We see that the answer is 1/5 which agrees with our solution to the probability of two aces given an ace. Now suppose you ask the more direct question "do you have the ace of hearts" and the answer is "yes". Then we have the tree analysis in Figure 9. We see that in this case the answer is 1/3 in agreement with the probability of two aces given the ace of hearts.

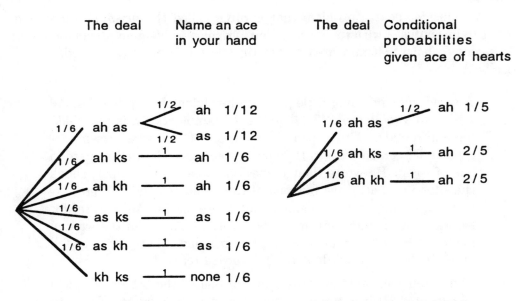

FIGURE 8. Probability of two aces given you answer "ace of hearts" to the request "name an ace in your hand".

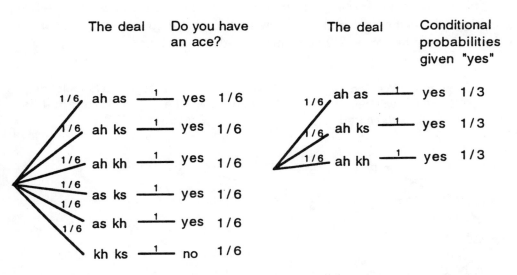

FIGURE 9. Probability of two aces given that you answer "yes" to the question "do you have the ace of hearts?".

Thus we can get either of the standard solutions by asking the appropriate question. Just how to figure out the correct question seems to be a bit of an art. Also, we have certainly not explained the paradoxical fact that very slight changes in the way you ask the question can give a completely different answer.

We consider next a problem that is often called the *prisoner's dilemma*. (Not to be confused with the more famous *prisoner's dilemma of game theory*.) It seems to have first appeared in Martin Gardner's book (see Gardner [6]). He writes:

> A wonderful confusing little problem involving three prisoners and a warden, even more difficult to state unambiguously, is now making the rounds. Three men — A, B, and C — were in separate cells under sentence of death when the governor decided to pardon one of them. He wrote their names on three slips of paper, shook the slips in a hat, drew out one of them and telephoned the warden, requesting that the name of the lucky man be kept secret for several days. Rumor of this reached prisoner A. When the warden made his morning rounds, A tried to persuade the warden to tell him who had been pardoned. The warden refused.
>
> "Then tell me," said A, "the name of one of the others who will be executed. If B is to be pardoned, give me C's name. If C is to be pardoned, give me B's name. And if I'm to be pardoned, flip a coin to decide whether to name B or C."

The warden tells A that B is to be executed and A assumes now that his probability of being executed has decreased from 2/3 to 1/2 by virtue of this information. Is he correct? The solution is given in Figure 10.

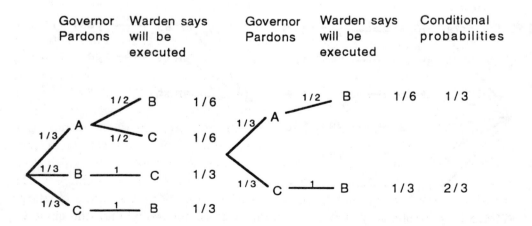

| Governor
Pardons | Warden says
will be
executed | | Governor
Pardons | Warden says
will be
executed | Conditional
probabilities |

FIGURE 10. Solution to the prisoner's dilemma.

We see that the probability that A was pardoned is still 1/3 and so he has not improved his chances of being executed by learning that it is between him and B. Had we not assumed that the warden tosses a coin when he has a choice, the probability would have changed and, as usual, would depend upon the probability that the guard chooses B when he has a choice between B and C.

We return now to the *Monty Hall* problem. That we have already solved. We want to now solve it by our standard tree diagram method. This problem has also been discussed at length in letters in *The American Statistician* (see Selvin [11]). The problem was revived by a letter from Craig Whitaker to Marilyn vos Savant for consideration in her column in *Parade Magazine* (see Marilyn vos Savant [8]). Craig wrote:

> Suppose you're on Monty Hall's *Let's Make a Deal !* You are given the choice of three doors, behind one door is a car, the others, goats.
>
> You pick a door, say 1, Monty opens another door, say 3 which has a goat. Monty says to you "Do you want to pick door 2?" Is it to your advantage to switch your choice of doors?

After posing the puzzle Craig goes on to say:

> I've worked out two different situations (based upon Monty's prior behavior, i.e., whether or not he knows what's behind the doors). In one situation it is to your advantage to switch, in the other there is no advantage to switch. What do you think?

Here we have a problem that is purported to be a real life problem and so we have to decide on the appropriate scenario. Craig in his letter already suggests that a basic question is whether Monty knows where the car is.

In her discussion of the problem, Marilyn vos Savant assumed that Monty did know where the car was and that he would open a door that did not have the car, but not the one tentatively chosen by the contestant. Thus, if the car is behind door 2 he must open door 3 and if it is behind door 3 he must open door 2. If the car is behind door 1, Monty has a choice and, as usual, we can either assume that he tosses a coin (as Marilyn vos Savant did) or more generally that he chooses door 3 with probability p and 2 with probability $1 - p$. Let's make the more general assumption. Now our unconditional tree is a bit larger than usual. The first step would show where the car is put. (We assume the choice is random.) The second step would show which door the contestant tentatively chooses. (Again we assume a random choice.) The third step would show Monty's choice. Rather than draw the somewhat large tree, we realize by now that to solve the conditional probability questions we need only draw the branches that are possible under the information given. Thus the answer is provided by the tree diagram in Figure 11.

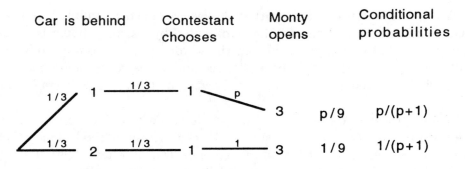

FIGURE 11. Solution to the Monty Hall problem.

Under our assumptions, given that the contestant chose door 1 and Monty chose door 3, the probability that the car is behind door 1 is $\frac{p}{1+p}$. From this we see that if $p = 1/2$ the probability the car is behind door 1 is 1/3 and that it is behind door 2 is 2/3, so the contestant should certainly switch. This was the solution of Marylin vos Savant, and was perfectly correct for her assumptions, but many readers found this solution difficult to believe. Since $\frac{p}{1+p} > 1/2$ for all p except $p = 1$, in all but this extreme case it is to your advantage to switch and even in this case you do not lose by switching. Thus you might as well switch.

If you assume that Monty does not know where the car is and just opens a door at random then, as Craig Whitaker remarked, there is no advantage to switching. Of course, this being a real life situation you can make lots of other assumptions. For example, it may well be that the host is out to trick you and sometimes offers you the choice (for example, if he sees that you have chosen the door with the car) and other times doesn't. When Monty Hall was interviewed by the *New York Times* (see Tierney [12]) he stated that

> "If the host is required to open a door all the time and offer you a switch, then you should take the switch. But if he has the choice whether to allow a switch or not, beware. *Caveat emptor.* It all depends on his mood."

5. COMMENTS ON THE TWO ENVELOPE PROBLEM

One of the tricks of making paradoxes is to make them slightly more difficult than is necessary to further befuddle us. As John Finn suggested to us, in this paradox we could just have well started with a simpler problem. Suppose Ali and Baba know that I am going to give then either an envelope with $5 or one with $10 and I am going to toss a coin to decide which to give to Ali, and then give the other to Baba. Then Ali can argue that Baba has $2x$ with probability 1/2 and $x/2$ with probability 1/2 and then the expected value would be $1.25x$. But now it is clear that this is nonsense, since, if Ali has the $5, Bala cannot possibly have 1/2 of this, namely, $2.50, since that was not even one of the choices. Similarly, if Ali has $10, Baba cannot have twice as much, $20. In fact, in this simpler problem the possibly outcomes are given by the tree diagram

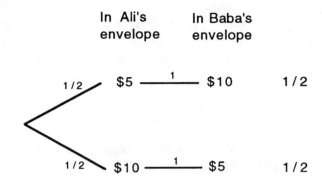

FIGURE 12. Tree diagram to compute Ali's expected winning.

From this we see that the expected amount received by either Ali or Baba is $1/2 \cdot 5 + 1/2 \cdot 10 = 7.5$.

6. COMMENTS ON THE CHOOSING THE BIGGER NUMBER PARADOX

This paradox shows that, in the first switching problem, there is an advantage to being allowed to switch. If Ali is allowed to switch or keep her number after looking at her envelope, he can improve her chances. Of course, we would have to assume that Baba must switch if Abi wants to. Let's try to see where this advantage really comes from. Assume that Ali were to know the distribution by which the two numbers were chosen. She will then know the probability $f(x)$ that the minimum of the two numbers is x. Then Ali can calculate the probability that she has the minimum given that she has an I.O.U worth a. That is,

$$P(a \text{ is the minimum} \mid \text{Ali has } a) = \frac{P(\text{Ali has } a \text{ and } a \text{ is the min})}{P(\text{Ali has } a)}$$
$$= \frac{1/2 f(a)}{1/2 f(a) + 1/2 f(a/2)}$$
$$= \frac{f(a)}{f(a) + f(a/2)}.$$

Thus the probability that Ali has the minimum of the two numbers is greater than one half only when $f(a/2) < f(a)$ and in this case, she should switch. Of course, Baba could reason in the same way and would not want to switch in precisely the cases where Ali would want to switch.

Now if we were picking the numbers and wanted to assure that Ali could not take advantage of the chance to switch, we would want to make the density $f(x)$ a constant so that she would never have an advantage. Unfortunately, there is no uniform density on the positive axis and so that is not possible. Thus one way to look at this paradox is to say that it comes about because we can't pick a real number with all possibilities equally likely.

References

[1] **Bar-Hillel, M. and Falk, R.,** Some teasers concerning conditional probabilities, *Cognition,* **11,** 109–122, 1982.

[2] **Barbeau ed.,** Fallacies, flaws and flimflam, *The College Mathematics Journal,* **24,** 149–159, 1993.

[3] **Bertrand, J.,** *Calcul des Probabilités,* Gauthier-Uillars, 1888.

[4] **Blackwell, David,** On the Translation Parameter Problem for Discrete Variables, *Annals of Mathematical Statistics,* **22,** 393–399, 1951.

[5] **Freund, J. E.,** Puzzle or paradox? *American Statistician,* **19,** 29, 44, 1965.

[6] **Gardner, M.,** *Mathematical Puzzles and Diversions,* Simon and Schuster, New York, l961.

[7] **Gridgeman, N. T.,** Letter, *American Statistician,* **21,** 38–39, 1967.

[8] **Marilyn vos Savant,** Ask Marilyn, Parade magazine, *Boston Globe,* 9 September; 2 December; 17 February 1990, reprinted in Marilyn vos Savant, *Ask Marilyn,* St. Martins, New York, 1992.

[9] **Marilyn vos Savant,** Ask Marilyn, Parade magazine, *Boston Globe,* 13 October 1991; 5 January 1992.

[10] **Paulos, John Allen,** *Innumeracy : Mathematical Illiteracy and its Consequences,* Hill and Wang, New York, 1988.

[11] **Selvin S.,** Letter to the Editor, *The American Statistician,* **29,** 134, 1975.

[12] **Tierney, John,** Behind Monty Hall's doors: puzzle, debate and answer? *New York Times,* 21 July 1991.

INDEX